The Electronics Dictionary For Technicians

THE ELECTRONICS DICTIONARY FOR TECHNICIANS

Tom Adamson

Merrill, an imprint of
Macmillan Publishing Company
New York

Maxwell Macmillan Canada
Toronto

Maxwell Macmillan International
New York Oxford Singapore Sydney

Acquisitions Editor: Dave Garza

Developmental Editor: Monica Ohlinger

Production Editor: bookworks

Production Manager: Aliza Greenblatt

Cover Designer: Thomas Mack

Illustrations: Publication Services

This book was set in Century Schoolbook by Monotype Composition Company, Inc., and was printed and bound by Arcata Graphics.

The cover was printed by New England Book Components.

Printed in the United States of America.

Macmillan Publishing Company
866 Third Avenue
New York, NY 10022

Macmillan Publishing Company is part of the
Maxwell Communication Group of Companies.

Maxwell Macmillan Canada, Inc.
1200 Eglington Avenue East, Suite 200
Don Mills, Ontario, M3C 3N1

Library of Congress Cataloging-in-Publication Data

Adamson, Thomas A.
The electronic dictionary for technicians / Tom Adamson.
p. cm.—(Merrill's international series in engineering technology)
ISBN 0-02-300820-2
1. Electronics—Dictionaries. I. Title. II. Series.
TK7804.A33 1992
621.381'03—dc20 91-45386
CIP

Printing: 1 2 3 4 5 6 7 8 9 Year: 2 3 4 5

MERRILL'S INTERNATIONAL SERIES IN ENGINEERING TECHNOLOGY

ADAMSON	*Applied Pascal for Technology,* 0-675-20771-1
	The Electronic Dictionary for Technicians, 0-02-300820-2
	Microcomputer Repair, 0-02-300825-3
	Structured BASIC Applied to Technology, 0-675-20772-X
	Structured C for Technology, 0-675-20993-5
	Structured C for Technology (w/ disks), 0-675-21289-8
ANTONAKOS	*The 68000 Microprocessor: Hardware and Software Principles and Applications,* 0-675-21043-7
ASSER/STIGLIANO/ BAHRENBURG	*Microcomputer Servicing: Practical Systems and Troubleshooting,* 0-675-20907-2
	Microcomputer Theory and Servicing, 0-675-20659-6
	Lab Manual to accompany *Microcomputer Theory and Servicing,* 0-675-21109-3
ASTON	*Principles of Biomedical Instrumentation and Measurement,* 0-675-20943-9
BATESON	*Introduction to Control System Technology,* 3rd ed., 0-675-21010-0
BEACH/JUSTICE	*DC/AC Circuit Essentials,* 0-675-20193-4
BERLIN	*Experiments in Electronic Device* to accompany *Floyd's Electronic Devices and Electronic Devices: Electron Flow Version,* 3rd. ed., 0-02-308422-7
	The Illustrated Electronics Dictionary, 0-675-20451-8
BERLIN/GETZ	*Experiments in Instrumentation and Measurement,* 0-675-20450-X
	Fundamentals of Operational Amplifiers and Linear Integrated Circuits, 0-675-21002-X
	Principles of Electronic Instrumentation and Measurement, 0-675-20449-6
BERUBE	*Electronic Devices and Circuits Using MICRO-CAP II,* 0-02-309160-6

Merrill's International Series in Engineering Technology

BOGART — *Electronic Devices and Circuits,* 2nd ed., 0-675-21150-6

BOGART/BROWN — *Experiments in Electronic Devices and Circuits,* 2nd ed., 0-675-21151-4

BOYLESTAD — *DC/AC: The Basics,* 0-675-20918-8
Introductory Circuit Analysis, 6th ed., 0-675-21181-6

BOYLESTAD/KOUSOUROU — *Experiments in Circuit Analysis,* 6th ed., 0-675-21182-4
Experiments in DC/AC Basics, 0-675-21131-X

BREY — *Microprocessors and Peripherals: Hardware, Software, Interfacing, and Applications,* 2nd ed., 0-675-20884-X
The Intel Microprocessors—8086/8088, 80186, 80286, 80386, and 80846—Architecture, Programming, and Interfacing, 2nd ed., 0-675-21309-6

BROGERG — *Lab Manual* to accompany *Electronic Communication Techniques,* 2nd ed., 0-675-21257-X

BUCHLA — *Digital Experiments: Emphasizing Systems and Design,* 2nd ed., 0-675-21180-8
Experiments in Electric Circuits Fundamentals, 2nd ed., 0-675-21409-2
Experiments in Electronics Fundamentals: Circuits, Devices, and Applications, 2nd ed., 0-675-21407-6

BUCHLA/MCLACHLAN — *Applied Electronic Instrumentation and Measurement,* 0-675-21162-X

CICCARELLI — *Circuit Modeling: Exercises and Software,* 2nd ed., 0-675-21152-2

COOPER — *Introduction to VersaCAD,* 0-675-21164-6

COX — *Digital Experiments: Emphasizing Troubleshooting,* 2nd ed., 0-675-21196-4

CROFT — *Getting a Job: Resume Writing, Job Application Letters, and Interview Strategies,* 0-675-20917-X

DAVIS — *Technical Mathematics,* 0-675-20338-4
Technical Mathematics with Calculus, 0-675-20965-X
Study Guide to Accompany Technical Mathematics, 0-675-20966-8
Study Guide to Accompany Technical Mathematics with Calculus, 0-675-20964-1

DELKER — *Experiments in 8085 Microprocessor Programming and Interfacing,* 0-675-20663-4

FLOYD — *Digital Fundamentals,* 4th ed., 0-675-21217-0
Electric Circuits Fundamentals, 2nd ed., 0-675-21408-4
Electronic Devices, 3rd ed., 0-675-22170-6
Electronic Devices: Electron Flow Version, 0-02-338540-5

	Electronics Fundamentals: Circuits, Devices, and Applications, 2nd ed., 0-675-21310-X
	Fundamentals of Linear Circuits, 0-02-338481-6
	Principles of Electric Circuits, Electron Flow Version, 2nd ed., 0-675-21292-8
	Principles of Electric Circuits, 3rd ed., 0-675-21062-3
FULLER	*Robotics: Introduction, Programming, and Projects*, 0-675-21078-X
GAONKAR	*Microprocessor Architecture, Programming, and Applications with the 8085/8080A*, 2nd ed., 0-675-20675-8
	The Z80 Microprocessor: Architecture, Interfacing, Programming, and Design, 0-675-20540-9
GILLIES	*Instrumentation and Measurements for Electronic Technicians*, 0-675-20432-1
GOETSCH	*Industrial Supervision: In the Age of High Technology*, 0-675-22137-4
GOETSCH/RICKMAN	*Computer-Aided Drafting with AutoCAD*, 0-675-20915-3
GOODY	*Programming and Interfacing the 8086/8088 Microprocessor*, 0-675-21312-6
HUBERT	*Electric Machines: Theory, Operation, Applications, Adjustment, and Control*, 0-675-21136-0
HUMPHRIES	*Motors and Controls*, 0-675-20235-3
HUTCHINS	*Introduction to Quality: Management, Assurance, and Control*, 0-675-20896-3
KEOWN	*PSpice and Circuit Analysis*, 0-675-22135-8
KEYSER	*Materials Science in Engineering*, 4th ed., 0-675-20401-1
KIRKPATRICK	*The AutoCAD Book: Drawing, Modeling, and Applications*, 2nd ed., 0-675-22288-5
	Industrial Blueprint Reading and Sketching, 0-675-20617-0
KRAUT	*Fluid Mechanics for Technicians*, 0-675-21330-4
KULATHINAL	*Transform Analysis and Electronic Networks with Applications*, 0-675-20765-7
LAMIT/LLOYD	*Drafting for Electronics*, 0-675-20200-0
LAMIT/WAHLER/HIGGINS	*Workbook in Drafting for Electronics*, 0-675-20417-8
LAMIT/PAIGE	*Computer-Aided Design and Drafting*, 0-675-20475-5
LAVIANA	*Basic Computer Numerical Control Programming*, 2nd ed., 0-675-21298-7
MACKENZIE	*The 8051 Microcontroller*, 0-02-373650-X
MARUGGI	*Technical Graphics: Electronics Worktext*, 2nd ed., 0-675-21378-9
	The Technology of Drafting, 0-675-20762-2
	Workbook for the Technology of Drafting, 0-675-21234-0
MCCALLA	*Digital Logic and Computer Design*, 0-675-21170-0

MacIntyre	*Study Guide to accompany Electronic Devices,* 3rd ed., 0-02-379296-5
	Study Guide to accompany Electronics Fundamentals, 2nd ed., 0-675-21406-8
Miller	*The 68000 Microprocessor Family: Architecture, Programming, and Applications,* 2nd ed., 0-02-381560-4
Monaco	*Essential Mathematics for Electronics Technicians,* 0-675-21172-7
	Introduction to Microwave Technology, 0-675-21030-5
	Laboratory Activities in Microwave Technology, 0-675-21031-3
	Preparing for the FCC General Radiotelephone Operator's License Examination, 0-675-21313-4
	Student Resource Manual to accompany *Essential Mathematics for Electronics Technicians,* 0-675-21173-5
Monseen	*PSPICE with Circuit Analysis,* 0-675-21376-2
Mott	*Applied Fluid Mechanics,* 3rd ed., 0-675-21026-7
	Machine Elements in Mechanical Design, 2nd ed., 0-675-22289-3
Nashelsky/Boylestad	*BASIC Applied to Circuit Analysis,* 0-675-20161-6
Panares	*A Handbook of English for Technical Students,* 0-675-20650-2
Pfeiffer	*Proposal Writing: The Art of Friendly Persuasion,* 0-675-20988-9
	Technical Writing: A Practical Approach, 0-675-21221-9
Pond	*Introduction to Engineering Technology,* 0-675-21003-8
Quinn	*The 6800 Microprocessor,* 0-675-20515-8
Reis	*Digital Electronics Through Project Analysis,* 0-675-21141-7
	Electronic Project Design and Fabrication, 2nd ed., 0-02-399230-1
	Laboratory Manual for Digital Electronics Through Project Analysis, 0-675-21254-5
Rolle	*Thermodynamics and Heat Power,* 3rd ed., 0-675-21016-X
Rosenblatt/Friedman	*Direct and Alternating Current Machinery,* 2nd ed., 0-675-20160-8
Roze	*Technical Communication: The Practical Craft,* 0-675-20641-3
Schoenbeck	*Electronic Communications: Modulation and Transmission,* 2nd ed., 0-675-21311-8
Schwartz	*Survey of Electronics,* 3rd ed., 0-675-20162-4
Sell	*Basic Technical Drawing,* 0-675-21001-1

SMITH	*Statistical Process Control and Quality Improvement,* 0-675-21160-3
SORAK	*Linear Integrated Circuits: Laboratory Experiments,* 0-675-20661-8
SPIEGEL/LIMBRUNNER	*Applied Statics and Strength of Materials,* 0-675-21123-9
STANLEY, B.H.	*Experiments in Electric Circuits,* 3rd ed., 0-675-21088-7
STANLEY, W.D.	*Operational Amplifiers with Linear Integrated Circuits,* 2nd ed., 0-675-20660-X
SUBBARAO	*16/32-Bit Microprocessors: 68000/68010/68020 Software, Hardware, and Design Applications,* 0-675-21119-0
TOCCI	*Electronic Devices: Conventional Flow Version,* 3rd ed., 0-675-20063-6
	Fundamentals of Pulse and Digital Circuits, 3rd ed., 0-675-20033-4
	Introduction to Electric Circuit Analysis, 2nd ed., 0-675-20002-4
TOCCI/OLIVER	*Fundamentals of Electronic Devices,* 4th ed., 0-675-21259-6
WEBB	*Programmable Logic Controllers: Principles and Applications,* 2nd ed., 0-02-424970-X
WEBB/GRESHOCK	*Industrial Control Electronics,* 0-675-20897-1
WEISMAN	*Basic Technical Writing,* 6th ed., 0-675-21256-1
WOLANSKY/AKERS	*Modern Hydraulics: The Basics at Work,* 0-675-20987-0
WOLF	*Statics and Strength of Materials: A Parallel Approach,* 0-675-20622-7

To my sister
Elizabeth A. Reardon

Preface

You are now holding a dictionary unlike any you may have seen before. All of its terms are defined using basic English and do not assume that you, the reader, have any background in electronics or computers.

This dictionary is intended for students of electronics; whether you be a full-time student enrolled in an electronics program or someone who needs to know electronic and computer terminology for their job. This is the dictionary you have been looking for. It contains all of the latest terms used by today's electronic technician. It does not contain old, outdated terms found in larger electronic dictionaries, where the terms you really need to know are hidden in pages of old technology.

The terms selected for this dictionary are the most commonly encountered terms by today's students majoring in electronics technology. This book also includes the essential terminology that the electronic technician needs from the field of computer science as well as electronic communications. A well selected set of appendices are included. Use these as a quick review of material before a test or job interview.

My special thanks to Monica Ohlinger, Developmental Editor; Dave Garza, Electronics Editor; Steve Helba, Executive Editor; and Jeff Smith, Director of Software Development. I would also like to thank the Production Editor, bookworks.

My special thanks to those who painstakingly reviewed the original manuscripts and whose recommendations helped to shape this dictionary: Ray Fleming, Edison State Community College; Robert J. Larsen, Texas State Technical Institute; Bill Medcalf, Hallmark Institute of Technology; Jim Predko, Lansing Community College; and Ronald F. Ravelle, Commonwealth Colleges.

The Electronics Dictionary For Technicians

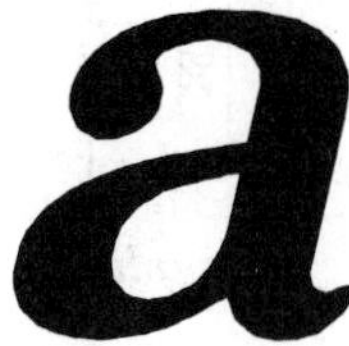

A Letter symbol for ampere. See ampere.

Å Symbol for angstrom. See angstrom.

above ground A voltage reading that is positive with respect to ground.

absolute maximum ratings Electrical specifications given by the manufacture of a device that must not be exceeded. If any of the absolute maximum ratings of a device are exceeded, the device may become permanently damaged.

absolute value The value of a number without regard to its sign. Signified by | |. For example, $|-5| = 5$.

absolute value amplifier An amplifier that will always produce a positive output signal regardless of the polarity of the input signal.

absolute zero The temperature at which all molecular motion ceases. Absolute zero is −273.15 °C.

ac Abbreviation for alternating current. See alternating current.

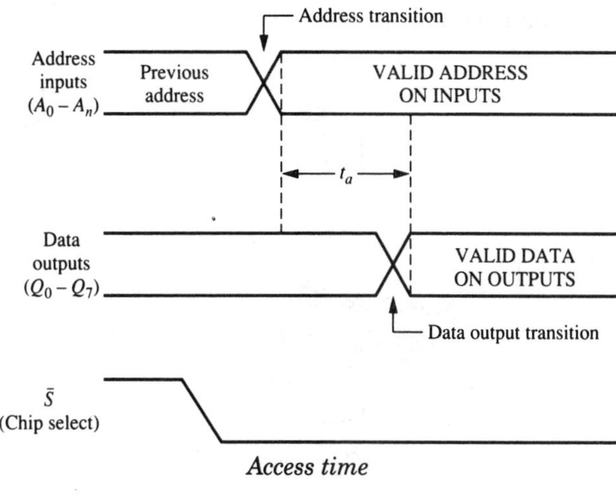

Access time

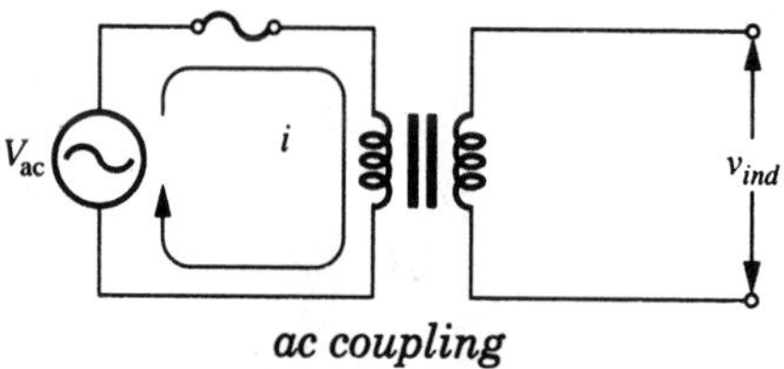

ac coupling

access time In microcomputer circuits, the amount of time it takes for valid data to appear once a valid address or similar instruction has been selected.

ac coupling The process of allowing the effects of ac to go from one point in a circuit to another (while usually blocking dc). A capacitor or a transformer may be used for ac coupling.

accumulator One of the major registers inside a microprocessor. Used for the intermediate storage of data. Also used to store temporarily the result of an arithmetic, logic or transfer process.

ac generator A device that produces an alternating voltage.

ACIA Abbreviation for asynchronous communications interface adapter. A digital device used to interface between a computer and another digital system.

A_{CL} Symbol for closed-loop gain. See closed-loop gain.

A_{CM} Symbol for common-mode gain. See common-mode gain.

acoustic coupler An electrical device that converts electrical impulses into sound and sound into electrical impulses. Used for the transmission and reception of digital data over the phone lines in older computer systems.

acquisition time The amount of time it takes for a system to respond to a given input value. In sample-and-hold circuits, the amount of time from when the sample command is given to

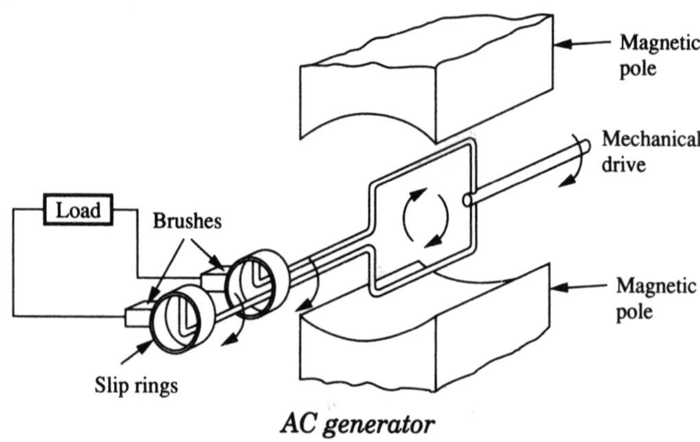

AC generator

when the circuit is able to acquire and hold the new sample value.

ac ripple In a power supply, the small variations in amplitude of the output voltage. In a half-wave rectifier, the ac ripple is equal to the line frequency; in a full-wave rectifier, the ac ripple is equal to double the line frequency.

ac stability In a power supply, the ability of the power supply to reject ac interference.

active filter An electronic circuit that is frequency selective and consists of devices such as transistors or operational amplifiers.

active logic level In an electrical digital circuit, the level (HIGH or LOW) that it takes to activate a device. As an example, in an active LOW LED circuit, the LED will be ON when its control input is LOW; in an active HIGH circiut, the LED will be ON when its control input is HIGH.

active low A logic condition that has an effect on the output

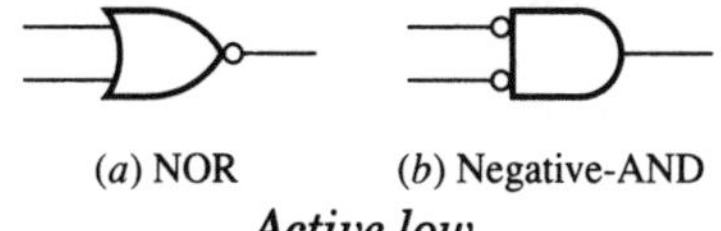

(*a*) NOR (*b*) Negative-AND

Active low

of a logic circuit when it is in the LOW state.

A/D Analog-to-digital (converter).

ADC Abbreviation for analog-to-digital converter. See A-to-D converter.

adder In digital circuits, a logic circuit used to add binary numbers. See half adder and full adder.

address A label, name, or number identifying a location where data are stored. Computer memory is organized in a column-row fashion. The columns contain the bit pattern that represents the data, while each row is distinguished from the next by a unique address. Addresses are usually sequentially assigned from zero up to the maximum number of rows of memory contained in the computer.

address bus A group of conductors from a microprocessor that carries the information for the activation of a specific memory or device location. The information is usually in the form of a binary number. See address register.

address modes The methods used by microprocessors for transferring information between memory and the microprocessor.

address multiplexing The process of splitting the address into two or more parts so that fewer address lines are required by the microprocessor.

address register A register that is used inside a microprocessor that stores a binary value that represents the memory location (address) that will be activated by the microprocessor.

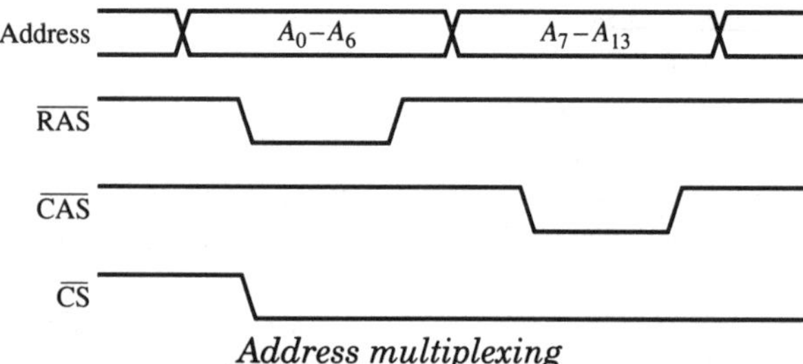

Address multiplexing

adjustable voltage regulator A voltage regulator whose output voltage is adjustable. This is usually accomplished by the addition of a few external components to a fixed-voltage regulator.

admittance (*Y*) The reciprocal of impedance. Mathematically, $Y = 1/Z$, where Z is the impedance of the circuit. Admittance is measured in siemens (S). Admittance means how well an ac circuit will admit the flow of current.

AFC Abbreviation for automatic frequency control. See automatic frequency control.

AFT Abbreviation for automatic fine tuning. See automatic fine tuning.

AGC Abbreviation for automatic gain control. See automatic gain control.

algorithm A sequence of instructions that explains how to solve a given problem. As an example, a computer program is an algorithm written in programming code understood by the computer system using the algorithm.

aliasing distortion When the sampling rate is less than the sampling theorem allows (see sampling rate), all the correct information will not be transmitted. An example of aliasing distortion is the wagon wheel effect, where in old Western movies the wheels of wagons appeared to turn in the wrong direction relative to the motion of the wagon. This is caused by the low sampling rate of the film compared to the rotational frequency of the wagon wheel.

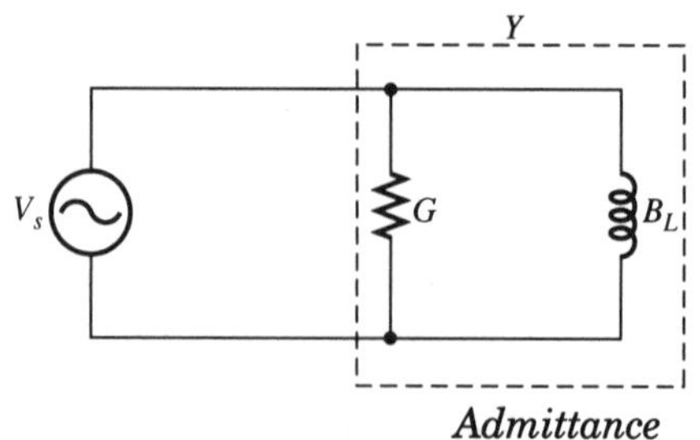

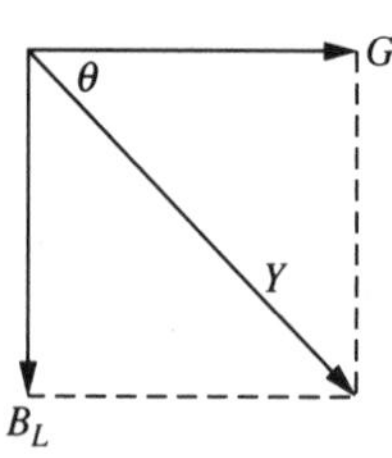

Admittance

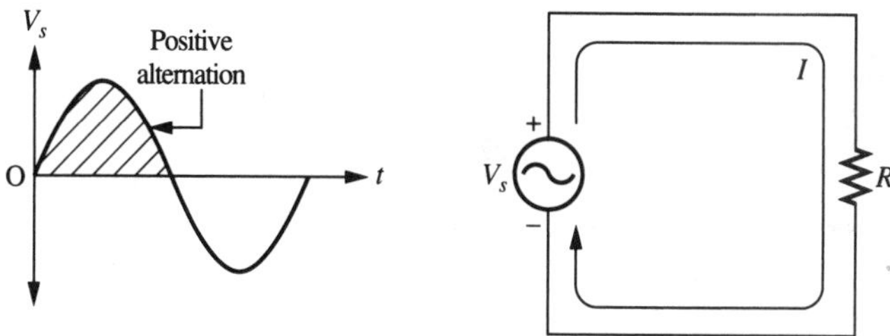

(*a*) Positive voltage: current direction as shown

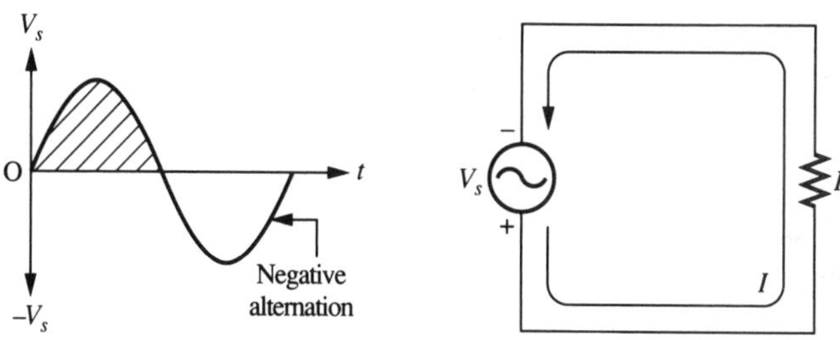

(*b*) Negative voltage: current reverses direction

Alternating current

alpha (α) The ratio of transistor collector current to emitter current. Mathematically, $\alpha = I_C/I_E$. Because $I_E > I_C$, α is always less than one.

alphanumeric characters Codes that represent letters of the alphabet, numerals, and other symbols such as punctuation marks and mathematical operation symbols, such as "+" and "=."

alterable memory Computer memory that is capable of having the storage of digital information changed.

alternating current Current that changes its direction in response to a change in voltage polarity.

alternating voltage Voltage that changes its polarity.

ALU Abbreviation for arithmetic logic unit. A digital circuit capable of performing arithmetic and logic operations.

AM Abbreviation for amplitude modulation. See amplitude modulation.

AM receiver Amplitude modulation receiver. A communications receiver capable of receiving an AM transmission, removing the intelligence from the wave, and reproducing it in a usable form.

ambient Surrounding. For example, the ambient air temperature means the temperature of the surrounding air.

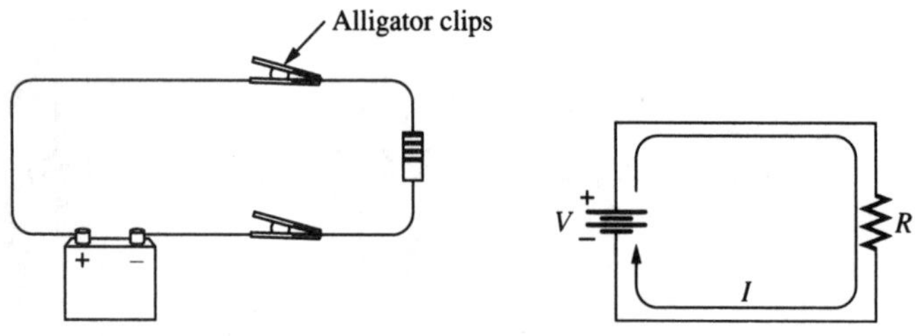

(*a*) Circuit in which the current is to be measured

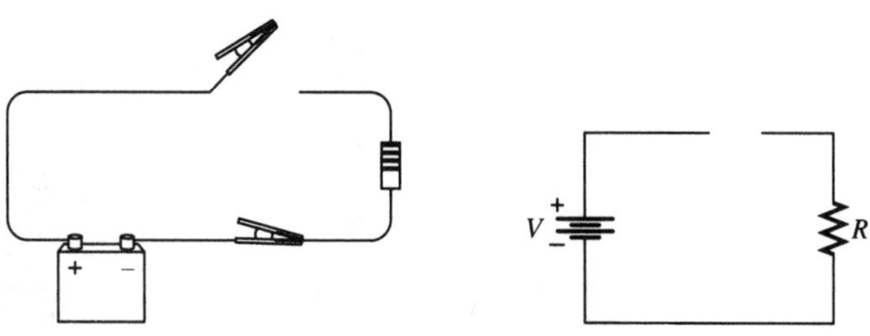

(*b*) Open the circuit either between the resistor and the positive terminal or between the resistor and the negative terminal of source.

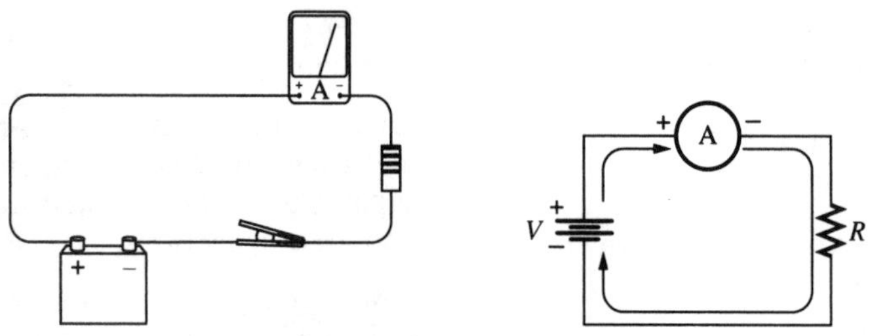

(*c*) Install the ammeter with polarity as shown (negative to negative–positive to positive).

Ammeter

American Standard Code for Information Interchange See ASCII.

ammeter An electrical instrument used for the purpose of measuring the amount of electrical current.

amp-hr rating See ampere-hour rating.

ampere The unit of measurement for current. One ampere is equal to a current flow of one columb per second in the SI system.

ampere-hour rating Rating for a source of energy that tells how long a given level of current

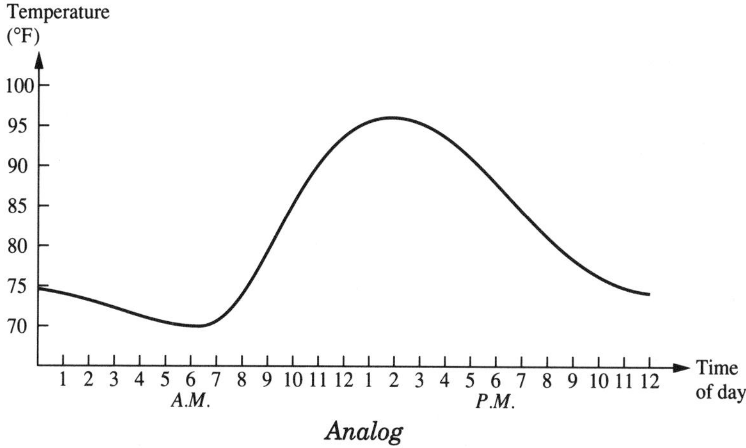

Analog

can be sustained from a given power source.

amplification Process of increasing the current, voltage, or power of an electrical signal.

amplifier An electronic circuit designed for the purpose of amplification. See amplification.

amplitude The amount of voltage or current as measured from its zero level.

amplitude modulation The process of causing information to change the strength of a high-frequency carrier wave.

amplitude response Of an active filter. It is the gain represented by the relationship between the output voltage and input voltage at different frequencies. Amplitude response is usually expressed in decibels.

analog In electronics a continuous, noninterrupted smooth change in an electrical process. A sine wave is an analog signal because it is a continuous, noninterrupted smooth change. A square wave is not an analog signal because it represents abrupt changes that are digital rather than analog.

analog bilateral switch A solid-state integrated circuit that acts as a single-pole, single-throw switch. A small control voltage (usually +5 volts) controls the open or closed condition of the switch.

analog computer A device capable of receiving information, processing the information, and displaying the results through the use of electrical circuits that mathematically simulate the problem being solved. A simple analog computer consists of two resistors in series where the total resistance represents the sum of

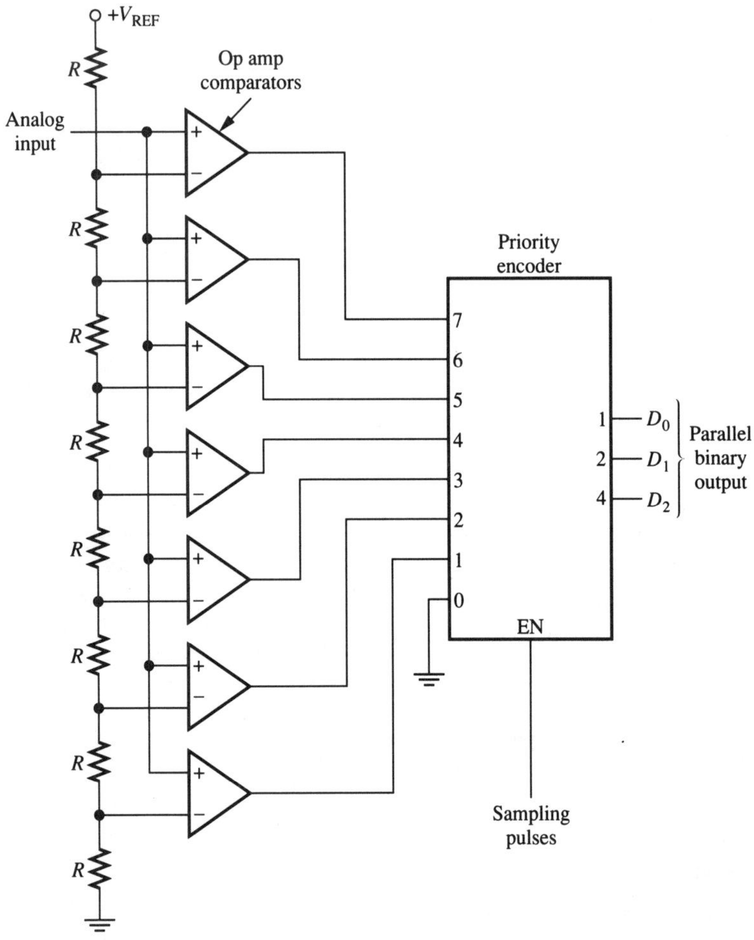

Analog-to-digital conversion

two numbers, each of which is the value of the individual resistor.

analog multimeter An electrical measuring instrument used to measure voltage, current, and resistance that utilizes an analog indicator for indicating values.

analog signal A smooth and continuous change of a voltage or current that represents some kind of desirable information.

analog-to-digital converter An electrical circuit that takes an analog signal and converts it to some form of a digital code. Analog-to-digital converters are

A	B	AB = X
0	0	0 • 0 = 0
0	1	0 • 1 = 0
1	0	1 • 0 = 0
1	1	1 • 1 = 1

AND function

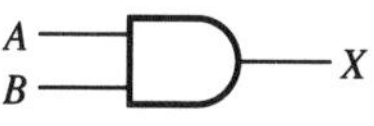

(*a*) Distinctive shape

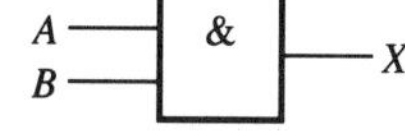

(*b*) Rectangular outline with AND (&) qualifying symbol

AND gate

used in computer systems to convert temperature readings into a digital code that is understood by the computer.

AND function A logical function whose output condition is TRUE only if all of its input conditions are TRUE. Expressed as $X = A■B$, where X is TRUE (or "1") only if both A and B are TRUE (a "1"). Otherwise, the output condition is FALSE, (a "0").

AND gate A digital circuit that performs the logical AND function.

AND/OR Combinational logic circuit consisting of 2 AND gates feeding into a single OR gate. Each AND gate consists of 2 inputs; thus this entire combination has a total of 4 inputs, allowing for 16 possible input combinations.

AND/OR/INVERT Combination logic circuit consisting of 2 AND gates feeding into a single OR gate, the output of which feeds into a single INVERTER. Each AND gate consists of 2 inputs; thus, this entire combination has a total of 4 inputs,

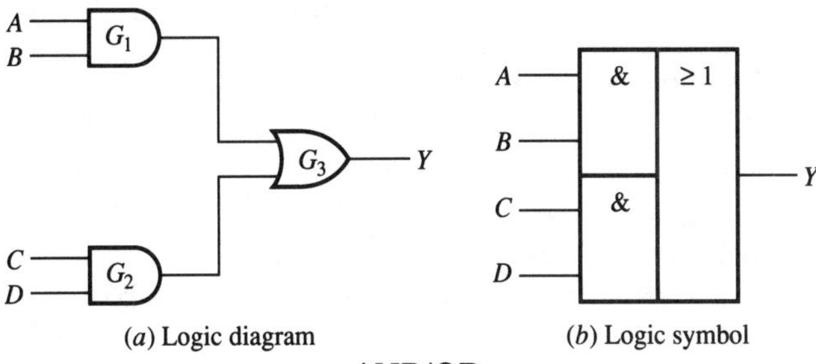

(*a*) Logic diagram

(*b*) Logic symbol

AND/OR

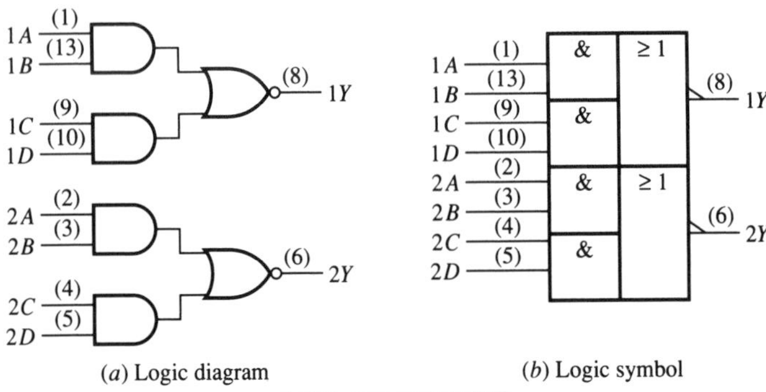

(*a*) Logic diagram (*b*) Logic symbol

AND OR INVERT

allowing for 16 possible input combinations.

angle modulation A general classification for both FM and PM. Angle modulation is changing the angle of the carrier wave from a reference value by a modulating signal.

angstrom A unit of measurement of 1×10^{-10} meters.

anode That part of an electrical device that is to have a positive voltage applied in relation to its cathode in order to have current flow. See cathode.

ANSI American National Standard Institute.

ANSI/IEEE logic symbols Logic symbols used to represent logic circuits approved by ANSI and the IEEE.

antenna A device that converts electromagnetic radiation into electrical energy or produces an electromagnetic radiation from an applied electrical energy. Almost all conductors will serve

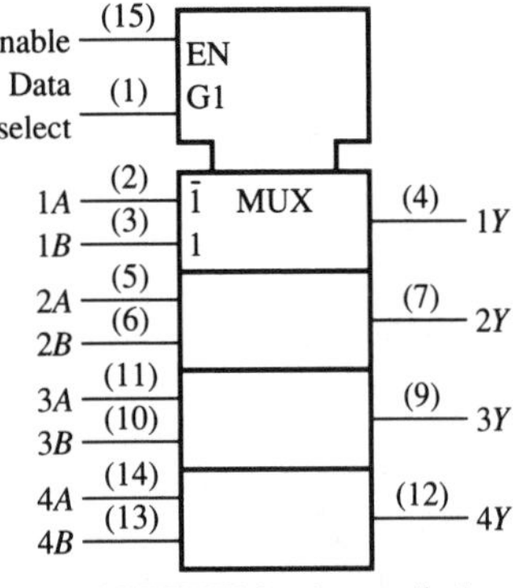

ANSI/IEEE logic symbol

as antennas. Devices used to transmit and receive electromagnetic radiation.

antenna coupling The method used to connect the antenna signals electrically to the rest of the circuit. One common form of antenna coupling is through the use of a transformer.

antenna gain The comparison of the output strength of the antenna radiation in a particular direction and a reference an-

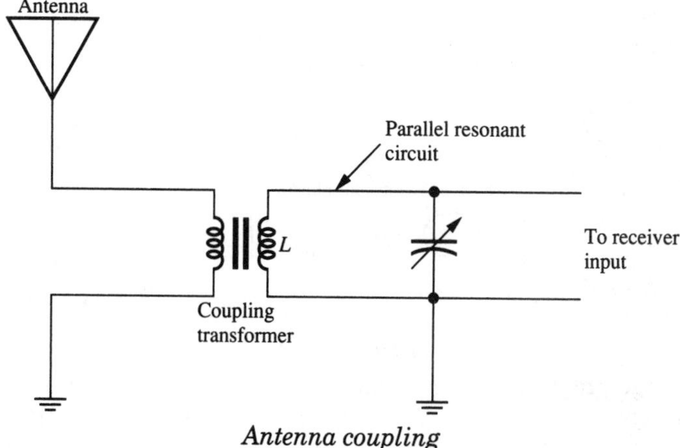

Antenna coupling

tenna. Expressed as A (dB) = 10 $\log_{10}(P_2/P_1)$, where A is the antenna gain in decibels, P_2 is the power of a reference antenna, and P_1 is the power of the actual antenna. The reference antenna is a hertz antenna that radiates an omnidirectional wave.

antenna *Q* The quality of an antenna defined as $Q = f_r/BW$, where f_r is the resonant frequency of the antenna and BW is the antenna bandwidth, both measured in hertz. The higher the Q of the antenna, the sharper its response curve and the more selective the antenna will be in choosing one frequency (f_r) over others.

A_P Letter symbol for difference gain. See difference gain.

applied current Current from an external source of electrical energy.

applied power Electrical power from an external source of electrical energy.

applied voltage Voltage from an external source of electrical energy.

Armstrong oscillator A circuit that generates its own signal. Uses a transformer to feed back the output signal to the input. Secondary winding is sometimes called a tickler coil. Value of the circuit capacitor and transformer inductance determines the resonant frequency.

ASCII Abbreviation for American Standard Code for Information Interchange. Pronounced "askee." The standard data transmission code used to achieve compatibility between data devices. Has a total of 128 unique characters.

assembler A computer program that translates an assembly

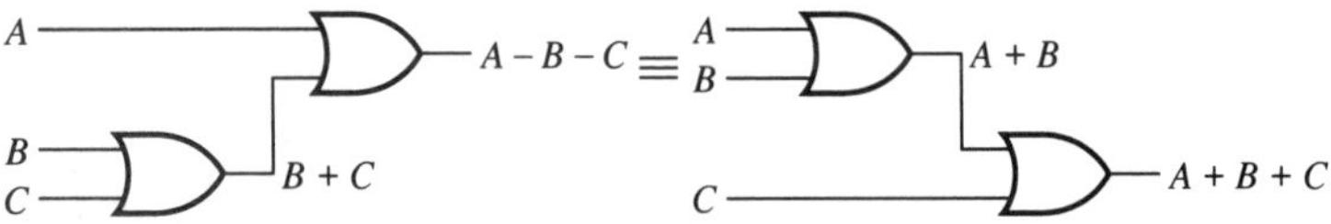

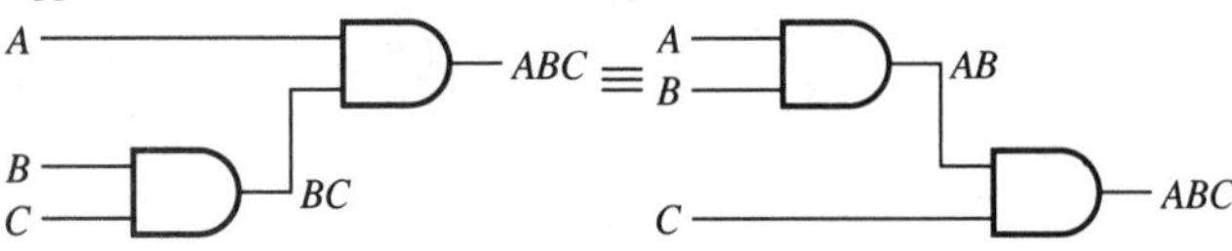

Associative laws

language program into machine language.

assembly language A computer programming language that uses mnemonics to represent the processes of a specific microprocessor. Each statement corresponds to one machine language statement.

assembly language programming The process of developing a computer program in assembly language. See assembly language.

associative law In Boolean algebra, states that it makes no difference how variables are grouped when they are ORed; the results will be the same. Thus, $A + (B + C) = (A + B) + C$. Also applies to Boolean multiplication: $A \cdot (B \cdot C) = (A \cdot B) \cdot C$.

astable multivibrator A multivibrator that switches constantly between two states (usually 0 volts and +5 volts). Used to produce a square-wave output.

asynchronous In electronics, a waveform that is not synchronized with any other quantity. See synchronous.

asynchronous counter A digital circuit acting as a counter that does not have all of its flip-flops synchronized directly by the clock.

asynchronous inputs Control inputs to a digital device that

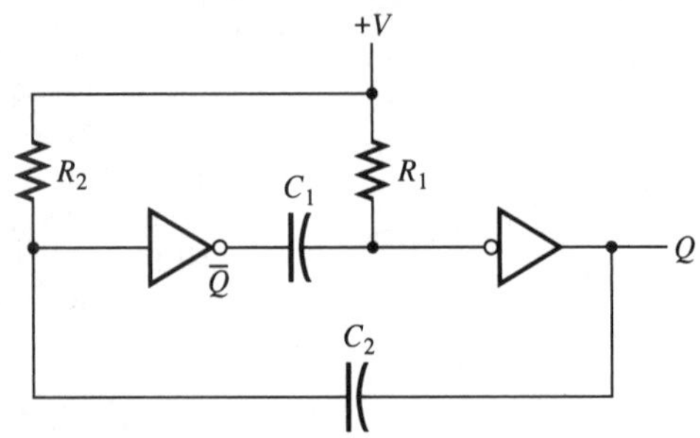

Astable multivibrator

Logic symbol for a J-K flip-flop with active-low preset and clear inputs.

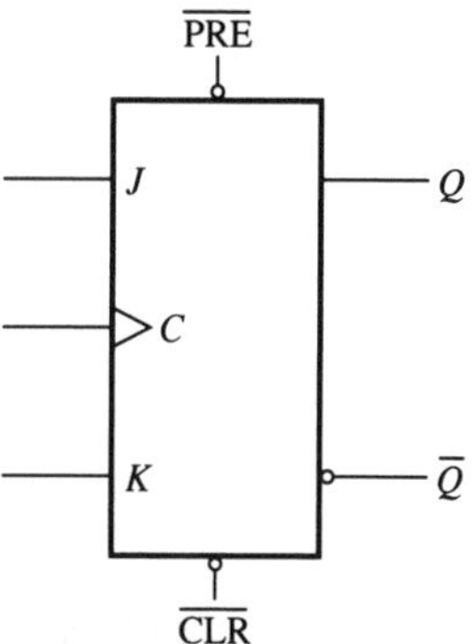

Logic diagram for a basic J-K flip-flop with active-LOW preset and clear.

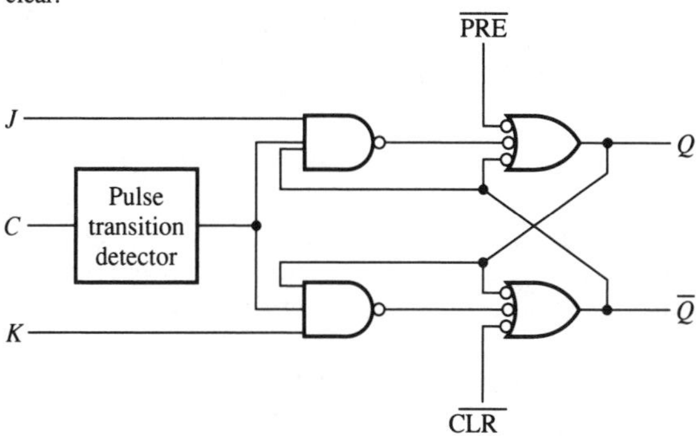

Asynchronous inputs

are independent of any clock or timing pulse. Flip-flops will use asynchronous inputs.

ATM Abbreviation for automatic teller machines. See automatic teller machine.

A-to-D converter Abbreviation for analog-to-digital converter. An electrical device that converts analog information into a digital code.

atom The smallest particle of an element that still possesses the characteristics of that element. In electronics, the atom is visualized as a miniature solar system, where the central nucleus is orbited by tiny particles called electrons. The central nucleus consists of positively charged protons, while the orbiting electrons are negatively charged.

atomic number The value of the number of electrons in a neutral atom.

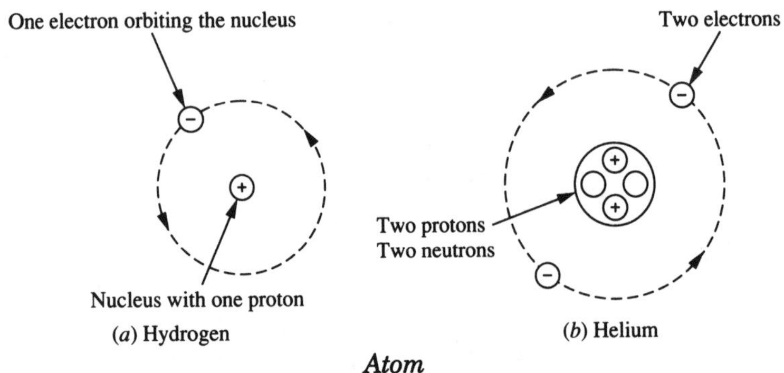

Atom

atomic weight A measure of the number of protons and neutrons in the nucleus of an atom.

attenuation A reduction of the current, voltage, or power of a signal.

attenuator An electrical circuit capable of reducing the strength of a signal. Attenuators may be found on the output circuits of signal generators for the purpose of controlling the strength of the output signal.

audio Pertaining to the frequencies within the range of human hearing. Usually considered in the range of 20 Hz to 20 kHz.

audio amplifier An electronic circuit capable of amplifying frequencies within the range of human hearing. See audio.

automatic fine tuning Circuit that locks in on a received signal without user adjustment. Automatic fine-tuning circuits are usually found in the RF amplifier section of television receivers.

automatic frequency control A method of automatically controlling the frequency of an oscillator. Automatic frequency control is used in communication receivers such as FM and AM radio as well as television to help maintain a stable reception.

automatic gain control An electronic circuit that automatically controls the strength of a signal. In a communications receiver, the automatic gain control controls the gain of the IF amplifier for the purpose of maintaining a given signal strength to the detector.

automatic teller machine A computer terminal that provides 24-hour banking service for customers of the bank. Through the use of a coded plastic card, bank customers may make deposits to and withdrawals from their accounts.

autoranging The ability of an instrument automatically to select the correct scale for reading an electrical variable. For example, an autoranging multimeter will automatically select the best range for reading a resistance, voltage, or current.

autotransformer A transformer with one tapped winding that is used both as the primary and the secondary. See transformer.

avalanche breakdown See reverse breakdown.

average value The area under the curve of a sine wave. The average value of half the sine wave produced from half-wave rectification is given by $V_{avg} = V_P/\pi$. The average value for a full-wave rectifier is twice that of the half-wave.

axis Lines of a graph used to indicate the magnitude of variables represented by the graph. A two-dimensional graph uses two mutually perpendicular lines as the axes. The vertical line is referred to as the *Y*-axis and the horizontal line as the *X*-axis.

backbone In communications, the part of a network that handles the major traffic. It may interconnect multiple locations, and smaller networks may be attached to it.

back current The current that flows when a pn junction is reverse biased. Also called reverse current. Current flow caused by minority carriers.

back diode A tunnel diode that is used in its reverse-current characteristic curve.

backfilling Assigning EMS memory to conventional memory, less than 1 MB in PCs using 8086 and 286 processors. The original motherboard chips are disabled, and the EMS chips are assigned the low-memory addresses. Allows multitasking programs to run more programs concurrently.

background (1) Noninteractive processing in the computer. (2) The base, or background, color on screen, the default of which is black.

background ink A highly reflective ink in optical character recognition systems used to print parts of a form that are not going to be detected by a scanner.

background noise In communications, undesirable random signals present when not receiving any transmitted signal. Background noise is usually in the form of static on a communications receiver.

backing storage With a computer, any external peripheral storage, such as disk or tape. Same as secondary storage or auxiliary storage.

backlit An LCD screen that provides its own light source from the back of the screen. Causes the display to be viewed without the need for ambient light.

backplane 1. The reverse side of a panel or board that contains interconnecting wires. 2. A printed circuit board containing slots, or sockets, for plugging in boards or cables.

back porch In a transmitted television signal, the area of the signal that is between the trailing edge of the horizontal-synchronous pulse and the trailing edge of the blanking pulse. In color transmission, the color synchronous burst is located here.

backup and recovery The use of manual and machine procedures that can help restore lost data in the event of hardware or software loss. The use of backup files, databases, and programs and system logs that keep track of the computer's operations are all part of a backup and recovery program.

backup copy 1. An extra copy of computer data for the purpose of preserving data if the original data should be lost. It is considered good programming practice to make frequent backup copies of the program as it is being developed. This helps to minimize data loss from power failures or lost or damaged storage media. 2. Any disk, tape, or other machine-readable copy of a data or program file.

backup disk The disk that is used to hold duplicate copies of files.

Backus-Naur form Also known as Backus normal form, it was the first metalanguage to define programming languages, developed by John Backus and Peter Naur in 1959.

backward chaining In artificial intelligence, a form of reasoning that starts with the conclusion and works backward. The goal is broken into many subgoals, or sub-subgoals, that can be solved more easily. Also known as the top-down approach.

backward compatible Same as downward compatible.

backward diode A heavily doped germanium pn junction that has a negative-resistance region. Called backward because its easy-current direction is in the negative-voltage region of its characteristic curve. Similar to the tunnel diode.

bad sector A segment of disk that cannot be read or written to because of a physical problem in the disk. Bad sectors on hard disks are marked by the operating system and bypassed. If data are recorded in a sector that becomes bad, special software, and sometimes special hardware, must be used to recover it.

BAK file (BAcKup file) In Microsoft DOS and IBM OS/2, a file extension for backup files.

BAL Abbreviation for basic assembly language; assembly language for the IBM 370/3000/4000 mainframe series.

balanced In electronics, circuits that are electrically alike and are symmetrical with respect to ground.

balanced bridge A bridge circuit where its various components are adjusted so that it has zero output volts.

balanced line A transmission line where two identical conductors are used and operated in such a manner so that there are identical voltages and currents on each line equal in magnitude but opposite in polarity. A balanced line reduces the effects of crosstalk and noise.

balanced modulator An electronic circuit that mixes two frequencies and produces a resultant sum and difference while eliminating the original frequencies on its output. Called a balanced modulator because the result of the carrier frequency is balanced out. Used in single- or double-sideband transmitters.

balanced output A three-conductor output where the signal voltage changes around the third neutral conductor. A balanced output reduces the effects of noise pickup.

ballast lamp A lamp that keeps an almost constant current by increasing its resistance when its applied voltage increases.

balun Acronym meaning *bal*anced to *un*balanced. Usually consisting of a transformer for matching an unbalanced coaxial transmission line to a balanced two-wire system. In television systems a balun coil (transformer) is used to accept a 75-ohm balanced input and deliver the signal to the 300-ohm twin lead conductor.

banana jack A small connector designed in the shape of a banana where its metal sides are under tension to fit tightly into a mating receptacle. Usually used in panel mounted connections.

band Any range of frequencies that lie between two defined limits.

bandpass An electrical filter that allows a continuous group of frequencies to pass and stops the passage of all lower and all higher frequencies.

bandpass amplifier An amplifier especially designed to amplify a specified range of frequencies.

bandpass filter An electrical filter used to pass a specified range of frequencies.

band reject Defined as not passing a specified range of frequencies.

band-reject filter An electrical circuit designed to prevent the passage of a given range of frequencies.

bandstop An electrical filter that prevents a continuous group of frequencies from passing and allows the passage of all lower and all higher frequencies.

bandstop filter An electrical filter used to reject a specified range of frequencies.

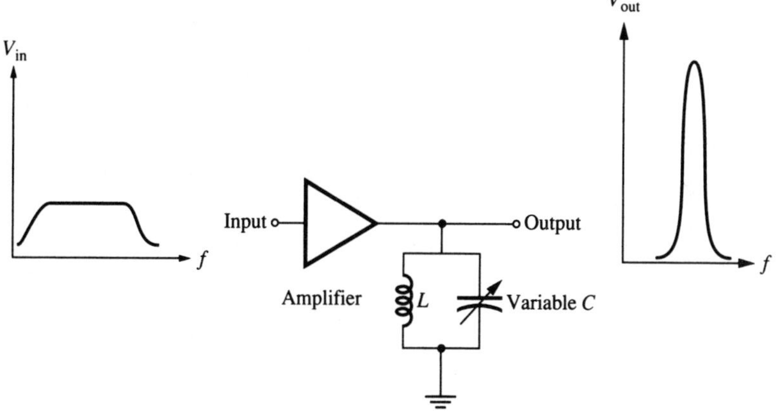

Bandpass amplifier

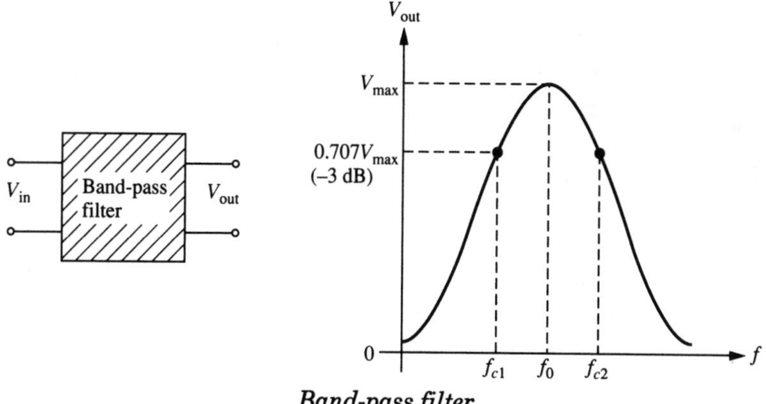

Band-pass filter

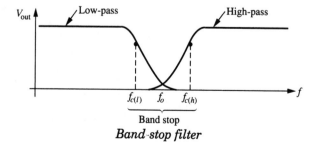

V_{out}
Low-pass
High-pass
f
$f_{c(l)}$ f_o $f_{c(h)}$
Band stop

Band-stop filter

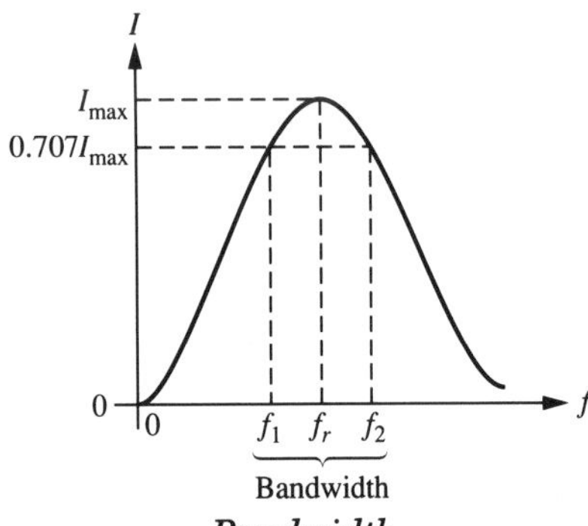

Bandwidth

bandwidth The measurement of a frequency selective circuit where the range of frequencies is measured from 70.7% of its maximum value. $BW = f_2 - f_1$, where f_1 and f_2 are the frequencies and the output is 70.7% of the maximum output. The 70.7% represents the half power or 3-db points.

bank switching Activating and deactivating electronic circuits. Bank switching is used when the design of a system does not allow all circuits from being addressed or activated at the same time, requiring that one unit be turned on while the others are turned off.

bar chart A graphical representation of information in the form of bars. Primarily used in business graphics.

bar code 1. A pattern of vertical bars whose width and separation contain information about a particular item. Bar codes are frequently used on consumer products and contain pricing as well as other information. Usually read with a laser beam. 2. A code used for the rapid identification of items with an optical scanner. Coding of the bar depends upon the width of the bar, not the height. The extended height allows tolerance for the different scanning systems.

Barkhausen criterion As applied to an oscillator. The Barkhausen criterion states that for an oscillator to oscillate, the product of the amplifier gain (A_V) and the amount of signal fed back from the output in phase with the input (B_V) must be equal to unity. Mathematically, this is $A_V B_V = 1$.

Barkhausen effect When the magnetizing force acts on a piece of magnetic material, a series of abrupt changes may occur. These changes are known as the Barkhausen effect.

base 1. The central part of a bipolar transistor. Used to help control the current between the emitter and the collector. The base of a number system is the number of digits used in that number system. For example, the base of the binary number system is 2 because it uses only two symbols, a 0 and a 1. 2. The number of characters used in each position of a number system. For example, the binary number system is to the base 2, because it uses only two symbols, 0 and 1, to represent any value.

base address The location in memory where the beginning of a program is stored. An address relative from the instruction in the program that is added to the base address to derive the absolute address. See base displacement.

base alignment The alignment of different font sizes on a baseline.

base displacement A method used for running programs from any location in memory. The addresses in the machine language program are displacement addresses (relative to the beginning of the program). As the program is running, the hardware adds the displacement address to the base address (where the beginning of the program is currently stored), and from this, the absolute address is determined.

base subscript notation Subscript used to indicate the base of a number. For example, 101_2 indicates that the number is a binary number since the subscript is 2 and is equal to the decimal number 5. The decimal number system is usually written without a subscript; thus, 101 means one hundred one.

BASIC Beginners All-purpose Symbolic Instruction Code. One of the first and most popular computer languages used with microcomputers. Designed by John Kemeny and Thomas Kurtz in

Instruction Word	Description	Example
CLS	Clears video screen	CLS
PRINT	Causes the computer to display or print out any message that follows in quotes or the value of a designated variable or the result of a specified calculation.	PRINT "HELLO" → HELLO PRINT X → Value of X PRINT 2 + 3 → 5
LET	Assigns a value to a variable. Can be omitted in some versions of BASIC for simplicity.	LET X = 8 or simply X = 8 LET Y = X/2 or simply Y = X/2
INPUT	Allows data to be entered into computer, such as variable values.	INPUT X (The computer stops and waits for you to enter a value for *X* from the keyboard.)
GOTO	Causes the computer to branch from its place in the program to a specified line number and skip everything in between.	GOTO 50
FOR/NEXT	These two instruction words are used together to set up loops.	A value for *Y* is calculated for each of three values of *X* (1, 2, and 3). 10 FOR X = 1 TO 3 20 Y = X*2 30 NEXT X
IF/THEN	These two instruction words are used together to set up conditional statements.	IF X = 6 THEN GOTO 50 or IF X = 6 THEN 50

Mathematical Operator	Description	Example
=	Equals	X = 3 (*X* equals 3)
+	Addition	Y = X + 5 (*X* plus 5)
–	Subtraction	A = 10 – X (10 minus *X*)
*	Multiplication	Z = 2*X (2 times *X*)
/	Division	W = Y/4 (*Y* divided by 4)
↑([ ) (**)	Exponentiation	X↑2 (*X* squared)

BASIC

the 1960s as an easy-to-learn interactive computer language.

bass Sounds in the low-audio-frequency range. As measured by the standard piano keyboard, they are all the notes below middle C (261.63 Hz).

bass boost Circuits that cause the audio frequency response of an audio amplifier to give increased emphasis to the bass frequencies (low audio frequencies).

bass control A manual control for emphasizing the amount of low frequencies reproduced by an audio system.

bass-reflex enclosure A container for speakers designed in such a manner that bass notes (lower frequencies) are emphasized.

batch process In electronics, the process of fabricating monolithic resistors, diodes, and capacitors with the same process and at the same time.

batch processing In computers, processing information in such a manner where no interactive communication between program and program user is possible.

battery A voltage source consisting of two or more fundamental energy converting units called cells.

battery changer Normally, an electronic device used to allow house current (120 VAC) to store chemical energy in a battery. Later, this stored chemical energy may then be converted to dc electrical energy.

baud The unit that measures the speed information is transferred. Equal to the reciprocal of the pulse width of the shortest pulse measured in seconds. When using an RS-232 interface, the baud rate is equal to the data rate in bits per second.

BBS Abbreviation for bulletin board system. An interconnection of computers through telephone lines for the purpose of sharing information.

BC Abbreviation for base-collector junction in a transistor.

BCD Abbreviation for binary coded decimal. See binary-coded decimal.

BCD decade counter A digital counter that goes through the counting sequence from 0 to 9. Its binary sequence is 0000_2 to 1001_2.

BCD to decimal A logic circuit that converts each BCD code word (8241 code) into 1 of the 10 possible decimal-digit values. This type of decoder has 4 inputs and 10 outputs.

BCS 1. The Boston Computer Society, one of the largest computer associations in the world. Address: One Center Plaza, Boston, MA 01108. 2. The British Computer Society. Address: BCS, 13 Mansfield St., London, England W1M 0BP.

BE Abbreviation for base-emitter junction in a transistor.

A synchronous BCD decade counter.

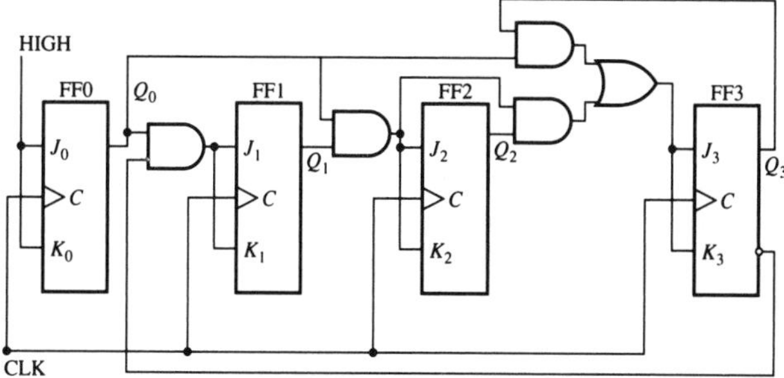

States of a BCD decade counter.

Clock Pulse	Q_3	Q_2	Q_1	Q_0
0	0	0	0	0
1	0	0	0	1
2	0	0	1	0
3	0	0	1	1
4	0	1	0	0
5	0	1	0	1
6	0	1	1	0
7	0	1	1	1
8	1	0	0	0
9	1	0	0	1

BCD decade counter

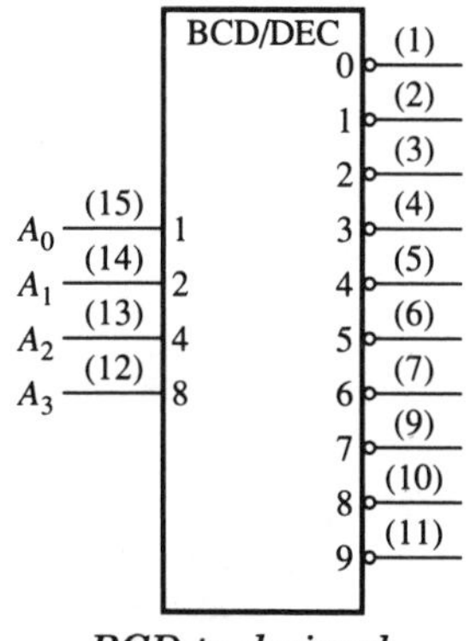

BCD to decimal

beam A parallel flow of electromagnetic or particle energy, such as a beam of light.

beat frequency oscillator Oscillator that produces a signal that is to be combined with another signal to produce the sum and difference frequencies of the two signals.

bell character A control code that is used to sound an audible bell or tone in order to alert the user. In ASCII, it has a numeric value of 7.

Bell Labs The research and development center of the AT&T Company.

below ground A voltage reading that is negative with respect to ground.

benchmark A method of testing the performance of a computer or peripheral equipment. Uses the actual application programs and files that the end user will use.

bench test Checking the performance of an electrical system with laboratory test equipment. A bench test may include testing the system under "normal" operating conditions or may also include testing the system under extreme conditions (such as temperature).

bending loss In fiber optics, signal loss that occurs because of a "kink" in the optical fiber.

Bernoulli box A floppy disk system that works on the Bernoulli principle. Unlike a hard disk on where the read-write head flies over a rigid disk, the Bernoulli disk is a flexible disk that flies up to a rigid head. This effect is named in honor of Daniel Bernoulli, an eighteenth-century Swiss scientist who discovered the fluid dynamic principle upon which this disk system depends.

Bessel filter An electrical frequency selective circuit. Usually used as a high-pass or a low-pass filter. Similar characteristics as in the Butterworth filter, but with a smaller decrease in amplitude beyond the break frequency.

Bessel function A mathematical formula that will indicate the relative strength of a given sideband of an FM signal when the modulation index is known.

Beta 1. The first home VCR format. Developed by Sony, it records video on ½″ tape cassettes. Beta hi-fi added CD-quality audio, and SuperBeta improved the visual image. Beta is no longer in popular production. 2. A television recording and playback system that uses magnetic tape for copying the signal. A system that produces better quality than VHS because of its higher recording speed of 6.9 meters per second.

Beta (β) Symbol for the current gain characteristics of a junction transistor. $\beta = I_C/I_B$, where I_C is the collector current and I_B is the base current.

beta test A test of hardware or software that is performed by users under normal operating conditions. Usually done prior to the release of the software or hardware for general use.

betaware Software that has been provided to a large number of users in advance of the formal release for the purpose of testing the software.

BFO Abbreviation for beat frequency oscillator. See beat frequency oscillator.

***B-H* curve** A curve showing the relationship between the flux density B and the applied magnetizing intensity H for a given magnetic material.

bias Influence. In electronics it is the application of a small dc voltage to an active device such as a diode or transistor to produce specific electrical characteristics. For example, it is the biasing of

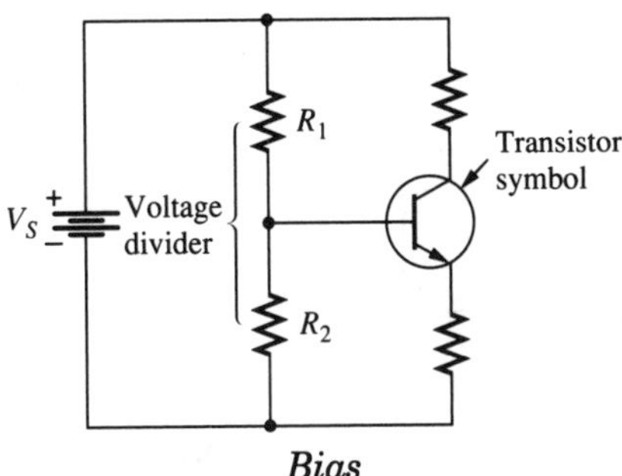

Bias

a transistor that determines its class of operation (class A, class B, etc).

bidirectional Transmission in either direction. An electrical connection is bidirectional if it can transmit information or conduct in either direction.

bidirectional bus A group of wires treated as a unit where data may flow in either direction within this group of wires. In most microprocessor-based digital systems, a bidirectional bus is used to control and communicate with other parts of the system.

bidirectional printer A printer capable of printing as the printer head is scanning the paper in both directions. Faster than a single-direction printer.

bidirectional shift register A shift register that has the capability of shifting its bits in either direction.

bidirectional tristate buffer Two tristate buffers connected in such a manner that the input of one is connected to the output of the other. In this fashion, data can flow in either direction (bidirectional) along a data bus line.

BIFET An electronic device that uses a combination of bipolar junction transistors and JFETs in its internal circuity.

bifilar A winding in which the effects of inductance are reduced by winding two wires that will carry current in opposite directions, tending to cancel out inductance effects.

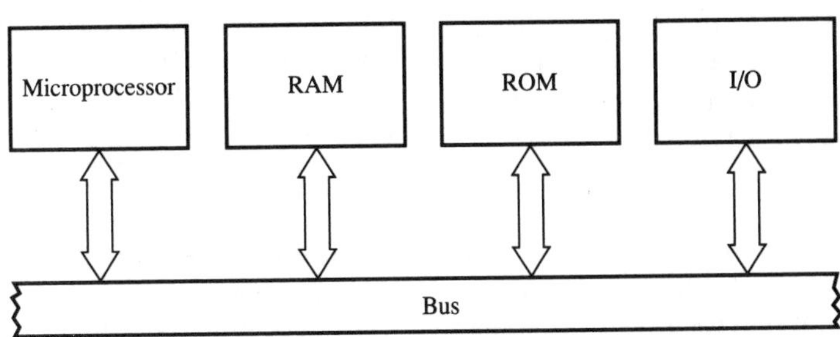

Bidirectional bus

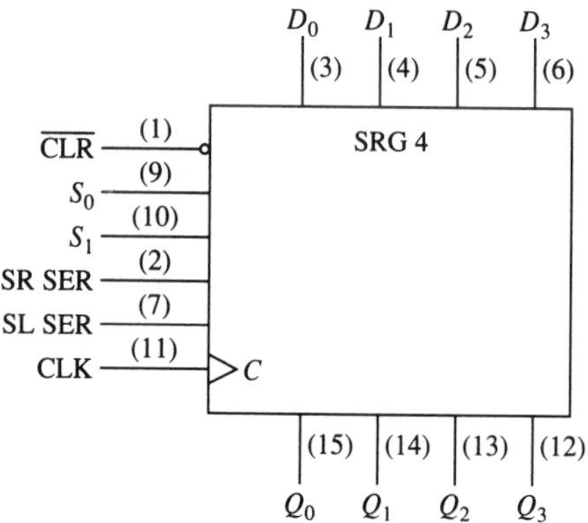

Bidirectional shift register

bilateral In both directions. An example is a bilateral analog switch that is a solid-state switch that may have current flow in either direction when closed.

bilateral switch A solid-state CMOS switch that can be operated in one of two directions and acts as an electrically controlled mechanical SPST switch.

binary addition The process of adding binary (base 2) numbers. Involves only two symbols, a 0 and a 1. In its simplest form 1 + 0 = 1, and 1 + 1 = 10_2. The most elementary form of addition used by a microprocessor.

binary-coded decimal A binary code that allows an easy conversion between itself and the decimal number system.

binary counter An electrical device that produces a binary count on its output for each input of a pulse. As an example, a 4-bit binary counter has a single input and four outputs. The outputs assume a binary count of 0000_2, 0001_2 to and including 1111_2, and then automatically resets back to 0000_2.

binary digit Also called a bit. A character that has the value of either 0 or 1.

binary division Division accomplished in computers by repeated subtraction of binary numbers. Essentially, division is the process of finding how many times one number can be subtracted from another; thus, 10/2 = 5 because 2 can be subtracted from 10 five times.

binary number system The number system to the base 2. Can represent any number value through the use of just two sym-

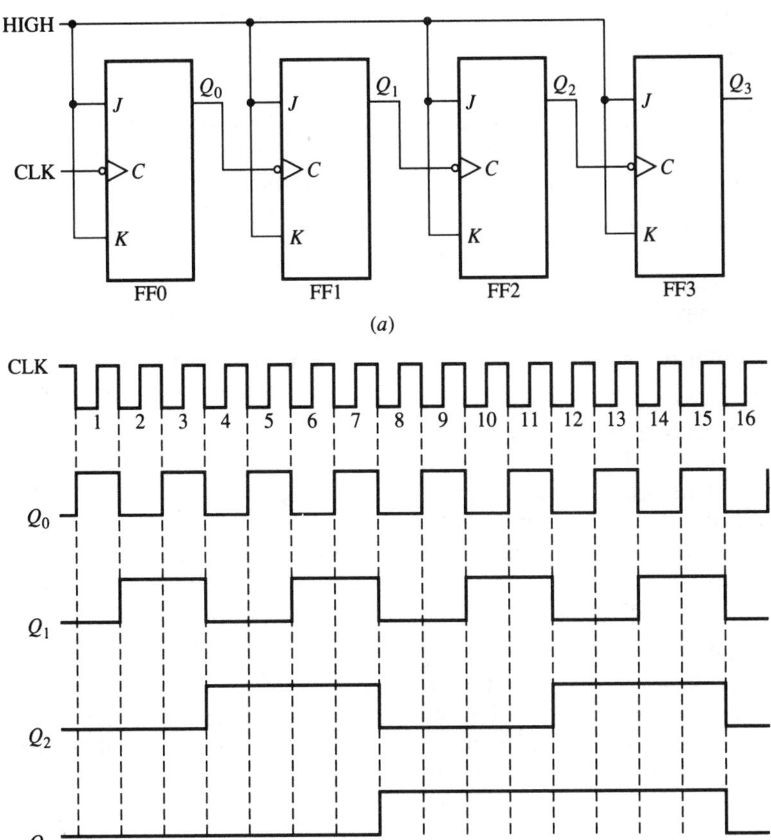

Binary counter

bols: 0 and 1. The most fundamental number system used by all digital computers.

binary search A programming method of locating an item of data from a group of organized data. Accomplished by starting the search in the middle of the data and then determining which half contains the desired data. The process is repeated where each "half" is broken into another "half" until the desired piece of data is found.

binary-to-decimal converter An electrical logic circuit that converts the binary representation of a number into the decimal representation of the same number.

binary-to-hexadecimal converter An electrical logic circuit that converts the binary representation of a number into the hexadecimal representation of the same number.

binary word A group of 1's and 0's treated as a unit. A binary word may be a byte, consisting of 8 bits or larger or smaller.

binding post A terminal for making temporary connections usually consisting of a bolt-and-nut type of structure.

bioengineering See bionics.

bionics The duplication of living systems in the form of mechanical and electrical hardware. Reducing life processes to mathematical terms so that hardware may simulate the process. The study that treats electronic simulation of biological phenomena.

BIOS Basic input-output system. In CP/M that part of the program that controls the input and output of data. In computers, a program in ROM that contains instructions for computer input and output.

bipolar Two pn junctions. As an example, a bipolar transistor has two pn junctions consisting of the emitter and base as one junction and the collector and base as the other junction.

bipolar memories Electrical memories manufactured from bipolar transistors.

bipolar transistor A semiconductor device consisting of three fused parts: the emitter, the base, and the collector. Consists of either an npn type or a pnp type.

bistable To have two stable states. See bistable multivibrator.

bistable multivibrator A multivibrator consisting of two stable states. It is the basic building block for flip-flops.

bit Abbreviation for binary digit. A single binary digit consisting of either a 0 or a 1.

bit density The number of bits of information contained in a given area, such as on a magnetic disk.

bit-mapped A method of displaying graphics on a computer where each picture element (pixel) is specified by the program. This includes the *X*- and *Y*-coordinates as well as the color.

bit rate The number of bits transmitted per unit time. A bit rate of 80 means that 80 bits are transmitted per second.

bits per second The number of binary bits transmitted each second.

blackbody An idealized solid that radiates or absorbs energy with no internal loss of the energy.

black box A term used in electronics to denote a complex circuit where the circuit details are not necessary for input and output analysis. When using a black box concept, you are interested only in the characteristics of the input and output terminals or the effect that the input has on the output.

black noise In a given electromagnetic spectrum, isolated noise spikes in a predominantly noise-free environment. What may be observed when scanning a black surface that may occasionally and randomly contain a few white specks (representing the noise).

blanking The process of canceling the effects of a portion of an electrical signal. For example, a *blanking* pulse is used to stop the retrace of the horizontal oscillator in a CRT display from appearing on the screen.

bleeder current Current that is intentionally caused to flow from a power source in order to help improve the voltage regulation of the power source.

bleeder resistor The resistor used to develop a bleeder current from a power source. May also be a resistor placed in parallel with a capacitor to discharge it safely when the circuit is disconnected from its power source.

block In computers, a continuous group of data stored in successive memory locations.

block diagram A drawing representing the major sections of an electrical system, where the essential details of each section are omitted and only the input and outputs are represented. Used to get an overall view of an electrical system without getting bogged down in the details of the system.

block move The process of moving a large group of data from one computer memory location to another.

board Any flat surface used for making electrical connections.

bode plot A graphical representation of the frequency response of an amplifier where the vertical axis represents the gain in dB and the horizontal axis is a logarithmic measure of the frequency being amplified.

Bohr model A model of the atom that states that it consists of negatively charged electrons orbiting around a stationary nucleus containing positively charged protons. Much the same as the model of the solar system with planets orbiting around the sun. Used as a model to help explain the nature of electricity.

bomb In programming, a computer program that fails completely in its stated purpose. A program that is written to "bomb" the computer system.

Boolean Having to do with the theories and concepts of George Boole, a nineteenth-century mathematician who developed, among other things, a two-valued system of logic. This system can be applied to help analyze the two-state circuits used by digital computers. See Boolean algebra.

Boolean addition Boolean representation of the OR function. Represented by the "+" sign, for example, $1 + 0 = 0$, $1 + 1 = 1$.

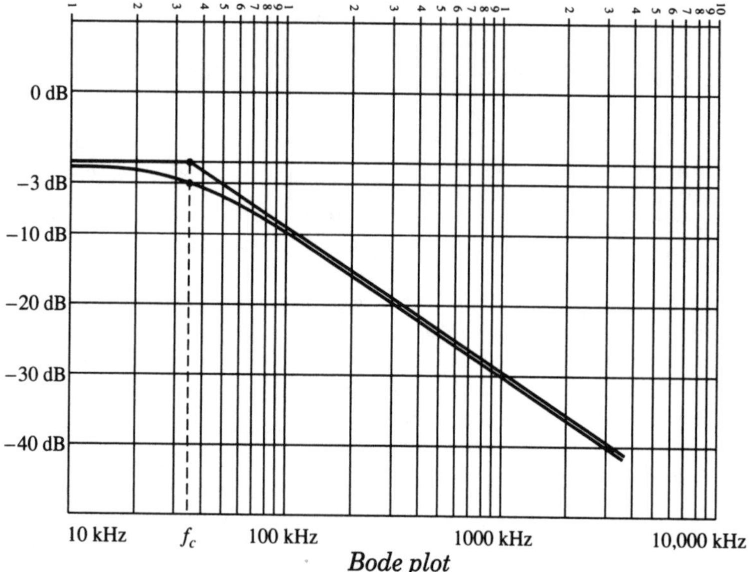

Bode plot

Boolean algebra An algebra developed by George Boole in the nineteenth century. Used to describe a two-level logic consisting of true and false. Used today to help analyze the two-state (high or low) logic circuits found in digital computers.

Boolean analysis The process of using Boolean algebra to predict the output of a logic network. The use of Boolean algebra to simplify a combinational logic network.

Boolean expression An algebraic expression that obeys the two-valued rules of Boolean algebra. In Boolean algebra any variable may have a value of only 1 or 0. Boolean algebra is useful when analyzing digital logic circuits.

Boolean logic The "mathematics of logic," developed by the English mathematician George Boole in the mid nineteenth century. Its rules and operations govern logical functions (TRUE/FALSE) rather than numbers.

Boolean multiplication Boolean representation of the AND function. Represented by the dot (·). Identical to the results achieved when multiplying any combination of 1 and 0.

Boolean search A searching method for specific data that uses the logic of AND, OR, and NOT.

boot 1. Term used to describe the process of turning on a computer so that a program is automatically loaded into it that will allow it to interact with its external environment. Derives from the concept of one "pulling your-

1. $A + 0 = A$
2. $A + 1 = 1$
3. $A \cdot 0 = 0$
4. $A \cdot 1 = A$
5. $A + A = A$
6. $A + \overline{A} = 1$
7. $A \cdot A = A$
8. $A \cdot \overline{A} = 0$
9. $\overline{\overline{A}} = A$
10. $A + AB = A$
11. $A + \overline{A}B = A + B$
12. $(A + B)(A + C) = A + BC$

Note: *A*, *B*, or *C* can represent a single variable or a combination of variables.

Boolean algebra

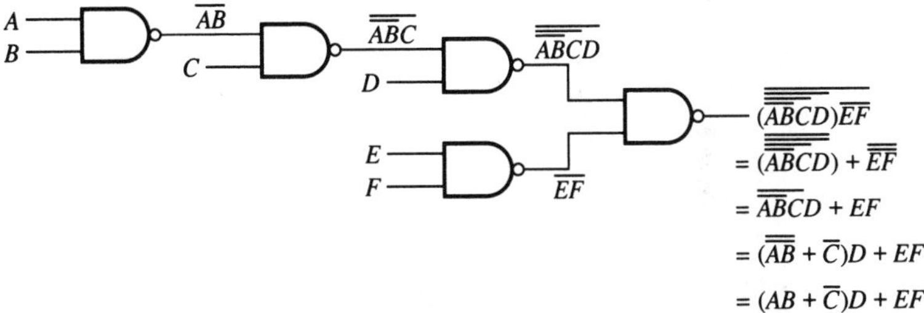

(*a*) Several Boolean steps are required to arrive at final output expression.

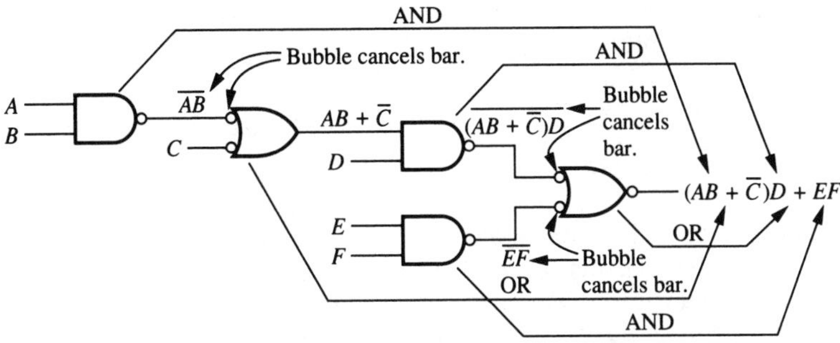

(*b*) Output expression is derived directly from diagram.

Boolean analysis

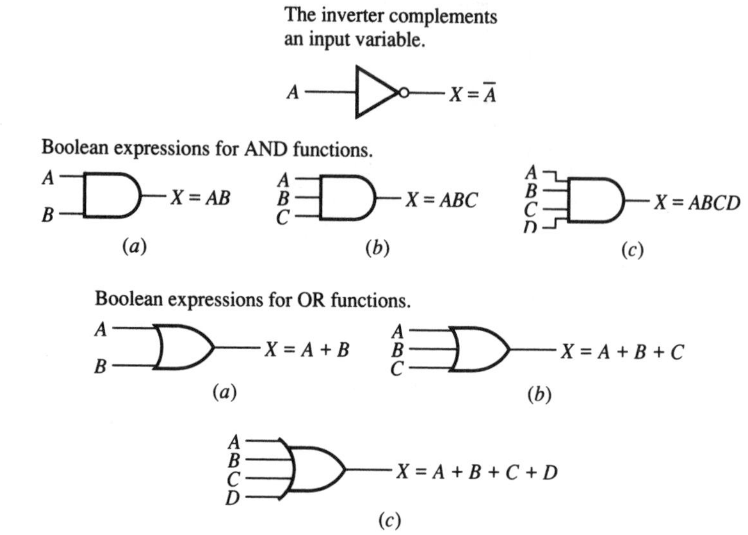

The inverter complements an input variable.

$A \rightarrow X = \overline{A}$

Boolean expressions for AND functions.

(*a*) $X = AB$ (*b*) $X = ABC$ (*c*) $X = ABCD$

Boolean expressions for OR functions.

(*a*) $X = A + B$ (*b*) $X = A + B + C$

(*c*) $X = A + B + C + D$

Boolean expression

self up by your bootstraps." Used to be called "bootstrapping."

bootable disk A disk that contains the operating system. Personal computers normally look for a bootable disk in the primary floppy drive. If a hard-disk system cannot find a bootable floppy upon start-up, it will boot from the hard disk.

boot drive The disk drive that contains the operating system.

boot failure The inability to locate and/or read the operating system from the designated disk.

boot sector The area on a disk that contains instructions and/or data that cause the computer to locate and load the operating system. Usually the first sector of a disk partition.

bootstrap See boot.

bootstrap loader The device for loading in the first initial instructions to a computer that is just activated. These instructions will then usually point to other instructions to complete the process.

Borland Borland International, Inc., a leading microcomputer software company. Developers of Turbo Pascal, C++, Quattro Pro, Paradox, Sidekick +, and many other software packages used by students and professionals the world over.

bounding In electronic amplifiers, bounding is the process of limiting the output range of the amplifier.

bpi Abbreviation for bits per inch. Used to measure the number of bits stored in one inch of a track on a recording surface.

bps 1. Abbreviation for bits per second. See bits per second. 2. Refers to the number of data bits transmitted each second. 3. Used to measure the speed of data transfer in a digital communication system.

braided wire A flexible wire made up of many different strands woven together.

branch 1. A part of a parallel circuit. The action of a computer program that causes execution of another part of the program. 2. An instruction that directs the computer to ignore its next sequential instruction and go elsewhere in the program. 3. A connection between two blocks in a flowchart or two nodes in a network.

branch current The current in part of a circuit that is in parallel with another part.

breadboard A term used to describe a temporary construction of electrical and electronic circuits for the purpose of testing and or modifying a design. Term originally came from the early days of radio construction where a wooden board (similar to wooden boards used for cooling and cutting oven-baked bread) would be used with metal nails for the construction of crystal radio receivers and similar types of electrical equipment.

break To stop temporarily or permanently executing, printing, or transmitting or other program execution.

break frequency In electrical filters, that frequency where the output is 3 dB less then the maximum output of the filter.

"Break" key A key that is pressed to stop the execution of the current program or transmission.

breakout box A device that is connected into a multiline cable and provides terminal connections for testing the signals in a transmission. Breakout boxes may also contain a small light for each line that glows when a signal is present on that line.

breakpoint A place in a computer program that causes the

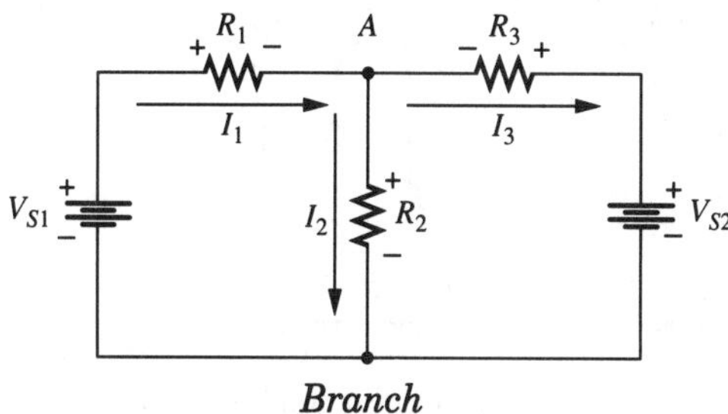

Branch

normal flow of the program to stop for the purpose of manual intervention or visual inspection. Often used in program development to help locate errors.

b-register In a microprocessor, a storage place for an instruction or data.

bridge 1. To cross purposefully or inadvertently from one circuit, channel, or element over to another. 2. A device that connects two networks of the same type together. 3. Any electrical circuit consisting of four components connected in such a fashion to produce some precise electrical measurement. A source of electrical energy is applied across one pair of junctions, and an electrical measurement is made across the other pair of junctions.

bridge circuit simplification A method of converting a bridge circuit into an equivalent series parallel circuit in order to simplify circuit analysis.

bridge network An electrical connection of components in such a manner that the resulting schematic generally has a diamond shaped appearance. A network in which no components are in series or in parallel.

bridge rectifier Four diodes connected in such a manner as to produce a pulsating dc output from an ac source input.

bridgeware Hardware or software that converts data or translates programs from one format into another.

briefcase computer A portable computer that fits inside a briefcase. See laptop computer.

brightness control In a television receiver an adjustment that determines the intensity of the display on the face of the

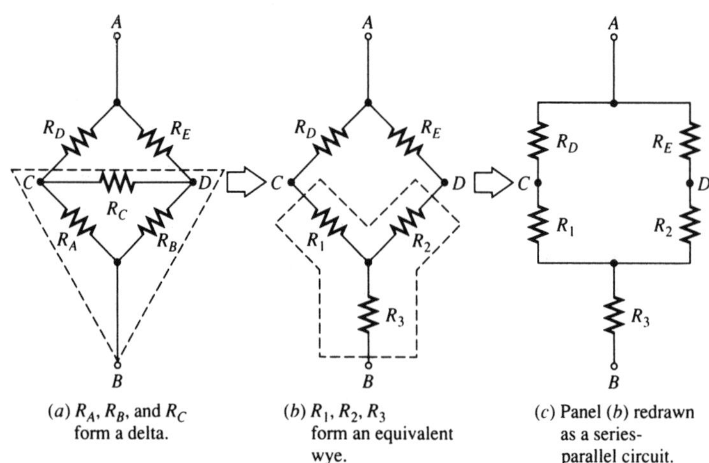

(*a*) R_A, R_B, and R_C form a delta.

(*b*) R_1, R_2, R_3 form an equivalent wye.

(*c*) Panel (*b*) redrawn as a series-parallel circuit.

Bridge circuit simplification

CRT. This control changes the strength of the electron beam; the stronger the beam, the brighter the image.

broadband Having a flat response over a wide range of frequencies.

broadband amplifier An amplifier having essentially flat frequency response over a wide range of frequencies.

broadband antenna An antenna capable of receiving and/or transmitting a wide range of frequencies.

broad tuning Tuned circuits that respond to a wide range of continuous frequencies.

brownout An extended period of insufficient power line voltage. May cause damage to computer equipment and other electrical equipment.

brush As applied to an electrical motor, the parts that make electrical contact between the stationary terminals and the rotating terminals.

brute force An approach to circuit or software design that uses seemingly inefficient methods to achieve an objective.

BTSC Abbreviation for broadcast television systems committee. The committee appointed by the FCC which recommended the first MTS television transmission system. The system allows for stereo audio as well as room for a second audio channel and a professional channel included within each commercial television channel.

bubble memory 1. Nonvolatile memory where data are represented by the presence or absence of magnetized areas (called

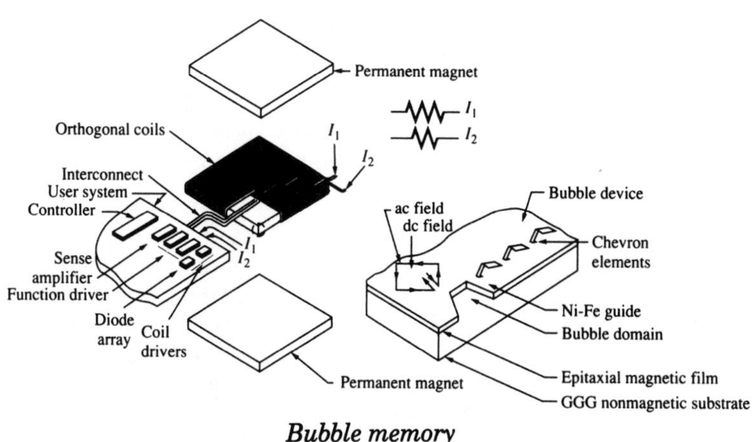

Bubble memory

bubbles) formed on a thin piece of garnet. 2. A form of computer memory that retains its information with the absence of electrical power. Data are represented by the presence or absence of magnetized areas (called bubbles) formed on a thin magnetic film of garnet.

bubble sort A programming method of arranging information in order. The simplest and slowest kind of sorting algorithm where two data samples at a time are compared and a switch is made according to their relative values. In this manner, data are sorted or "bubbled" up to the top of the list.

bubbled input-output A graphical method of showing a logic inversion.

buck To oppose electrically or magnetically, as two like magnetic fields "buck" each other.

buffer An electrical circuit that serves as a power amplifier or to help isolate one circuit from another. In computers a buffer is a place in memory assigned as a temporary holding place for data.

bug 1. In programming, an unintentional error in a program. In hardware, an unintentional error in hardware design. 2. A fault in an electrical circuit. An error in a computer program. Word derives from insects which were attracted to the glow of filaments in the first vacuum tube computers and would get stuck in the mechanical relays.

bulk eraser Electronic device for erasing information quickly from magnetic media.

bulk resistance In semiconductor material, the resistance at a pn junction that is not part of the potential barrier. For most semiconductor devices, the bulk resistance is only a few ohms and can usually be ignored.

burn-in The operation of electrical devices or circuits in order to help stabilize the failure rate. Burn-in is sometimes used to help weed out components that may experience early failure. Sometimes done at extreme temperatures and higher than normal voltages.

burst refresh A method of refreshing DRAM where each cell row is sequentially refreshed. See DRAM.

bus A group of separate conductors treated as a unit. One of the most common uses of a bus in a computer is for the transmission of digital information within the system.

bus driver A circuit used to amplify signals that will be communicated along a group of wires (the bus).

bus oriented Any system that uses groups of conductors as a unit. For example, a microprocessor is a bus-oriented system because it uses groups of wires as a unit, namely, three bus units called the address bus, data bus, and control bus. A fourth bus that supplies power to the microprocessor is usually counted separately from the other three buses because this fourth bus does not carry any data.

Butterworth filter 1. A filter that exhibits the flattest response in its passband. 2. An electrical circuit that is frequency selective. A Butterworth filter approaches a constant slope of 6 dB per octave.

bypass capacitor A capacitor connected in parallel with another circuit element to provide a low-impedance path for an electrical signal.

byte A group of 8 binary bits treated as a unit.

byte-organized memory Digital memory where the data size is 1 byte (8 bits).

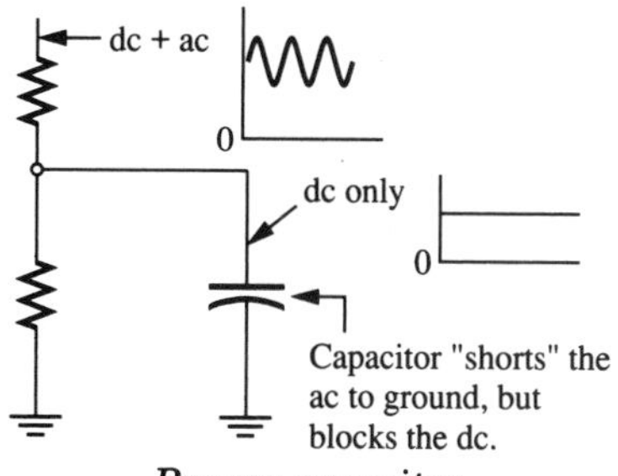

Bypass capacitor

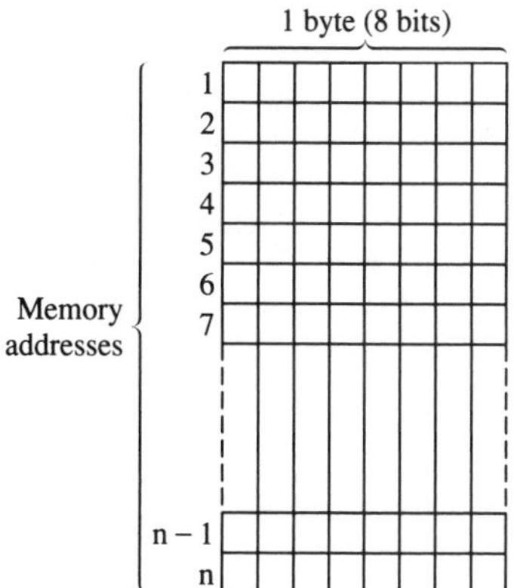

Byte-organized memory

C

C 1. A programming language developed in the 1970s at Bell Laboratories by Dennis Ritchie. For programmers where complete control of the computer is possible. Can also be given some of the characteristics and syntax of other languages. 2. Letter symbol for coloumb. 3. Abbreviation for capacitor.

C++ Programming language that is an object- oriented subset of the C programming language developed at AT&T Bell Laboratories by Bjarne Stroustrup. Used by programmers wishing to market programming tools that may be modified by the end user without direct access to the original source code. See object-oriented programming.

CAD Abbreviation for computer-aided design. See computer-aided design.

capacitance The ability to store an electrical charge. Measured in farads (F). Capacitance may be defined as $I = C(dE/dt)$, where I is the charging current in amps, dE is the change in applied voltage across the capacitor, and dt is the change in time during which the applied voltage is changing.

capacitive susceptance (B_C) The reciprocal of capacitive reactance, measured in siemens (S). Mathematically, capacitive susceptance is $B_c = 1/X_c$, where X_c is the capacitive reactance (in ohms).

capacitor An electrical element used to store electrical energy. Unit of measurement is the farad (F). Abbreviated as *C*. In its simplest form it can be thought of as two metal plates separated by an insulator called the dielectric.

capacitor charge The storage of electricity in the plates of a capacitor.

capacitor charging The process of placing a charge on the plates of a capacitor.

capacitor checking A method of testing a capacitor using the ohmmeter.

capacitor discharging The process of removing the charge from the plates of a capacitor.

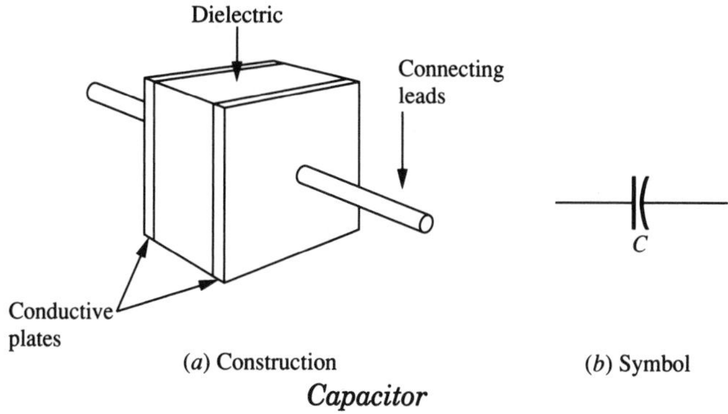

(*a*) Construction (*b*) Symbol

Capacitor

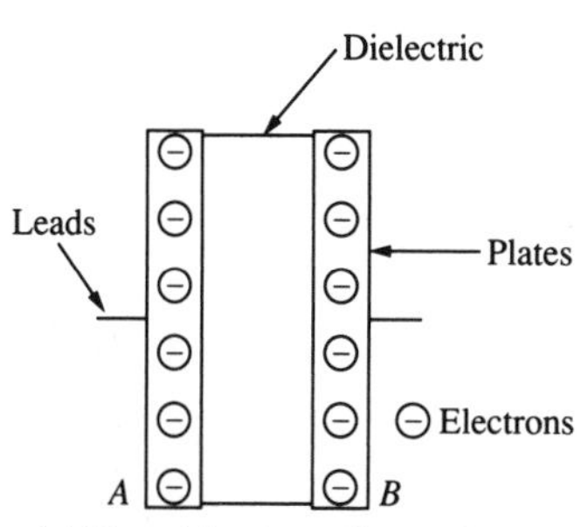

(*a*) Neutral (uncharged) capacitor (same charge on both plates).

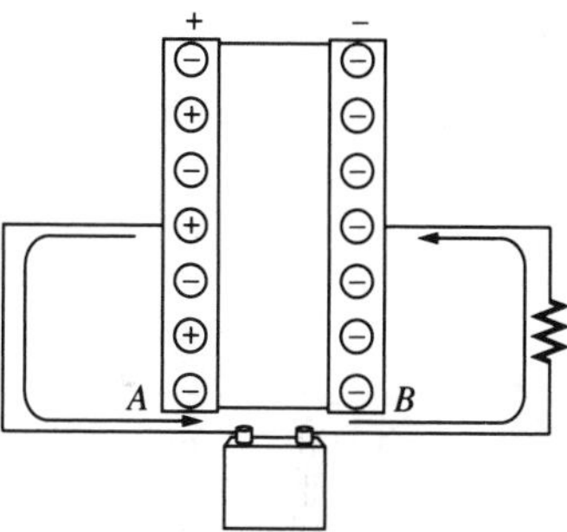

(*b*) Electrons flow from plate *A* to plate *B* as capacitor charges.

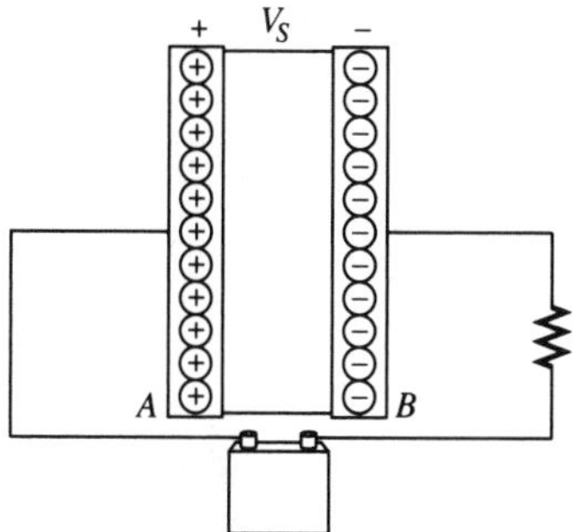

(*c*) Capacitor charged to V_S. No more electrons flow.

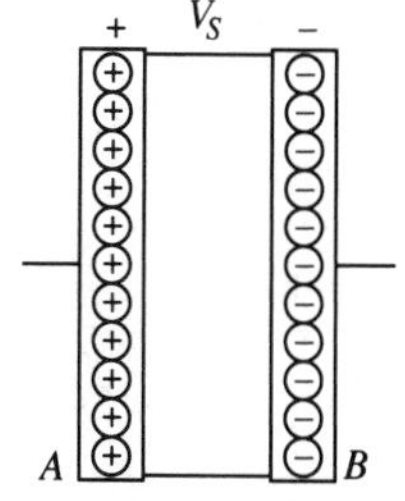

(*d*) Capacitor retains charge when disconnected from source.

Capacitor charge

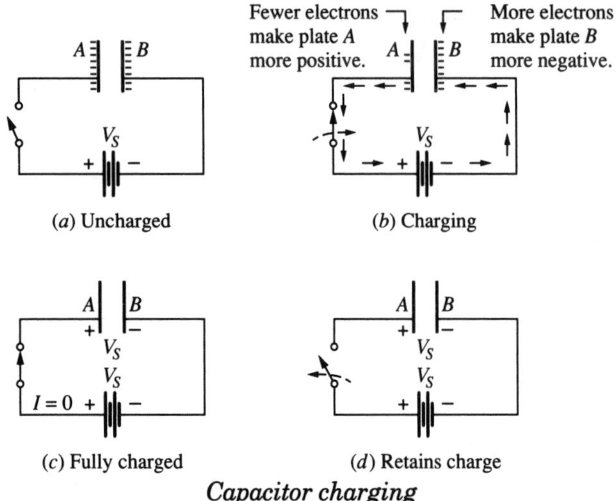

(*a*) Uncharged (*b*) Charging

(*c*) Fully charged (*d*) Retains charge

Capacitor charging

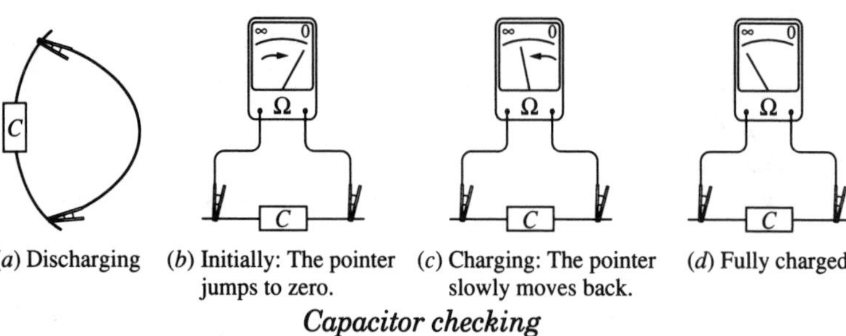

(*a*) Discharging (*b*) Initially: The pointer jumps to zero. (*c*) Charging: The pointer slowly moves back. (*d*) Fully charged

Capacitor checking

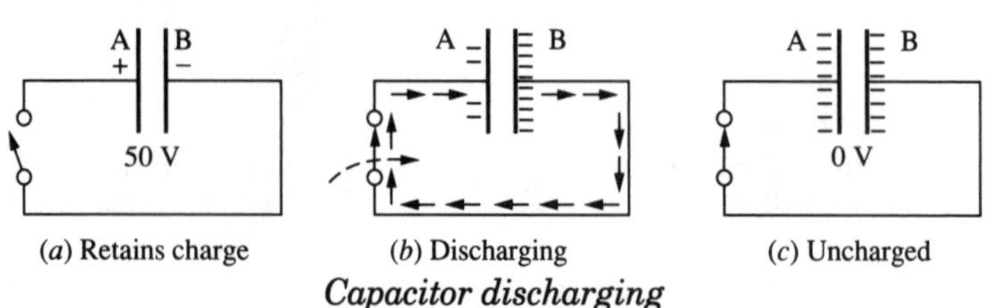

(*a*) Retains charge (*b*) Discharging (*c*) Uncharged

Capacitor discharging

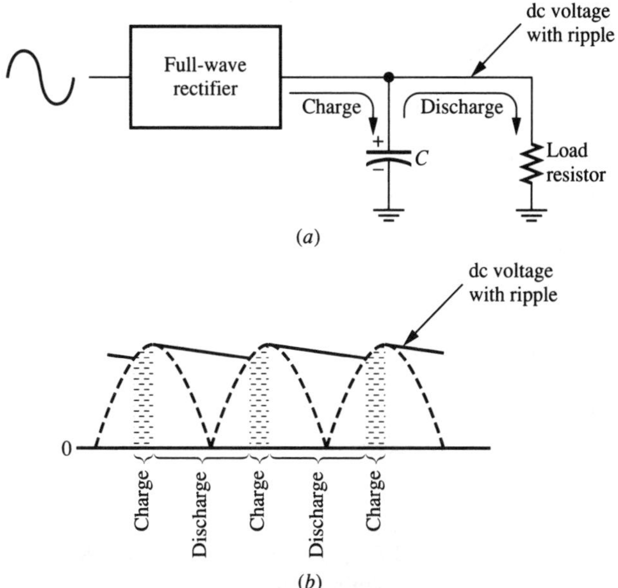

Capacitor filter

capacitor filter See capacitor-input filter.

capacitor-input filter An electrical circuit designed to remove fluctuations in voltage level consisting of capacitors and resistors and/or inductors. The input to a capacitor-input filter consists of a capacitor in parallel with the filter. See pi filter.

capacitor phase In a capacitor, the current leads the voltage by 90 degrees.

carbon composition resistor See carbon resistor.

carbon resistor The most common kind of fixed resistors. Carbon resistors are constructed from a combination of granulated carbon mixed with a ceramic binder. This process results in a low-cost, reliable resistor used for most all electronic applications.

card reader A device used to read cards. In the early days of computers, card readers were electromechanical devices that read holes punched into envelope-sized cards. The patterns represented by these holes represented data as well as computer programming instructions.

cardioid microphone A microphone that has a heart-shaped pickup pattern. A cardioid microphone will have a uniform pickup pattern for about 180 degrees and minimum response in the remaining direction.

carrier In electronics, the waveform or signal that is modu-

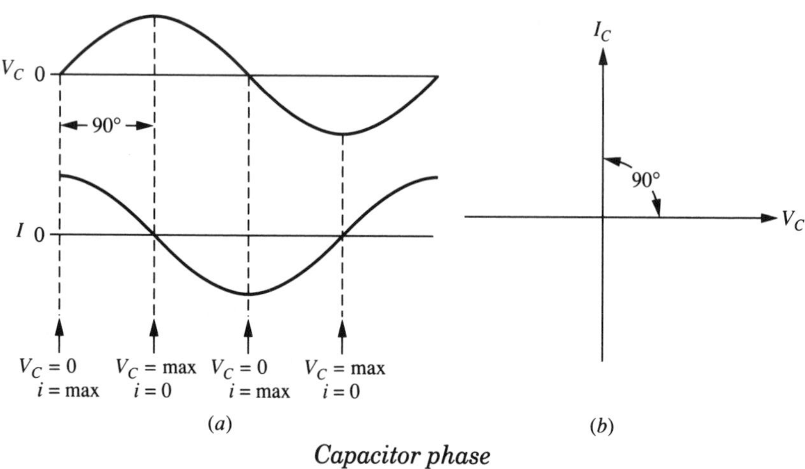

Capacitor phase

lated. The signal that will *carry* the information to be transmitted. For example, in FM transmission, the carrier is a radio frequency sine wave that will have its frequency characteristics changed according to the amplitude and frequency of the lower-frequency information signal (usually music or voice).

carrier reinsertion The process of reinserting the carrier frequency in a single-sideband (or double-sideband) receiver. Since the carrier is not transmitted in a sideband system, it must be reinserted at the carrier. Doing the reinsertion is the job of the carrier reinsertion oscillator.

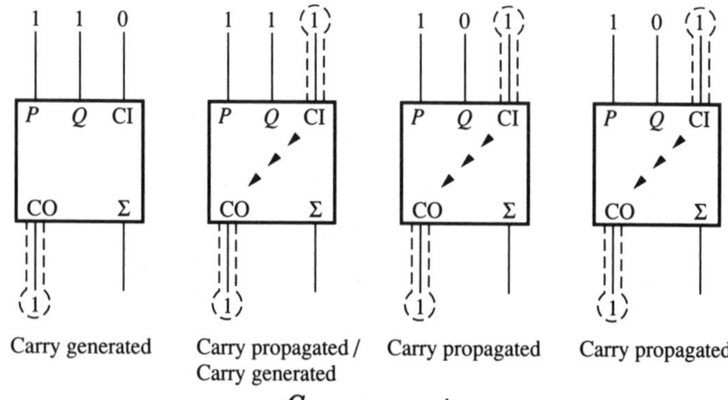

Carry generate

carry generate The process of having a logic circuit capable of adding binary numbers generate a carry bit throughout the logic adding circuits.

cascade Arranging two or more circuits so that an electrical signal travels serially from one circuit to the other. Also called tandem.

cascade amplifier Arranging two or more amplifiers so that an electrical signal travels serially from one amplifier to the other.

cascaded counter Connecting two or more counters in order to increase the full count. To cascade a counter, the output of one is connected to the input of the other.

cascode Arranging two or more circuits so that an electrical signal is applied to all circuits in parallel (at the same time).

cascode amplifier Arranging two or more amplifiers so that an electrical signal is applied to all amplifiers in parallel (at the same time).

CAT 1. Abbreviation for computer-aided troubleshooting. See computer-aided troubleshooting.

Two cascaded counters (all J, K inputs are HIGH).

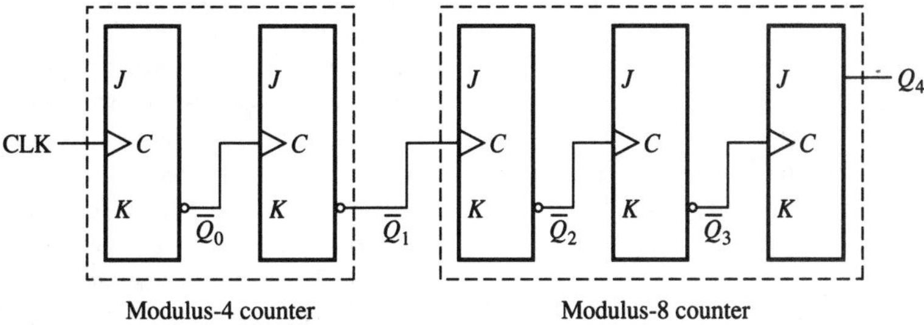

Timing diagram for a counter configuration.

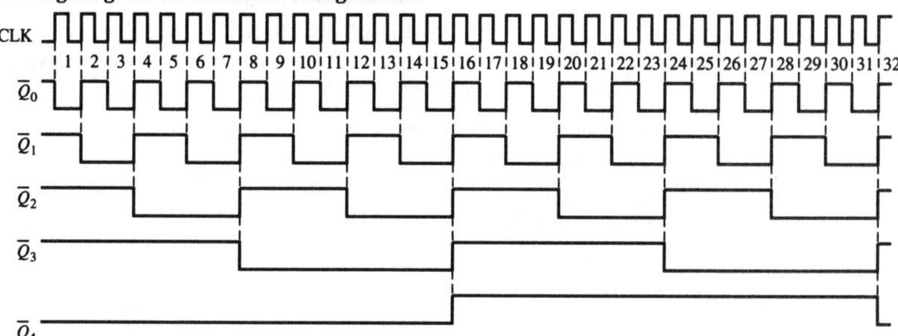

Cascaded counter

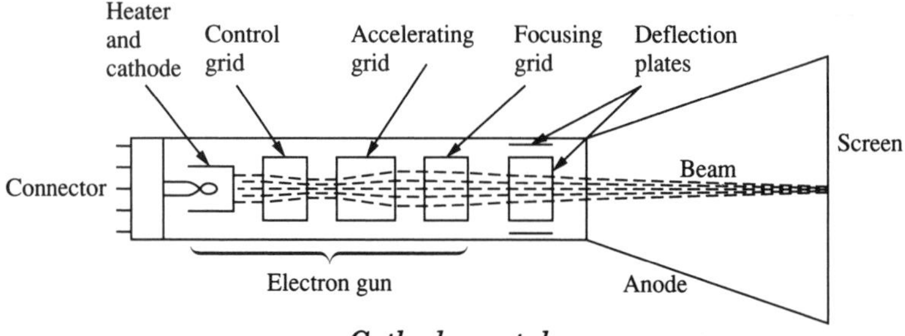

Cathode ray tube

2. Abbreviation for computer-aided tomography or computer-assisted tomography or computerized axial tomography. Means sectional graphics. Produces a graphical display of cross sections of the human body.

catastrophic failure A sudden failure of an electrical system. One of the causes of catastrophic failure is a power outage.

cathode That part of an electrical device that is to have a negative voltage applied in respect to its anode in order to have current flow.

cathode bias In a vacuum tube amplifier, using a resistor between the cathode and ground to be called the cathode resistor. The voltage developed across this resistor causes the cathode to be more positive than ground. If the control grid is tied to ground potential, this keeps the dc operating characteristics of the vacuum tube so that the control grid is more negative than the slightly positive cathode.

cathode follower See common plate.

cathode ray tube An electrical device that emits a stream of electrons for the purpose of displaying visual information on a glass faceplate. The glass faceplate is an extension of a glass tube that houses the source of electrons from a device called the electron gun. The whole assembly is placed under a vacuum, and the inside of the glass faceplate is coated with a phosphor that glows when struck by the electron stream. An electronic deflection system is used to control the position of the electron stream. Since the electrons are emitted from the cathode, they are referred to as cathode rays.

cavity wavemeter An adjustable waveguide whose physical dimensions may be changed to adjust its resonant frequency. Used to measure the frequency of waves inside the waveguide.

CB amplifier Abbreviation for common-base amplifier. See common base.

CC amplifier Abbreviation for common-collector amplifier. See common collector.

CCCS Abbreviation for current-controlled current source. See current-controlled current source.

CCD memory Abbreviation for charge-coupled device memory. Memory where charges are stored between tiny electrical plates fabricated in a high-density integrated circuit.

CCVS Abbreviation for current-controlled voltage source. See current-controlled voltage source.

CD Abbreviation for compact disk. See compact disk.

CE amplifier Abbreviation for common-emitter amplifier. See common emitter.

cell A basic physical unit that can supply electrical energy such as a chemical battery cell or a solar cell. Cells can be placed in series to increase the total available voltage. (See pg. 51 for illustration.)

cemf Abbreviation for counter-electromotive force. Produced by an inductance that opposes the applied voltage, as found in electrical motors.

center tap An electrical connection in the center of a transformer winding. An electrical connection that causes an equal division of electrical properties; a center-tapped resistor is a resistor with two halves of equal resistance.

center-tapped transformer A transformer that has either or both of its primary and secondary windings with an extra connection (tap) in the electrical center of the winding. Usually, a center tapped transformer will have only its secondary winding tapped in this manner.

central processing unit That part of a computer that performs all the arithmetic and logic functions. The central processing unit consists of the microprocessor and its supporting chips. It is connected to the rest of the microcomputer through three buses: the data bus, the address bus, and the control bus.

ceramic capacitor A capacitor constructed by using a ceramic dielectric.

CGS system The system of units that uses the *centimeter*, *gram*, and the *second* as the fundamental units of measurement.

channel 1. In television systems an assigned transmitting frequency. Each television channel has a bandwidth of 6 MHz. For example, the bandwidth of channel 2 is from 54 to 60 MHz. 2. On a field-effect transistor. See field-effect transistor.

characteristic impedance The impedance presented by a transmission line that is of theoretically infinite length. All transmission lines have a characteristic impedance. For example, the characteristic impedance of a standard TV lean-in is 300 Ω.

charge 1. The measurable phenomena of electricity. Produced

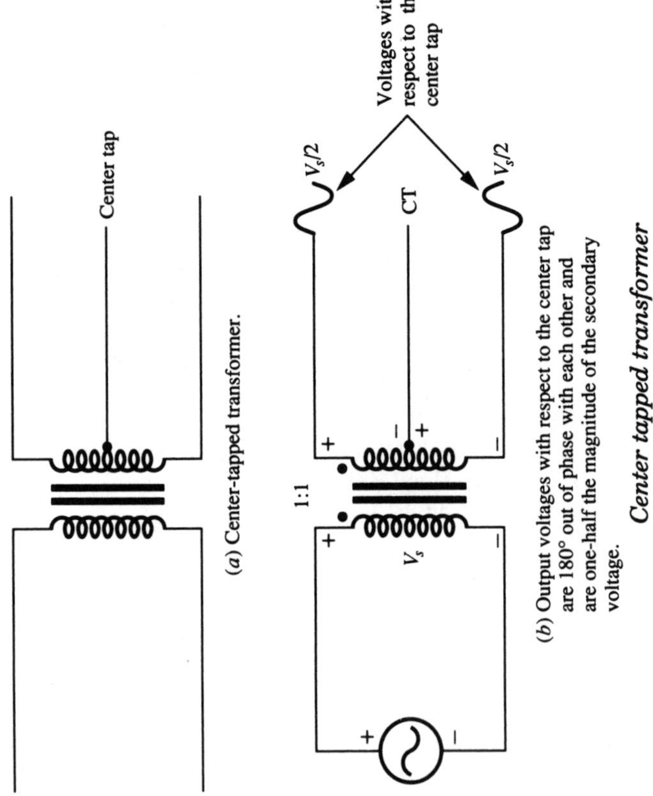

(*a*) Center-tapped transformer.

(*b*) Output voltages with respect to the center tap are 180° out of phase with each other and are one-half the magnitude of the secondary voltage.

Center tapped transformer

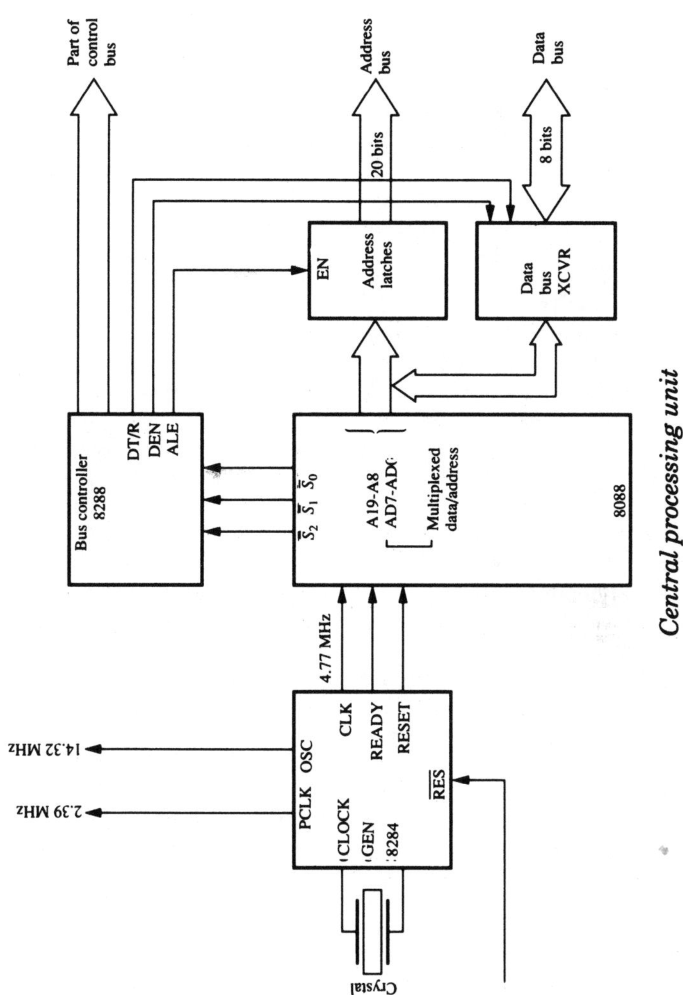

Central processing unit

by the absence of electrons (+ charge) or an excess of electrons (− charge). 2. To cause such an absence or excess to occur, such as to charge a capacitor. See coloumb.

charging lead-acid cell The process of converting electrical energy into chemical energy in a lead-acid cell. The process converts external electrical charges into chemical ions contained within the electrolyte.

Chebyshev filter An electrical frequency selective circuit. Commonly used as a high-pass or a low-pass filter. It exhibits more of a ripple in the range of frequencies it passes than do other types of filters.

chemical cell The smallest element of a chemical voltage source. As an example, a copper-zinc chemical where one electrode is made of copper and the other of zinc. These electrodes are immersed in a solution of water and hydrochloric acid which is the electrolyte.

chip A common term used to describe an integrated circuit. See integrated circuit.

choke A name given to an inductor used to reduce variations in a dc voltage source. A choke is used in power supplies to help reduce any line frequency fluctuations on the dc output.

choke-input filter See *LC* filter.

chroma bandpass amplifier In a color television receiver, the circuit that separates the high-frequency color (chrominance) signal from the rest of the composite video signal received from the video amplifier and amplifies it.

chroma oscillator In a color television receiver, a crystal-controlled 3.58-MHz oscillator used as the subcarrier that was suppressed at the transmitter. Its phase acts as the reference for the incoming color signals.

chroma phase detector In a color television receiver, a circuit that ensures that the 3.58-MHz oscillator is in phase with the transmitted color reference signal.

chrominance signals In a color television receiver, the resulting signals that contain color information. The phase relationships determine the color being presented.

circuit A connection of two or more electrical elements with the intention of serving some useful purpose.

circuit breaker A safety device similar to a fuse that will open when a specified current value is exceeded. A circuit breaker, unlike a fuse, can be reset to be used again.

circular mil The cross-sectional area of a wire having a diameter of one mil.

circular waveguide A waveguide in the form of a tube. Circular waveguides are used whenever a rotating element, such as a radar antenna, must be attached to the communications system using the waveguide.

clamper An electrical circuit designed to restore a specified dc level to a given signal. Sometimes called dc restorers.

Clapp oscillator A circuit that generates its own signal. Uses a capacitive voltage divider and an additional capacitor in series with an inductor. Resonant frequency is determined by the equivalent capacitance and value of the inductor.

class A amplifier An amplifier that is biased in such a manner as to allow a continuous current flow during all portions of the input signal. Class A amplifiers are commonly used as audio amplifiers.

class AB amplifier An amplifier that is biased halfway between a class A and a class B amplifier. See class A amplifier. See class B amplifier.

class B amplifier An amplifier that ideally is biased at cutoff. In an ideal class B amplifier, current flows for only 180 degrees of the input signal. Class B amplifiers are sometimes used in a push-pull arrangement on the outputs of audio amplifiers.

class C amplifier An amplifier that is biased below cutoff. A class C amplifier will conduct for less than 180 degrees of the input signal.

clipper See limiter.

clock In computer circuits the timing waveforms or circuits used to control the timing of the computer or other such similar system.

clocked sequential circuits In digital circuits, consists of a combinational logic section and a memory section. A clocked input is available to the memory section. At any given time, the memory is in a state called the "present state" and will advance to the "next state" on the next clock pulse as determined by specified conditions.

clock pulse In computer circuits a periodic waveform used to time the operations of a digital computer or other such device.

closed circuit A circuit in which the current has a complete path through which to flow.

closed loop 1. An electrical configuration where the output is connected back to the input. 2. In an electrical circuit, a continuous connection of circuit branches that allows tracing of a path that leaves a point in one direction and returns to the same point from another direction within the same circuit.

closed-loop gain The gain of an amplifier when feedback is present.

CM Abbreviation for circular mil. See circular mil.

CMOS 1. Abbreviation for complementary metal-oxide semiconductor. Digital circuits that have low-power dissipation compared to TTL. Manufactured as n-channel and p-channel enhancement mode devices on a silicon chip in a push-pull configuration. 2. Where TTL circuits use bipolar transistors, CMOS use field-effect transistors. The logic functions available are the same as for TTL. The main differences are their performance characteristics.

cmrr Abbreviation for common-mode rejection ratio. See common-mode rejection ratio.

coaxial cable A conductor consisting of a single solid wire surrounded by a polyethylene dielectric that is covered by a braided wire shield. The whole assembly is enclosed in a plastic jacket. A coaxial cable is used to help shield signals from stray interference.

coercivity A measure of the magnetic field that must be applied in order to cause the magnetization in the medium to reverse direction.

coherence When light waves are in phase with each other.

coherent light Light having a single wavelength.

collector That part of a transistor that is designed for power dissipation. A transistor has three leads: the emitter, the base, and the collector. The collector is drawn opposite the emitter and without an arrow (the emitter contains an arrow symbol, the direction of which determines the type of the transistor).

collector feedback bias In a transistor amplifier, a method of establishing the dc operating point by connecting the base through a resistor to the collector. This arrangement provides negative feedback to the amplifier.

collinear array An antenna constructed in such a manner that half-wave dipoles are placed in line with each other. The resulting radiation pattern is very directional.

color code See color coding.

color coding A method used to indicate the value of a device through the use of colors to represent the value and other information.

color control In a color television receiver, controls the amplitude of the chrominance signal from the bandpass amplifier. Determines the amount of color intensity (saturation) presented on the screen.

color killer In a color television receiver, the circuit that

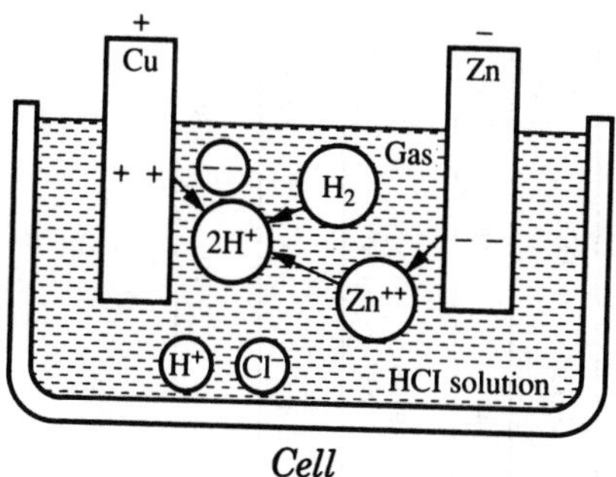

Cell

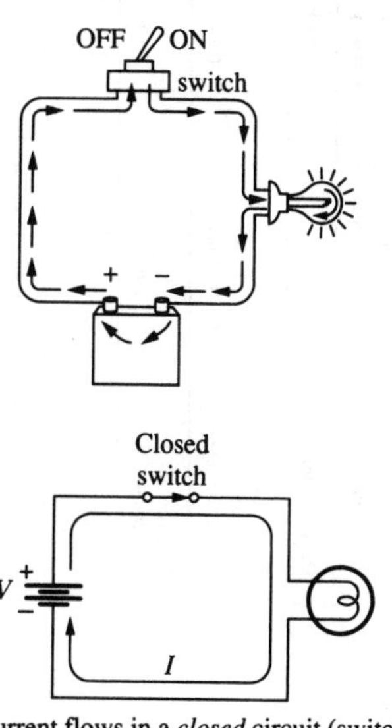

Current flows in a *closed* circuit (switch is ON or in the *closed* position).

Closed circuit

Technology	CMOS* (silicon-gate)	CMOS* (metal-gate)	TTL Std.	TTL LS	TTL S	TTL ALS	TTL AS
Device series	74 HC	4000B	74	74LS	74S	74ALS	74AS
Power dissipation							
Static	2.5 nW	1 μW	10 mW	2mW	19 mW	1 mW	8.5 mW
@ 100 kHz	0.17 mW	0.1 mW	10 mW	2mW	19 mW	1 mW	8.5 mW
Propagation delay time	8 ns	50 ns	10 ns	10 ns	3 ns	4 ns	1.5 ns
Fan-out (same series)			10	20	20	20	40

*Propagation delay is dependent on V_{CC}. Power dissipation and fan-out are a function of frequency.

CMOS

Color-code bands on a resistor.

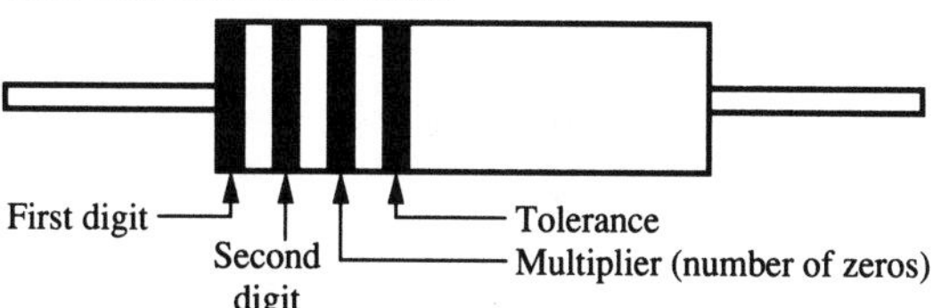

Resistor color code.

	Digit	Color
Resistance value, first three bands	0	Black
	1	Brown
	2	Red
	3	Orange
	4	Yellow
	5	Green
	6	Blue
	7	Violet
	8	Gray
	9	White
Tolerance, fourth band	5%	Gold
	10%	Silver
	20%	No band

Color code

ensures that no color appears on the screen while viewing a black and white signal. The color killer turns off the bandpass amplifier when there is no color signal.

color phase detector See chroma phase detector.

color sync burst In a television signal a short signal sent during the horizontal blanking pulse to provide color information to the television receiver. The color sync burst rides on the back porch of each horizontal blanking pulse and has a frequency of about 3.58 MHz. The signal is phase modulated to contain color information.

Colpitts oscillator A circuit that generates its own signal. Uses a capacitive voltage divider in parallel with an inductor. Resonant frequency is determined by equivalent capacitance value and inductor value.

combinational logic In a digital logic circuit, a digital logic circuit made up of two or more separate logic gates.

common base A unijunction transistor amplifier where the input signal is applied between the emitter and the base and the output signal is taken from between the collector and the base. Input signal and output signal are in phase, with a small input

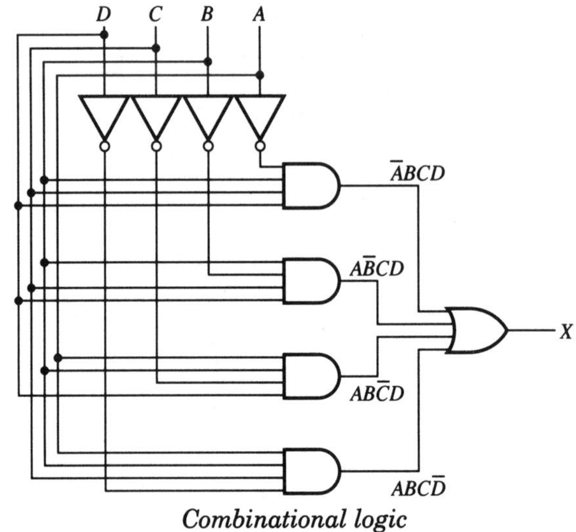

Combinational logic

impedance and large output impedance compared to other amplifier configurations.

common cathode A vacuum tube amplifier where the input signal is applied between the grid and the cathode and the output signal is taken from between the plate and the cathode. Similar signal characteristics to that of the common emitter.

common collector A unijunction transistor amplifier where the collector is electrically common to both the input and output signals. Input signal and output signal are in phase. Has a large input impedance and small output impedance compared to other amplifier configurations. Voltage gain is less than 1, but because of large current gain, configuration produces a large power gain. Usually referred to as an emitter follower.

common drain A field-effect transistor amplifier where the drain is electrically common to both the input and output signals. Similar signal characteristics to that of the common collector. Usually called a source follower.

common emitter A unijunction transistor amplifier where the input signal is applied between the base and emitter and the output signal is taken from between the collector and emitter. Input signal and output signal are 180 degrees out of phase, with a medium input and output impedance when compared to other amplifier configurations.

common gate A field-effect transistor amplifier where the input signal is applied between the source and the gate and the output signal is taken from between the drain and the gate. Similar signal characteristics to that of the common base.

common grid A vacuum tube amplifier where the input signal is applied between the cathode and the grid and the output signal is taken from between the plate and the grid. Signal characteristics similar to those of the common base.

common-mode gain Measure of the ratio of change in output voltage to a change in common-mode voltage. See common-mode voltage.

common-mode rejection ratio In operational amplifiers, the ratio of the open-loop gain to the common-mode gain. Common-mode rejection is a measure of the op amp's ability to reject common-mode signals (noise).

common-mode voltage Voltage applied to both inputs of a differential amplifier. The applied voltages are in phase and of equal frequency and amplitude.

common plate A vacuum tube amplifier where the plate is electrically common to both the input signal and the output signals. Similar signal characteristics to those of the common collector. Usually called a cathode follower.

common source A field-effect transistor amplifier where the input signal is applied between the gate and the source and the output signal is taken from between the drain and the source. Similar signal characteristics to that of the common emitter.

common source amplifier For a field-effect transistor, similar in characteristics to the common emitter transistor amplifier. See common emitter.

communications 1. Pertaining to electronics it is the general field of electronics concerning the equipment and devices used for the transmission and reception of information by electromagnetic radiation. 2. In electronics, the conveying of information from one point to another using electrical means. The most common forms of electronic communications are the radio, television, and telephone. The radio and television use electromagnetic radiation to transmit information (such as voice, images, and/or music), whereas the telephone uses both electromagnetic radiation and electrical wires or fiber optics.

commutative law In Boolean algebra states that it makes no difference in which order variables are ORed; the result is the same. Thus, $A + B = B + A$. Also applies to Boolean multiplication: $A \cdot B = B \cdot A$.

commutator In electronics, that part of the armature to which the coils of an electrical motor are attached.

compact disk An optically reflective disk that contains digital

Application of commutative laws of addition.

A, B → $A + B$ = B, A → $B + A$

Application of commutative law of multiplication.

A, B → AB = B, A → BA

Commutative laws

information stored as surface charges that change the reflection of a laser beam focused on the rotating disk.

companding To increase the number of quantization steps for small signals and decrease the number of steps for large signals in PCM. Companding reduces the number of coding bits that may otherwise be required for signals with large dynamic ranges.

comparator A circuit that compares an input voltage to a reference and produces an output voltage indicating the comparison.

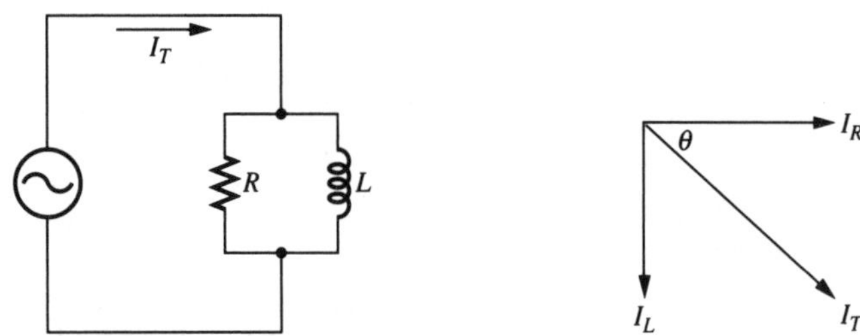

(a) Total current is the resultant of I_R and I_L

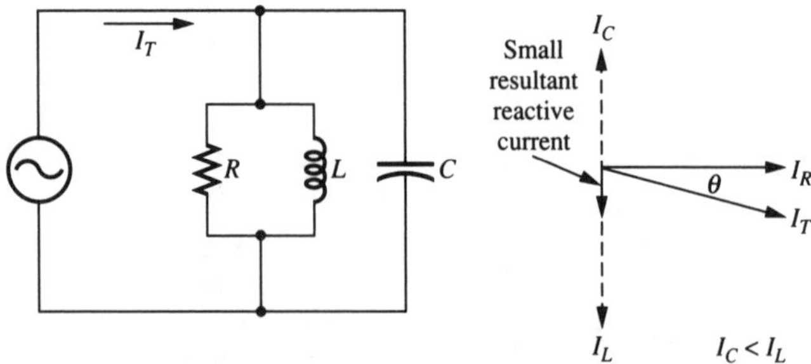

(b) I_C subtracts from I_L, leaving only a small reactive current, thus decreasing I_T and the phase angle.

Compensating capacitor

compensating capacitor A capacitor that is used in an ac circuit to help correct the power factor of the load.

compile To cause a binary-coded program, understood by the microprocessor, to be generated from a symbolic-coded program (such as C or Pascal) understood by the programmer. When a Pascal program, called the source code, is compiled, another program is generated called the object code. The object code contains the binary instructions understood by the microprocessor and required to execute the program.

compiler 1. In computers a program that converts the instructions of a high-level language into the instructions understood by the microprocessor. This new program is then the program that is executed by the computer. See interpreter. 2. A program used for converting an object code (such as Pascal or C) into a source code (a binary code understood by the computer). See compile.

complement A reversion of the digital state. The complement of OFF is ON, the complement of TRUE is FALSE, the complement of HIGH is LOW. Likewise, the complement of 0 is 1, and the complement of 1 is 0.

complementary pair Two transistors, one npn the other pnp that have similar electrical operating characteristics. A complementary pair would be found in a complementary push-pull amplifier. See complementary push-pull.

complementary push-pull A two-transistor amplifier consisting of one pnp and one npn transistor connected in such a manner so that one amplifies the positive half of the signal while the other amplifies the negative half.

complex number A number that contains a real and imaginary part. Complex numbers can be used in the solutions to ac circuit problems where the real part of the number is represented on the vertical axis and the imaginary along the horizontal axis. In this fashion, they become a mathematical representation of phasors.

complex plane A method of representing complex numbers where the horizontal axis represents real numbers and the vertical axis represents imaginary numbers. In electronics, the *j* pre-

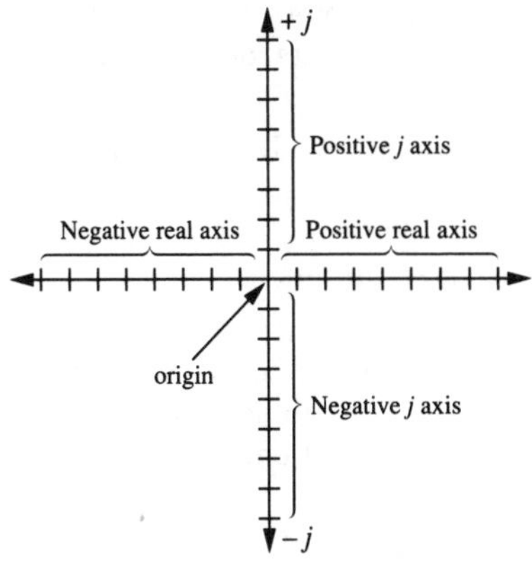

Complex plane

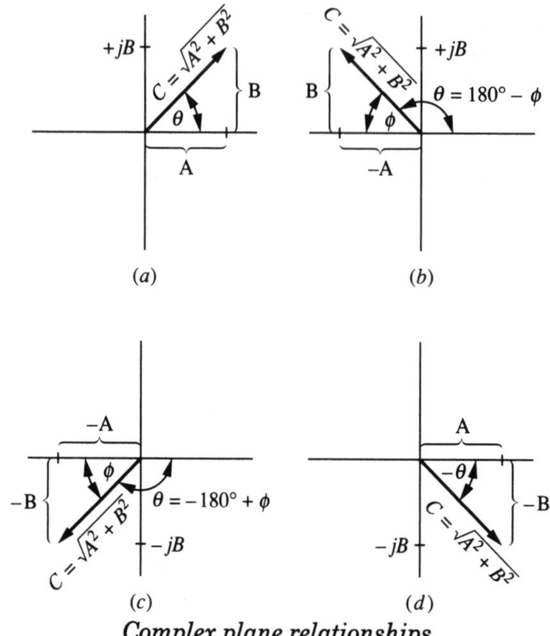

Complex plane relationships

fix is used to designate numbers that lie on the imaginary axis.

complex plane relationships In complex number notation, the relationship between the complex phasor and the real and imaginary axis of the complex plane.

component In electronics, a self-contained part such as a resistor, capacitor, inductor, or integrated circuit.

component side Concerning a printed circuit board, the side of a printed circuit board containing the electrical components. See component.

composite color signal That part of a television transmission that contains all the necessary information for proper color reception. A composite color signal will contain the color picture signal as well as all necessary synchronizing information.

computer A programmable processing machine, consisting of an input, output, processing unit, and memory, that is capable of accepting information, applying prescribed processes to the information, and supplying the results of that process.

computer-aided design Using the graphics and the computational capabilities of the computer itself to substitute for the traditional use of pencil, paper, and calculator. Thus the design of an office building or piece of machinery is done graphically while necessary calculations of costs, amount of materials, and other items are done while the design is in process or being modified.

computer-aided troubleshooting An automated system of troubleshooting that uses a computer program to control the steps and record the results of the troubleshooting process.

computer architecture Dealing with the physical (hardware) aspects of a computer. The relationships among the hardware components of a computer.

computer game A computer simulation used for the purpose of entertainment. Similar to watching television with the added advantage of user participation.

computer-supported cooperative work Electronic and computer technology that allows people in remote places to interact with each other and with the same documents and files through voice, data, and video interconnections.

conductance The reciprocal of resistance. A measurement of the relative ease for which current can flow. Conductance (*G*) is measured in siemens (S).

conduction band The energy level of an electron where it is now a free electron. When an electron in the valence band requires sufficient energy, it can be set free from the confines of the atom and is said to now be in the conduction band.

conductor Any material that easily allows the flow of current. Having a low resistance. See insulator.

contact bounce A condition where the pole of a switch strikes the contact upon the closure of a switch, the contacts will physically bounce (like rapid vibrations) several times before actually closing. A circuit such as an S-R latch can be used to eliminate the effects of contact bounce.

continuity tester An inexpensive and simply constructed electrical device for testing circuit continuity. A continuity tester can be used in checking for shorts or opens.

contrast control In a television receiver an adjustment that controls the difference in levels of brightness and darkness on the face of the CRT. This is accomplished by controlling the gain of the video amplifier.

control bus A group of conductors from a microprocessor that carries the information used to control the operation of devices external to the microprocessor. Also used to control the operation of the microprocessor from external devices.

conventional current The flow of current from the positive

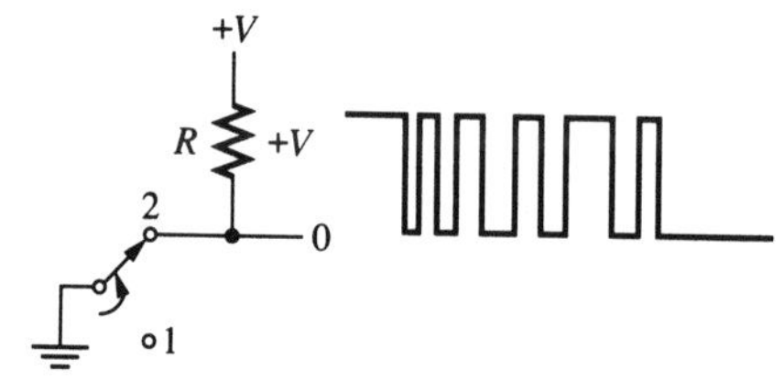

(a) Switch contact bounce

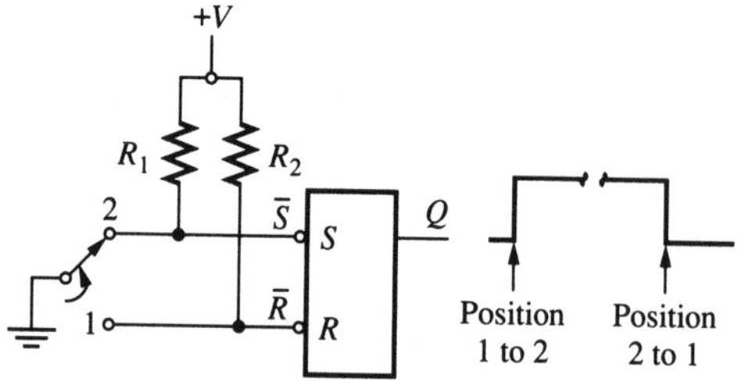

(b) Bounce eliminator

Contact bounce

terminal of the voltage source to the negative terminal of the same source. Opposite in the direction of electron flow. Any circuit may be analyzed using either conventional or electron flow and the same results and polarities obtained. Some prefer conventional current flow when analyzing solid-state circuits and electron flow when analyzing vacuum tube circuits.

conventional current flow A defined direction of current flow that is in the opposite direction of electron flow. See electron flow.

conversion errors Errors made during a conversion process. For example, in analog to digital (A/D) converters, there are conversion errors made between the value of the analog quantity and the resulting digital code that represents the value of that quantity.

conversion frequency In sideband transmitters, the conversion frequency is used to separate the sidebands so it is easier to filter out one of the sidebands. For example, when a low-frequency audio signal is being transmitted, the separation between its sidebands may be less than 100 Hz. A conversion oscillator with a frequency of 100 kHz is used to separate further the sidebands and ease the filtering requirements of the system.

conversion time The amount of time it takes to convert an analog voltage reading into a digital code. Term is used with analog-to-digital converters.

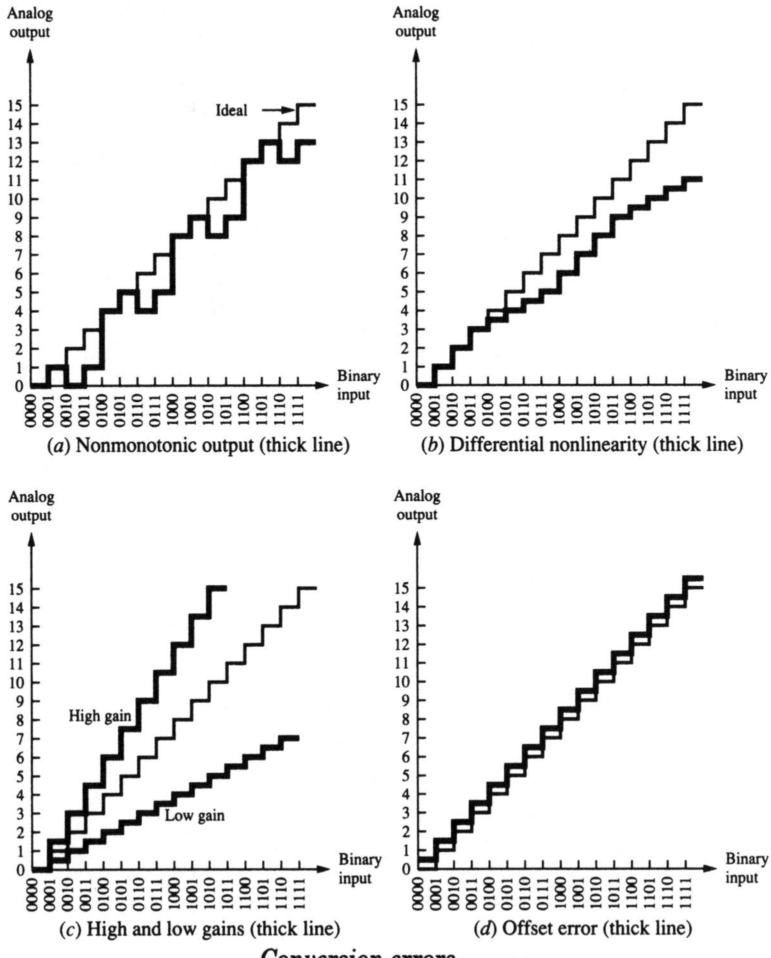

(*a*) Nonmonotonic output (thick line)

(*b*) Differential nonlinearity (thick line)

(*c*) High and low gains (thick line)

(*d*) Offset error (thick line)

Conversion errors

Cooper effect The tendency of electrons to pair as they travel in a medium.

copper An element that is a good conductor of electricity. Copper is one of the most common materials used to interconnect electrical parts to produce a desired circuit.

coprocessor A digital processing circuit that is made to be used in conjunction with a microprocessor. For example, many microprocessors have available a math coprocessor that will take over the math functions of the microprocessor, with the intent of making a faster overall system.

coulomb 1. The unit of measurement for electrical charge. One coulomb equals the amount of charge on 6.25×10^{18} electrons. 2. The basic unit of charge in the SI system.

Coulomb's law The relationship between the forces of charged particles. Mathematically expressed as $F = (kQ_1Q_2)/r^2$, where F is the force in newtons, k a constant $= 9.0 \times 10^9$, Q_1 and Q_2 are the charges in coulombs, and r the distance in meters between the two charges.

counter A digital circuit that counts events that are represented by changing levels or pulses or generating a particular code sequence. In order to

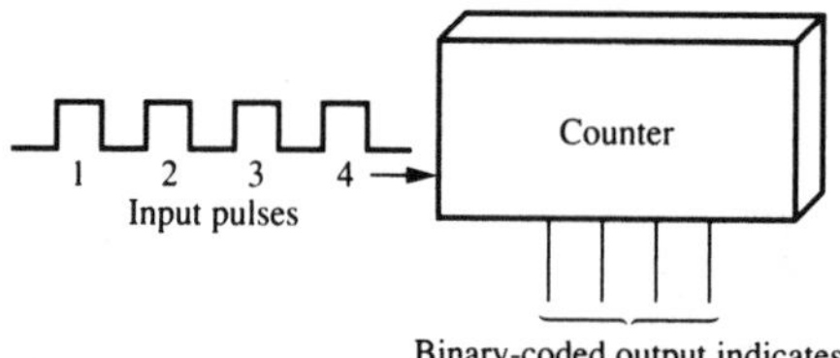

(*a*) Counter used to count events indicated by the occurrences of pulses.

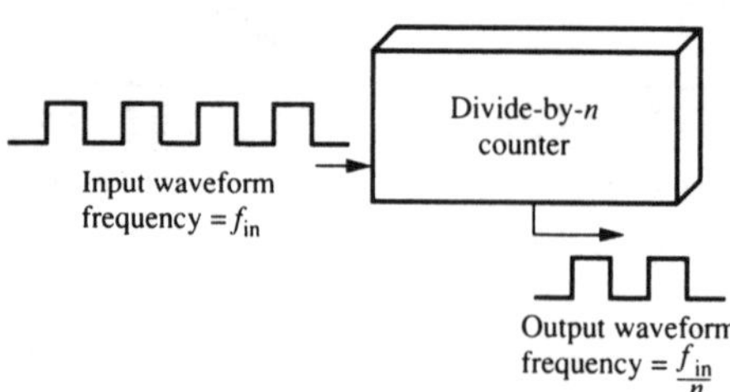

(*b*) Counter used to divide an input frequency by a factor *n*. For example, if $f_{in} = 100$ kHz and $n = 10$, $f_{out} = 10$ kHz.

Counter

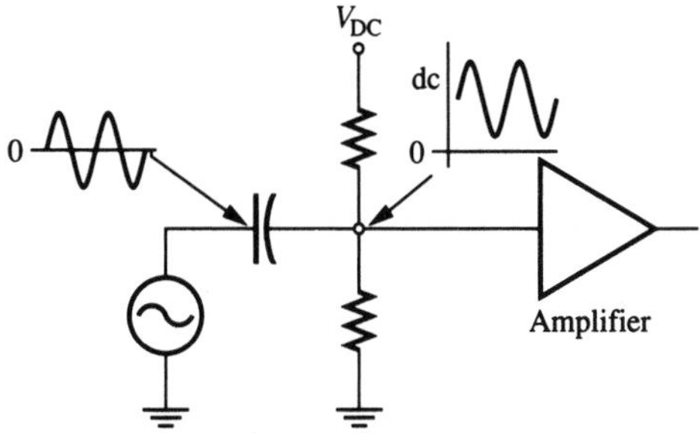

Coupling capacitor

"count," the counter must have some method of "remembering" its present value so it can count from that point to the next number in its sequence. Because of the memory requirement of counters, they are usually constructed from flip-flops.

coupling capacitor The arrangement of a capacitor in a circuit used to pass (couple) voltage changes but block dc.

coupling mismatch In optical fibers, a source of signal loss when attempting to connect on fiber optic to another. A coupling mismatch can result from improper alignment of the fibers, connecting ends of the fibers not being in parallel with each other or a separation of the two connecting fibers, resulting in light scattering.

covalent The bonding of two or more atoms by the interaction of their valence electrons.

CP/M Control program for microcomputers. An operating system produced by Digital Research.

CPU Abbreviation for central processing unit. See central processing unit.

CPU memory organization The way in which the CPU interfaces with memory in a microprocessor-based system. Generally, the CPU will interface with RAM (read-write memory) and ROM (read-only memory).

critical angle The angle of incidence that causes the angle of refraction to be exactly 90 degrees. As the angle of incidence is increased, so is the angle of refraction—up to a point where the refracted ray travels parallel to the surface of the reflecting medium. As the angle of the incident ray increases beyond the critical angle, the reflected ray no longer penetrates the surface of the medium but is now reflected from it. Below the critical angle, the refracted ray penetrates the material.

critical frequency The frequency at which the response of a circuit is 3 dB down from its maximum output.

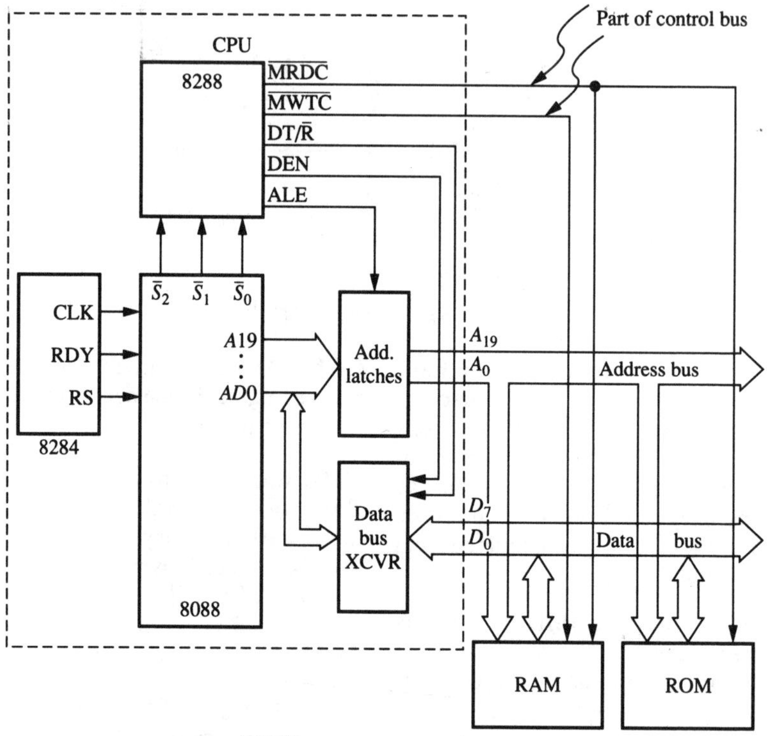

CPU memory organization

crossover distortion In a class B push-pull amplifier, the nonlinearity that occurs in the output signal when it changes from a positively going signal to a negatively going signal.

CRT Abbreviation for cathode ray tube. See cathode ray tube.

crystal An electrical device made from a crystal substance used to produce a stable frequency in an oscillator circuit. Crystals exhibit the piezoelectric effect. When an electrical voltage is placed across them, they vibrate at a very specific frequency that is determined by the physical characteristics of the crystal.

crystal filter A frequency selective electrical circuit that uses the effects of crystals in order to achieve frequency selection. A crystal filter has the advantage of having a very high *Q* which results in very narrow bandwidths.

crystal oscillator A circuit capable of producing its own signal. A crystal oscillator uses a crystal to determine the frequency of the resulting signal. See crystal.

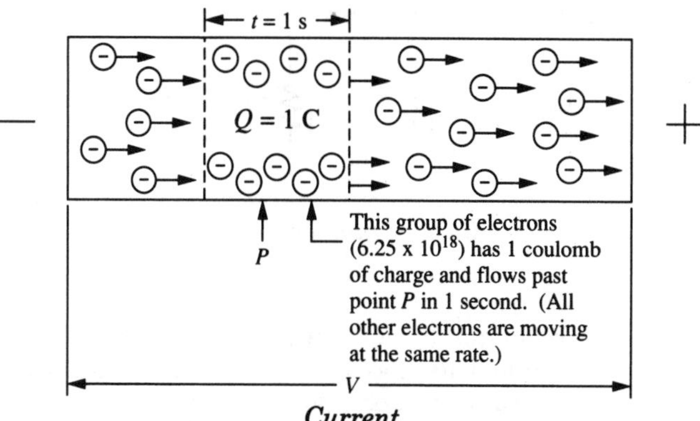

Current

CS amplifier Abbreviation for common-source amplifier. See common-source amplifier.

CSCW Abbreviation for computer-supported cooperative work. See computer-supported cooperative work.

CT Abbreviation for center tap. See center tap.

current The uniform movement of electrical charge. The measurement of one coulumb of charge per second. Unit of measurement is the ampere. Abbreviated as *I*.

current-controlled current source A current source whose output current is controlled by the current value of some other point in the circuit to which it is connected.

current-controlled voltage source A voltage source whose output current is controlled by the current value of some other point in the circuit to which it is connected.

current divider A parallel circuit is sometimes referred to as a current divider because there is more than one path for current flow and the value of the current in each path is determined by the value of the impedance of each parallel branch and the applied voltage.

current divider formula For resistors in parallel, $I_X = (R_T/R_X)\ I_T$, where I_X is the branch current, R_T is the total parallel resistance, R_X is the resistance of the branch for which the branch current is to be calculated, and I_T is the total circuit current.

current foldback Usually used in high-current-voltage regulators. Current foldback is a protection scheme where an overload or a short causes the output

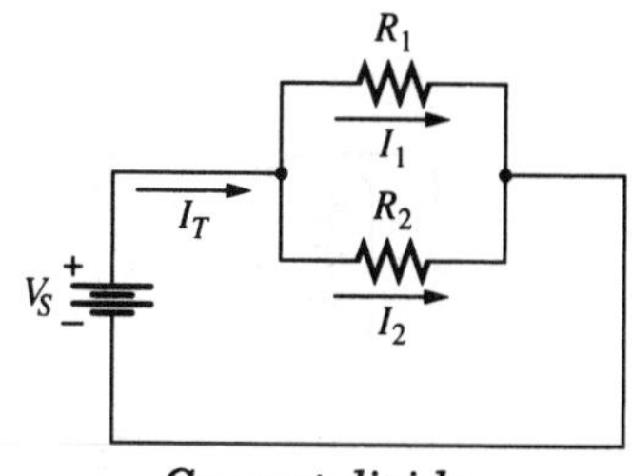

Current divider

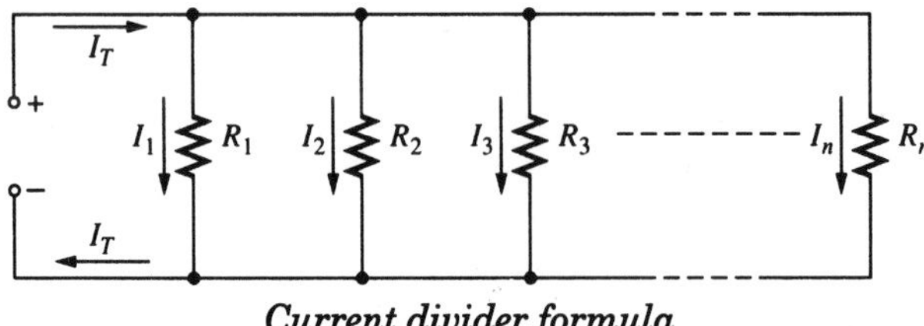

Current divider formula

current of the regulator to decrease to a small value. This process prevents the load or regulator from overheating.

current gain The ratio of the output current to the input current of a device; expressed as $I_{gain} = I_{out}/I_{in}$, where I_{gain} is the current gain (no units), I_{out} is the output current (in amps), and I_{in} is the input current (in amps). Note that if the output current is less than the input current, the current gain is less than one.

current limiting In regard to voltage regulators, when overloaded, the regulator current output will not exceed a certain amount. This applies even if the output of the regulator is short-circuited.

current sink A device that accepts current from an external source. Sometimes used to explain the electrical action of TTL digital logic circuits.

current source 1. A device that delivers current, usually in a fixed amount to a load. An ideal current source delivers the same amount of current to any value load. 2. Ideally supplies a constant amount of current to any circuit. In a practical manner it can be represented by an ideal current source in parallel with an equivalent internal resistance. See Norton equivalent circuit.

current standing-wave ratio The ratio of the maximum and minimum peak values of the current standing wave produced by a transmission line. See standing waves.

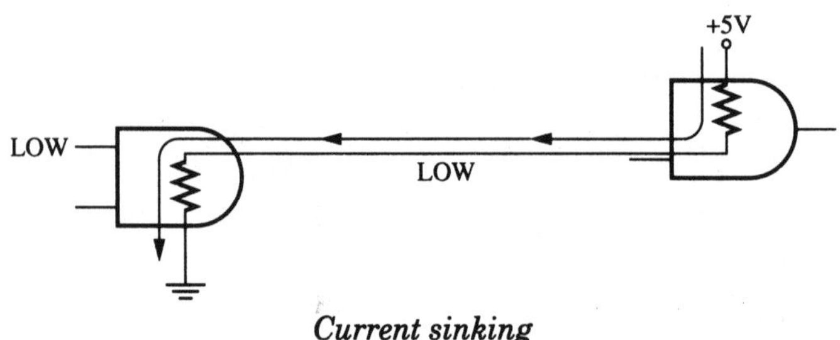

Current sinking

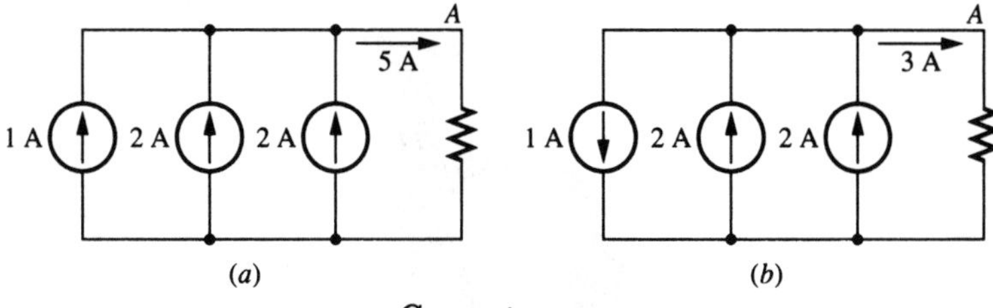

Current source

Current sourcing

current tracer A digital instrument that contains a lamp indicator that glows when the tip of the instrument is held over a logic circuit that is carrying a logic pulse or a series of pulses.

curve tracer An electronic instrument used to display the characteristic curves of an electrical device. Curve tracers are used, for example, to display the collector characteristic curves for junction transistors.

cutoff 1. The condition in a current controlling device such as a vacuum tube, transistor, or FET when the controlled current is prevented from flowing in the device. See saturation. 2. The nonconducting state of an electrical device. A transistor is said to be in cutoff when there is no longer any current between its emitter and collector.

cutoff wavelength In a waveguide, the lowest frequency that can be accommodated in a given waveguide; expressed mathematically as $l = 2a/m$, where l is the cutoff wavelength of the waveguide (in centimeters), a is the waveguide width (in centimeters), and m is the number of half-wavelength fields.

cycle The repetition of a waveform. For example, one cycle of a sine wave has occurred when the wave pattern begins to repeat itself. Comes from the word "circle."

DAC Abbreviation for digital-to-analog converter. See digital-to-analog converter.

D/A converter A device that converts a digital signal into an analog signal. See A/D converter.

D latch A type of latch that differs from the S-R latch in that it has only one input called the D (for data) input. This type of latch may also be gated.

dark current The current in a photosensitive electronic device when there is no light on the device. Ideally, the dark current is zero.

Darlington pair A transistor amplifier consisting of an arrangement of two transistors where the collector of one is connected directly to the base of the other. Gives a very high input impedance and high current gain.

d'Arsonval movement A meter movement. Consists of an iron-core coil that is mounted on bearings between a permanent magnet. A pointer is connected to the movable core to indicate the strength of the current in the coil.

DAT Digital audiotape. See digital audiotape.

data A general term used to identify fundamental elements of

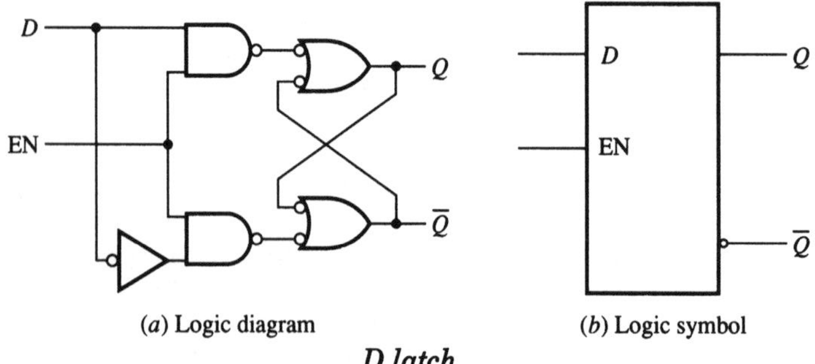

(*a*) Logic diagram

(*b*) Logic symbol

D latch

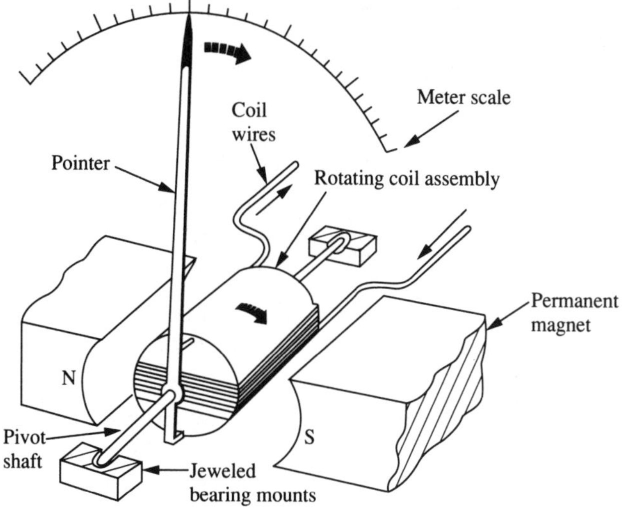

(*a*) Basic components of a d'Arsonval movement

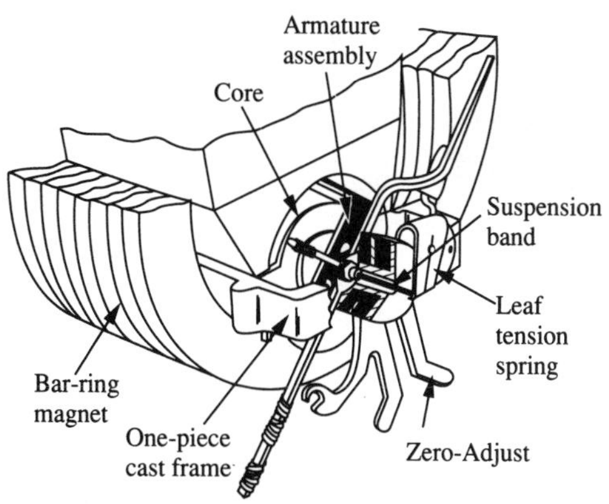

(*b*) Construction view of a typical d'Arsonval movement.

d' Arsonval movement

information that are capable of being processed or produced by a computer system.

data acquisition The gathering of information. Usually accomplished through the use of analog-to-digital and digital-to-analog converters.

data bus A group of conductors from a microprocessor that carries the information that will be stored in a specific memory location. The information is an ON/

OFF bit pattern represented by two different voltage levels, usually +5 volts and 0 volts.

data lock-out A circuit, such as a data lock-out flip-flop, intentionally designed to prevent data from influencing the circuit during specified times. See data lock-out flip-flop.

Logic symbol for a data lock-out J-K flip-flop.

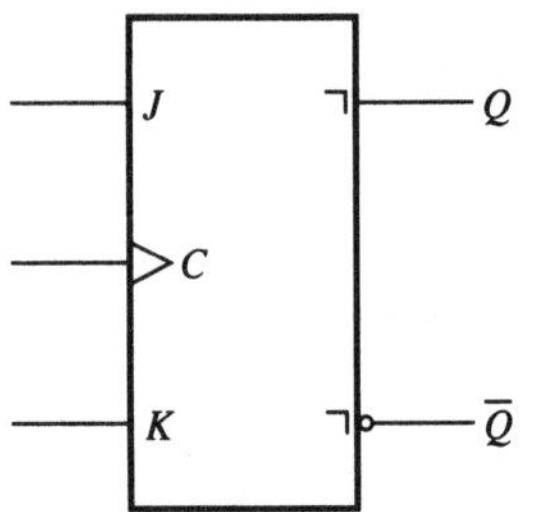

data lock-out flip-flop A master-slave flip-flop where the input is sensitive to changes only during the clock transition.

data selector A digital device that can select one of several different digital inputs and pass its condition to a single output. See multiplexer.

dB Letter symbol for decibel. See decibel.

dB gain Decibel gain. Expressed mathematically as $A\,(\text{dB}) = 10 \log_{10} A$, where A is the gain of the system.

dBm A measurement used to indicate a power gain with respect to one milliwatt. Mathematically expressed as P (dBm)

Logic symbols for the 74111 dual J-K master-slave flip-flop with data lock-out.

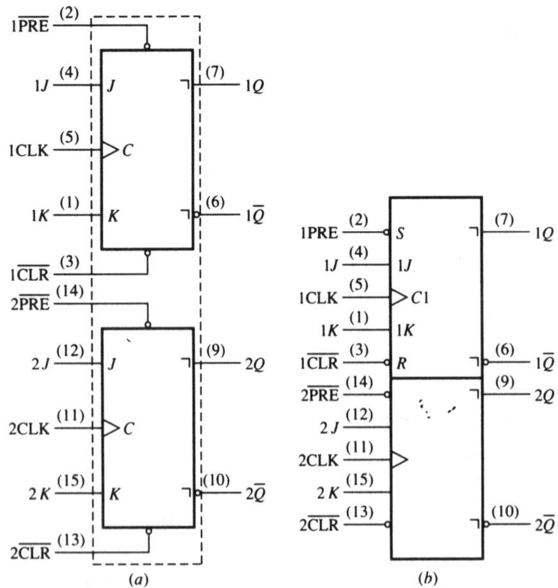

Data lock-out

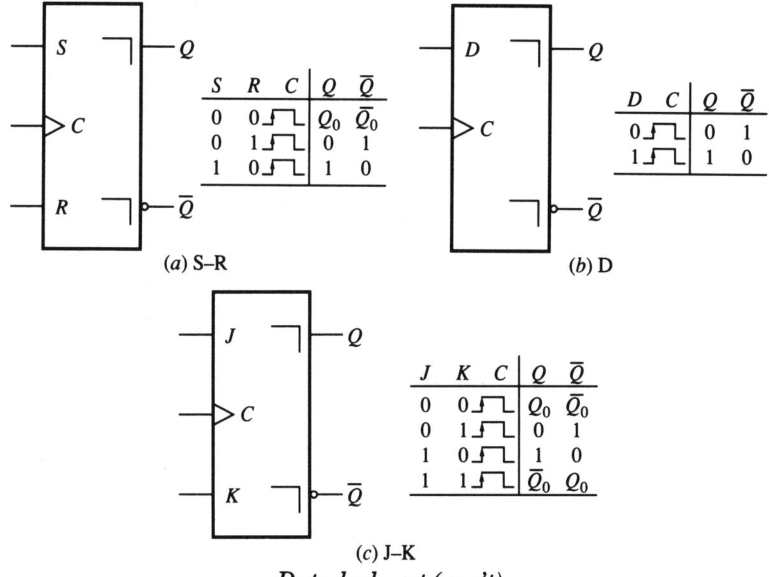

S	R	C	Q	$\bar{Q}$
0	0	⎍	Q_0	$\bar{Q}_0$
0	1	⎍	0	1
1	0	⎍	1	0

(*a*) S–R

D	C	Q	$\bar{Q}$
0	⎍	0	1
1	⎍	1	0

(*b*) D

J	K	C	Q	$\bar{Q}$
0	0	⎍	Q_0	$\bar{Q}_0$
0	1	⎍	0	1
1	0	⎍	1	0
1	1	⎍	$\bar{Q}_0$	Q_0

(*c*) J–K

Data lock-out (con't)

$= 10 \log_{10}(P/1 \text{ mW})$, where P (dBm) is the power gain with respect to 1 mW and P is the power gain (expressed in watts).

dc Abbreviation for direct current. See direct current.

dc amplifier See operational amplifier.

dc controller Operational amplifier used to control dc servomotors.

dc current source An electrical source of energy that will produce a constant voltage for all values of loads applied to it.

dc generator An electromechanical device that converts mechanical energy into a dc voltage and is capable of delivering electrical power to a load.

dc isolation Preventing dc voltages from being coupled from one part of a circuit to another. dc isolation may be achieved with a capacitor or with a transformer.

dc restorer See clamper.

dc-to-ac converter An electrical circuit that converts a given dc voltage to a specified ac voltage at a given frequency. An example of a dc to ac converter is a circuit that converts the 12 VDC of a car battery to 120 VAC at a frequency of 60 Hz.

dc-to-dc converter An electrical circuit that converts a given dc voltage to another dc voltage of a different value. An example

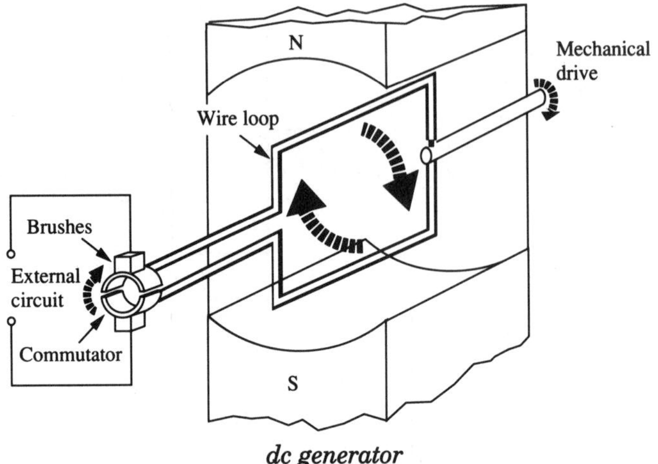

dc generator

of a dc-to-dc converter is a circuit that converts 12 VDC to 5 VDC.

dc value See average value.

debouncer A logic circuit used to remove unwanted changes in logic levels that are usually caused by mechanical switch contacts.

debugger A computer program used as an aid in finding programming errors.

decade In electronic frequency measurements, a change in frequency by a factor of 10.

decade counter A digital counter that has only 10 unique states.

decibel (dB) A ratio of the power output to the power input of a device or circuit. Defined as decibels = $10 \log_{10} P_{out}/P_{in}$.

decimal Having to do with 10. The decimal number system uses 10 separate symbols—0, 1, 2, 3, 4, 5, 6, 7, 8, and 9—to represent all possible numerical values. To represent fractional values, the decimal numbers are preceded by a decimal point (.).

decimal to BCD A logic encoder that converts a decimal

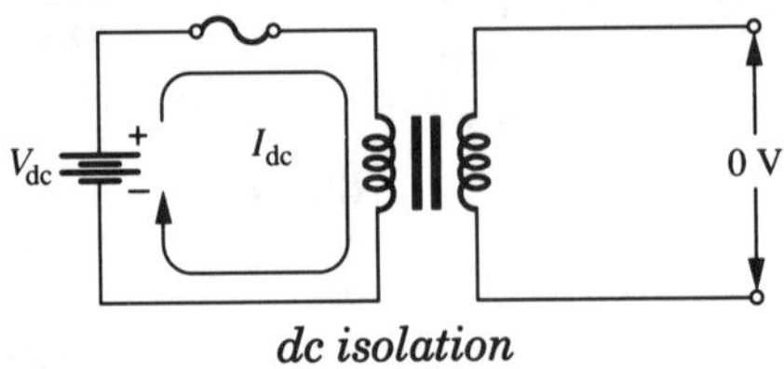

dc isolation

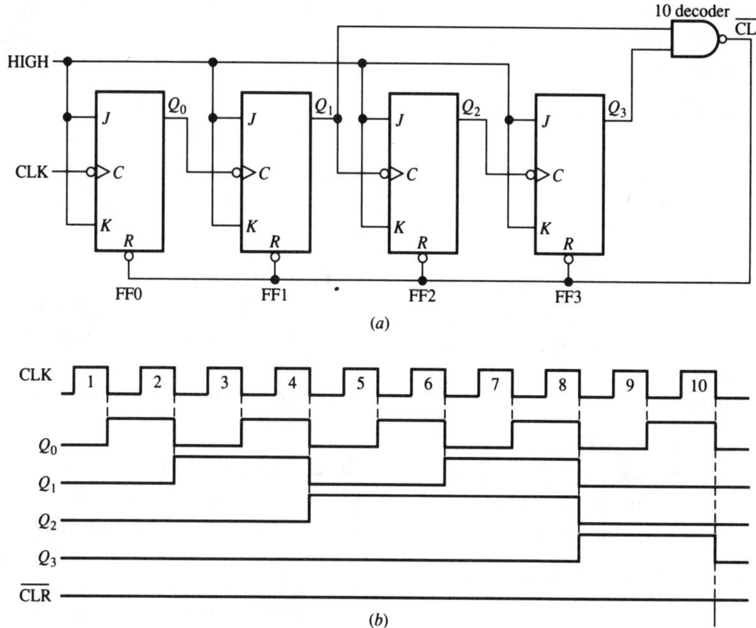

Decade counter

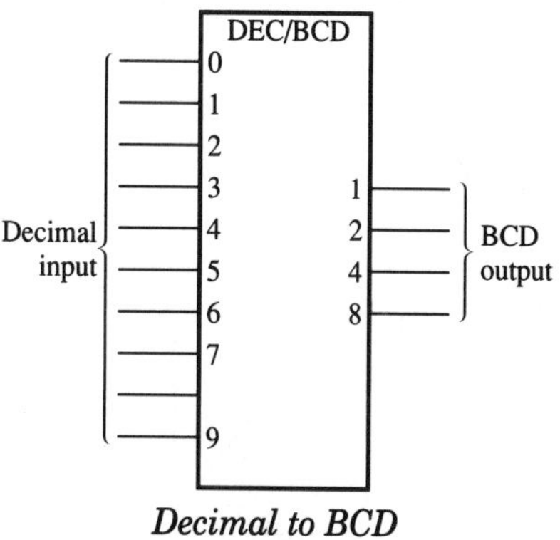

Decimal to BCD

number to its corresponding BCD (8241) value. This type of encoder has 10 inputs and 4 outputs.

decoder A logic network that will convert a specific bit pattern on its input to activate a specific line on its output. The most common form of a decoder is converting a binary input to an active line output such as in a two-line to four-line decoder where a 2-bit binary number on the input will activate one of the four lines of output.

decoding glitch An undesirable logic level of a short duration that is caused by time delays inherent in logic circuits.

decoupling network An electrical circuit used to present a low-impedance path to ground. Decoupling networks help prevent parasitic oscillations. See parasitic oscillations.

degrees One of the fundamental units of angular measurement. There are 360 degrees in a full circle.

delayed AGC A form of automatic gain control that acts only on signals above a certain strength. This permits reception of weaker signals.

delta-connected generator A generator made to operate from a delta-type line source.

delta-connection An electrical connection of components connected in such a fashion that their schematic representation appears as the Greek letter delta (Δ).

delta modulation A form of pulse code modulation where, in its simplest form, transmits just 1 bit per sample and the polarity of the bit indicates if the signal is larger or smaller than the previous sample.

delta-to-wye conversion A method of converting from a delta circuit to a wye circuit.

delta-wye A specific type of circuit connection. See delta-connection. See wye-connection.

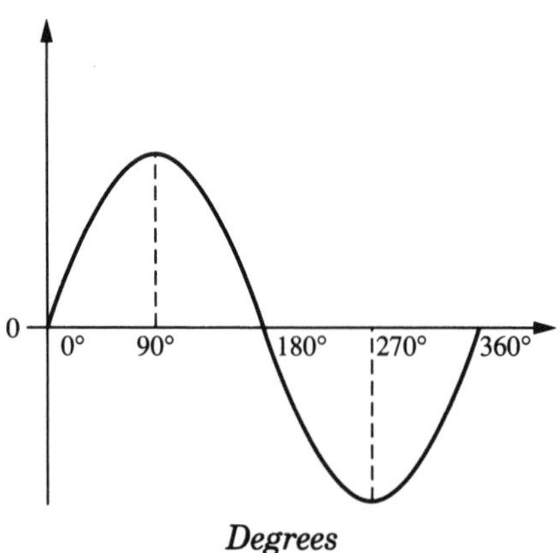

Degrees

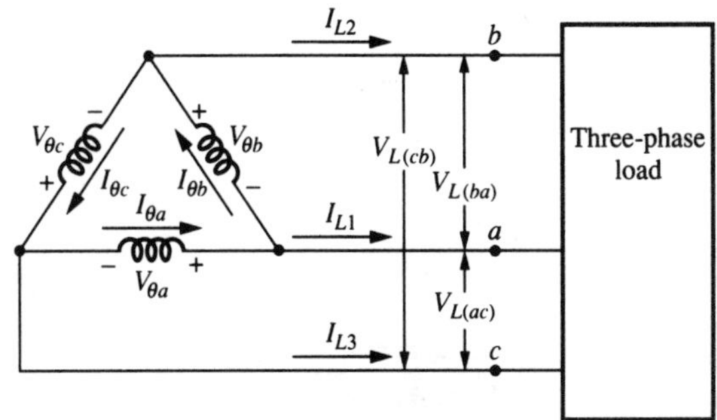

Delta-connected generator

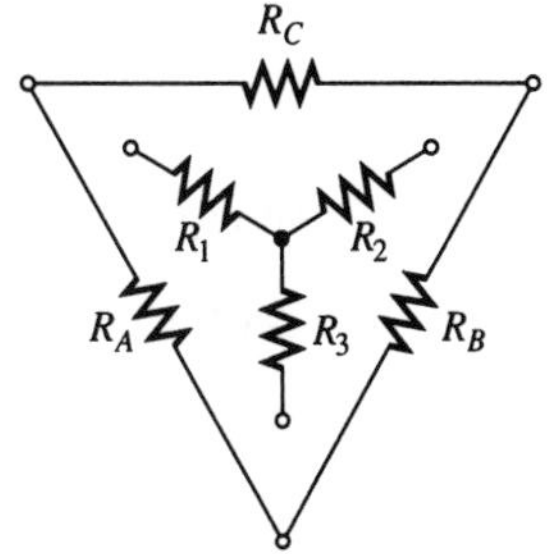

Delta-to-wye conversion

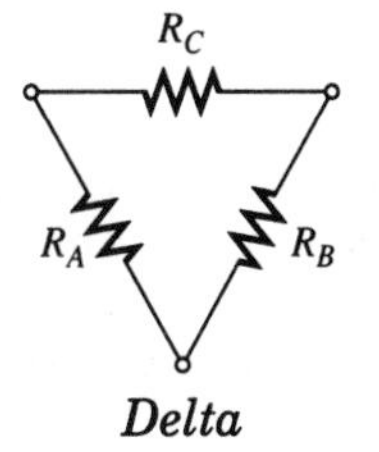

Delta

demodulate 1. The opposite of modulate. To remove electrical charges in a carrier wave that were caused by modulation for the purpose of extracting the intelligence represented by the modulation. See demodulator.

demodulator An electrical circuit or device that serves to demodulate. See demodulate.

DeMorgan's theorems There are two general theorems used in Boolean algebra developed by the logician and mathematican DeMorgan. These state that "The complement of a producet is equal to the sum of the complements." This is stated as $\overline{XY} = \overline{X} + \overline{Y}$. The other theorem states that "The complement of a sum is equal to the product of the complements." This is stated as $\overline{X + Y} = \overline{X}\,\overline{Y}$.

demultiplexer A digital network that converts the logic of one input line to one of several output lines. The active output line is selected by a set of select lines. See multiplexer.

denominator The value of the bottom part of a fraction. For example, in the fraction $\frac{2}{3}$, the 3 is the denominator. See numerator.

depletion area See pn junction.

depletion layer capacitance In semiconductor material, the capacitance that is formed as a result of the depletion region of

$$\overline{XY} \equiv \overline{X} + \overline{Y}$$

X	Y	$\overline{XY}$	$\overline{X}+\overline{Y}$
0	0	1	1
0	1	1	1
1	0	1	1
1	1	0	0

$$\overline{X+Y} \equiv \overline{X}\,\overline{Y}$$

X	Y	$\overline{X+Y}$	$\overline{X}\,\overline{Y}$
0	0	1	1
0	1	0	0
1	0	0	0
1	1	0	0

DeMorgan's theorems

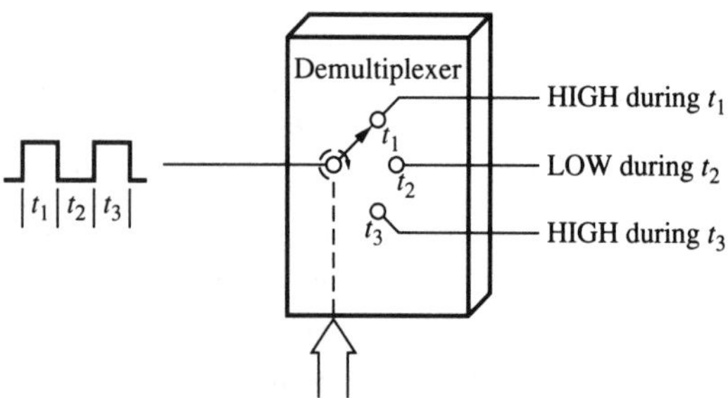

Selects the output to which the input is connected during each time interval

Demultiplexer

the pn junction. The depletion layer capacitance can be changed by changing the width of the depletion region with the application of an external dc voltage. It is this type of action that is utilized by the varactor diode. See varactor diode.

depletion mode A method of operating a MOSFET where a negative gate voltage is applied to the device.

derating factor The amount of change in the ratings of an electrical device according to another condition. For example, in solid-state devices the derating factor is used to lower some of the maximum current and voltage values as the temperature of the device is increased.

derivative A mathematical investigation of an instantaneous rate of change of a function.

desktop computer A personal computer designed to be used by a single person and be powered by conventional 120 VAC. Desktop computers find use in the home, school, and office.

detector The section of a communications receiver that removes (detects) the intelligence from the reproduced radio wave. As an example, the detector of an AM receiver removes the amplitude changes of the reproduced carrier wave and passes them on to be processed by the rest of the receiver. These amplitude changes represent the intelligence (voice or music in this case) transmitted by the radio wave.

determinants method A mathematical process for finding the unknown variables of two or more simultaneous linear equations.

deviation ratio Used in frequency modulation. Deviation ratio is a measure of the largest modulation index in which the maximum permitted frequency deviation and the maximum modulating frequency are used: $R_D = f_D/f_M$, where R_D is the deviation ratio (no units), f_D is the maximum deviation of FM carrier (in hertz), and f_M is the maximum modulating frequency (in hertz).

device An active electrical component other than a resistor, capacitor, or inductor.

diac A solid-state device that is capable of conducting current in one of two directions when properly activated.

diamagnetic materials Materials that possess permeabilities that are less than that of free space.

dielectric An insulator. Usually refers to the insulating material between the plates of a capacitor.

dielectric constant A measurement of the ratio of the capacitance of a capacitor with a specific dielectric material compared to the capacitance of the exact same capacitor where air is used as the dielectric.

difference amplifier An amplifier whose output voltage is proportional to the difference of two input voltages.

difference engine One of the first mechanical calculators.

difference gain In a differential amplifier, the ratio of a change in the output voltage to a change in the differential input voltage.

differential amplifier An electrical circuit whose output voltage is proportional to the difference between two input signals.

differential phase-shift keying A method of transmitting data between two digital systems that uses a synchronizing pulse along with the digital data. If there is a mark, the phase of the signal is shifted +90 degrees; if a *space,* the signal phase is shifted −90 degrees.

differential transmission A method of electrically connecting two systems through the use of differential line drivers. The action of the differential line drivers is to help cancel the effects of

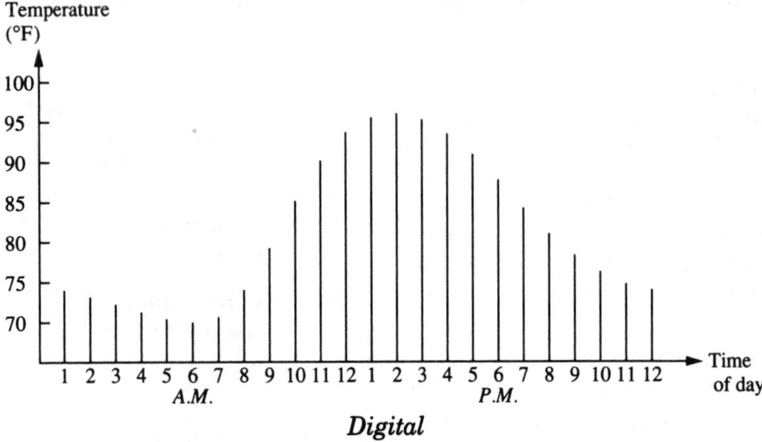

Digital

noise pickup between the two systems.

differentiator 1. An electrical circuit that produces an output that closely approximates the mathematical derivative of the input waveform. 2. An electrical circuit that is the opposite of an integrator. Consists of a high-pass filter.

diffusion The process of constructing electrical characteristics into silicon wafers. Diffusion is a step-by-step process that places dopants to the desired depth in a wafer by exposing it to very high temperature with specific chemicals.

digital Having to do with discrete values rather than continuous (analog) values. Digital electronic levels usually involve two possible states. These are represented by two voltage levels called HIGH and LOW.

digital adder A digital circuit that will add two numbers with a carry input. Its output consists of a sum and a carry output.

digital analysis The process of analyzing a digital circuit

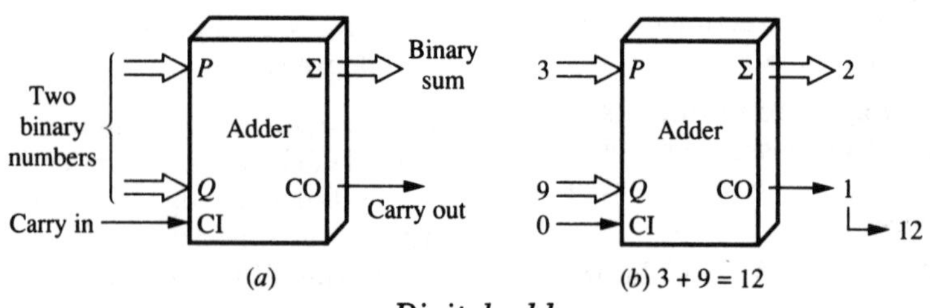

Digital adder

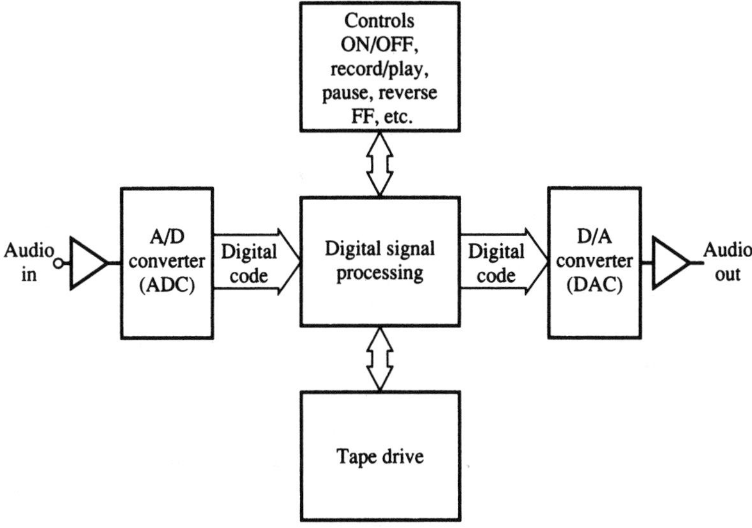

Digital audiotape

through the use of digital test equipment. Digital analysis is used in troubleshooting digital logic circuits and other similar electrical circuits.

digital audiotape A recording of sound by digitized information. Digital audiotape converts the analog sound quantity into a digital code that is recorded on the tape in the record mode. In the playback mode, the digital code is converted back into the original analog sound. Analog-to-digital and digital-to-analog converters are used in this process.

digital comparator A digital circuit that compares two quantities to indicate whether they are equal. If they are unequal, the digital comparator will indicate which of the two numbers is greater.

digital computer A machine that is capable of performing processes by using discrete levels. The most common kind of digital

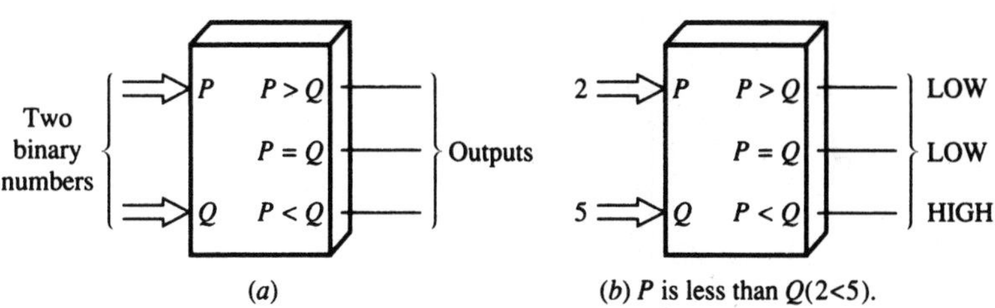

(*a*) (*b*) *P* is less than *Q*(2<5).

Digital comparator

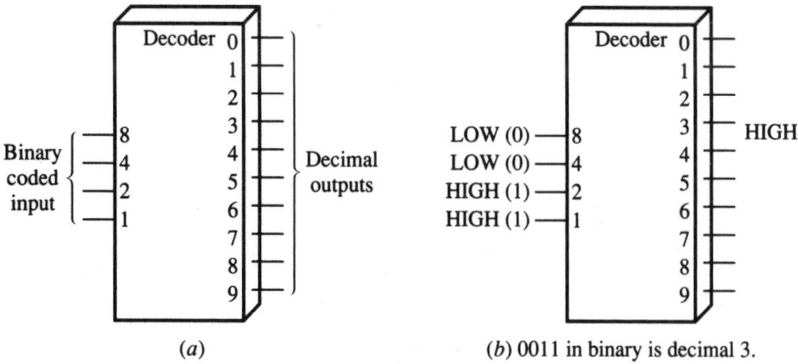

(*a*)

(*b*) 0011 in binary is decimal 3.

Digital decoder

computer is an electrical machine that uses two voltage levels to represent a HIGH (1 or TRUE) condition and a LOW (0 or FALSE) condition. An electrical digital computer is capable of performing mathematical and logical operations as well as combinations of these operations.

digital decoder A digital circuit that converts coded information, such as a binary number, into another form, such as a decimal number.

digital encoder A digital circuit that converts information, such as a decimal number or an alphabetic character, into some coded form.

digital integrated circuit An electronic circuit where all the components are produced on a single substance (such as silicon). Circuit is designed to respond to two-level electrical signal (usually 0 volts and +5 volts).

digital margin detector A circuit that produces a two-state

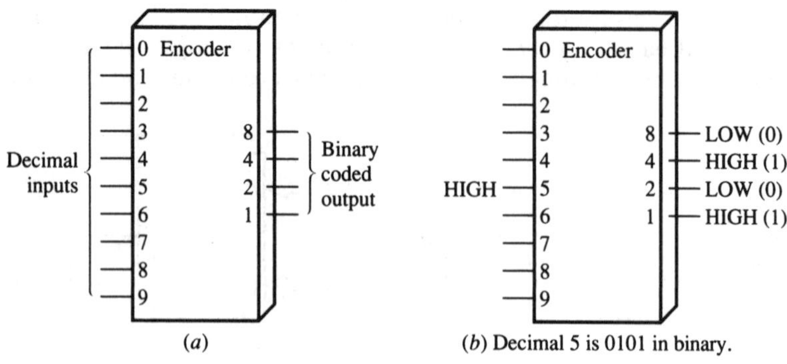

(*a*)

(*b*) Decimal 5 is 0101 in binary.

Digital encoder

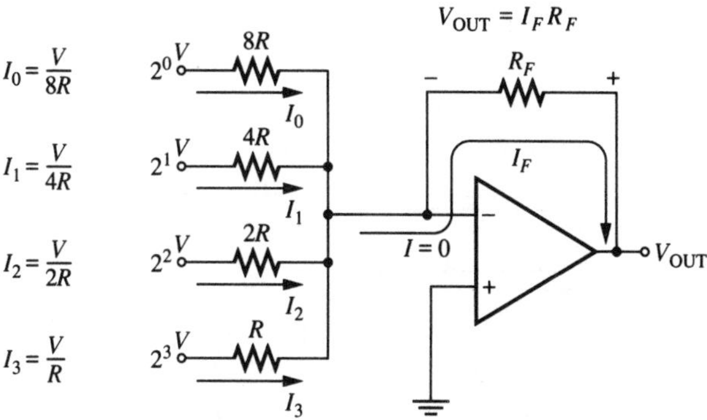

Digital-to-analog converter

output that depends upon a prescribed range of voltage inputs.

digital multimeter An electrical instrument with a digital readout capable of measuring voltage, current, and resistance.

digital phase detector A digital circuit consisting of an exclusive OR gate. The output of an exclusive OR gate is high only when the two inputs are different. Thus, when both digital inputs are in phase (happening at the same time), the output of the gate will be low.

digital probe An electrical instrument used to detect the presence of an ON or OFF signal. In its simplest form, consists of a test point electrically connected to a single LED. The test point is touched to the circuit in question, and the LED will be turned ON if the test point is ON or be OFF otherwise.

digital pulser An electrical instrument used to produce a digital pulse that is usually a brief transition from 0 volts to 5 volts and back to 0 volts. Consists of a metal probe that is placed against the connection where the digital pulse is required. Used for testing digital circuits.

digital-to-analog converter An electrical circuit that converts a digital code into an analog voltage.

digital toubleshooting The process of determining the problem in a nonfunctioning digital circuit. Usually consists of signal tracing and waveform as well as logic analysis.

digital voltmeter An instrument used to measure voltage with a digital readout.

diode An electrical device that allows current to flow through it in one direction only.

DIP Abbreviation for dual inline package.

DIP switch A small electrical switch made in the form of a dual

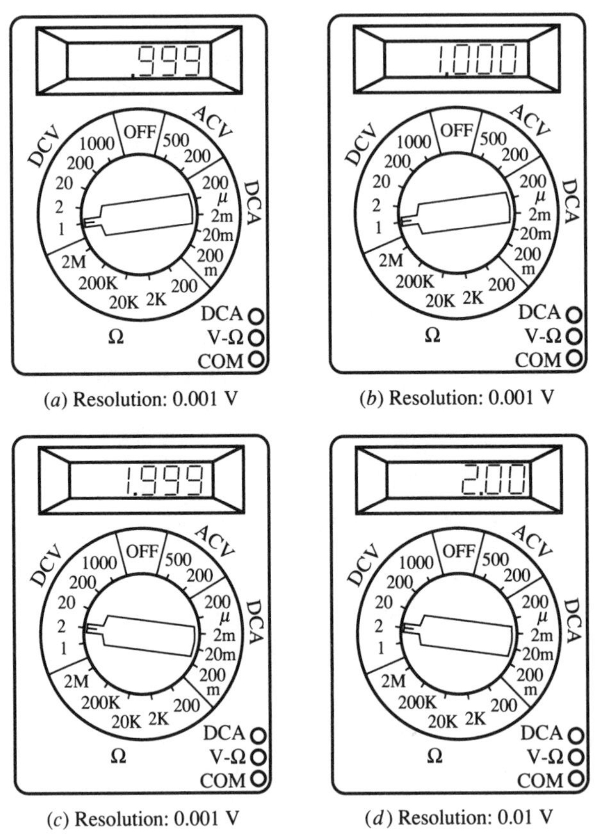

(*a*) Resolution: 0.001 V

(*b*) Resolution: 0.001 V

(*c*) Resolution: 0.001 V

(*d*) Resolution: 0.01 V

Digital multimeter

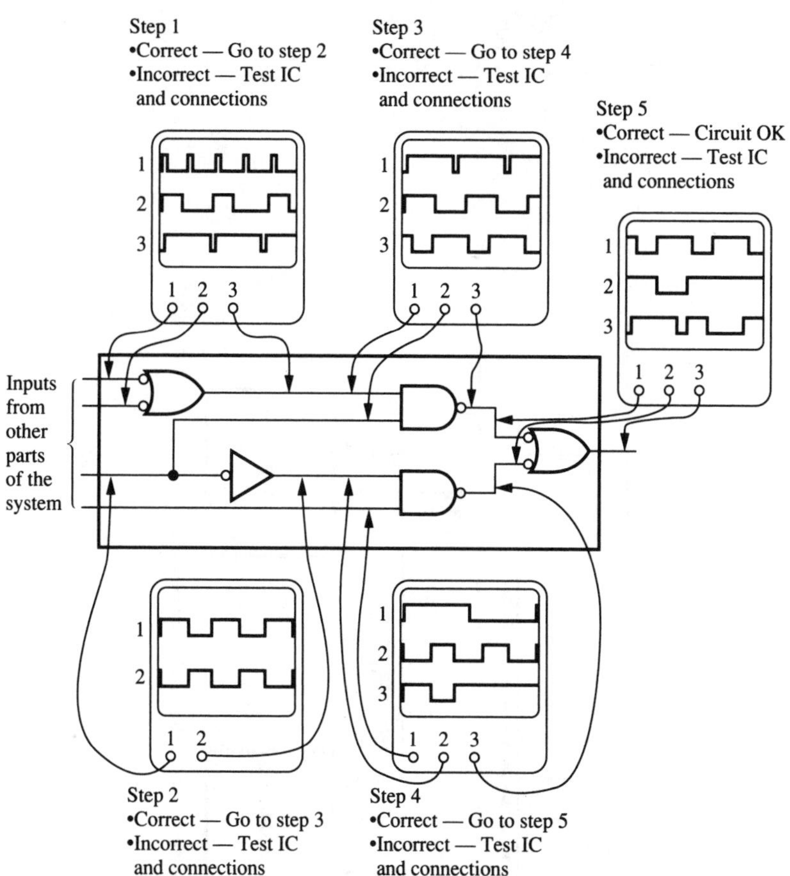

Digital troubleshooting

in-line package. Usually consists of two or more rows of independent SPST switches. Commonly found in digital equipment.

direct addressing An addressing mode where the process of transferring data between the microprocessor and the external system (usually memory) is done by designating the address for the location of the data.

direct current Current that flows in a circuit in the same direction and intensity. Referred to as dc.

direct memory access An I/O capable of transferring data to and from memory without intervention of the microprocessor.

discharge To remove any charge. To discharge a capacitor means to cause it to have all of the voltage across it (its charge) reduced to zero.

discharging lead-acid cell The process of converting chemi-

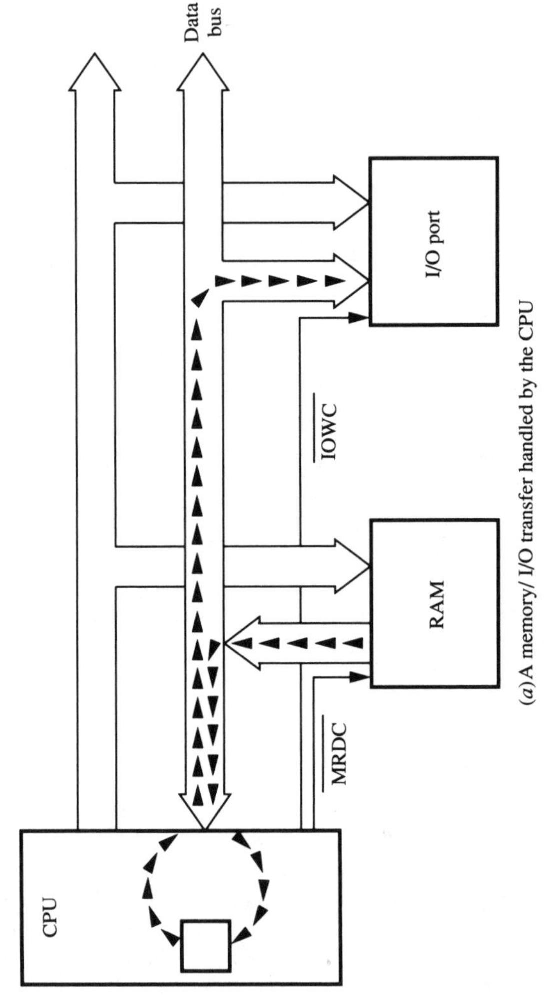

(*a*) A memory/ I/O transfer handled by the CPU

Direct memory access

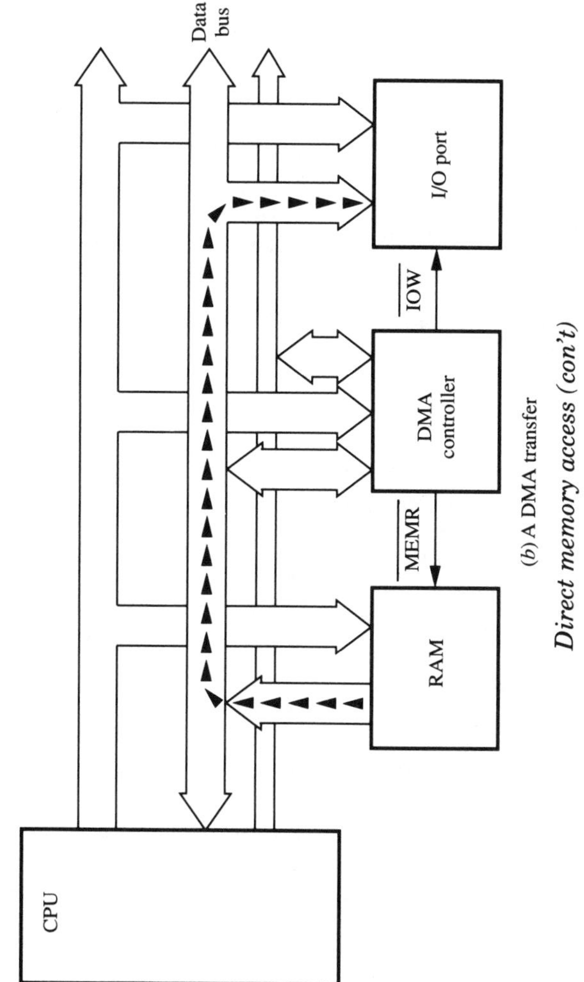

(*b*) A DMA transfer

Direct memory access (con't)

cal energy into electrical energy in a lead-acid cell. The process consists of leaving an excess of electrons on one electrode and a deficiency of electrons on the other.

discrete devices Individual electronic components that must be interconnected with other components to form a complete functional circuit.

discriminator An electrical circuit that produces an output voltage whose value is proportional to the frequency of an input signal. A discriminator does just the opposite of a voltage-controlled oscillator.

dish antenna See parabolic antenna.

disk operating system A program that instructs the computer on how to operate the disks as well as other necessary tasks.

distortion The action of an electrical device when the output signal is not a true representation of the input signal.

distributed refresh A method of refreshing DRAM where cell rows are refreshed between read and write operations.

distributive law In Boolean algebra, states that ORing variables that are then ANDed with a single variable is the same as ANDing each variable with the single variable and ORing the results. Thus, $A \cdot (B + C) = (A \cdot B) + (A \cdot C)$.

divergence Pertaining to fiber optics and lasers. The spreading of light as it leaves the light source.

DMA Abbreviation for direct memory access. See direct memory access.

DMA controller A device used to control the I/O transfer of data without microprocessor intervention.

DMM Digital multimeter.

dominant mode In a waveguide, the most natural mode of operation for a waveguide. This is the lowest possible frequency that can be propagated in a given waveguide.

don't cares A logic condition whose outcome has no effect on the rest of the system. Used in the analysis and design of logic circuits.

dopant A chemical that will cause an excess of positive or negative charges when added to a semiconductor material.

doping The process of adding impurities into pure semiconductor material in order to control the conduction characteristics of the material.

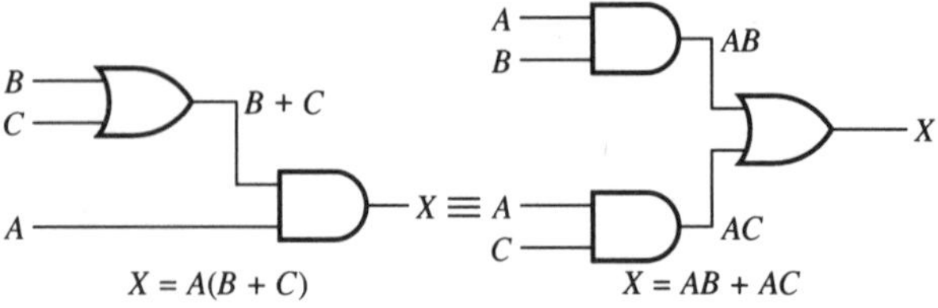

Distributive law

Doppler detector See phase-locked receiver.

DOS Abbreviation for disk operating system. See disk operating system.

double-pole, double-throw switch Two SPDT switches mechanically linked together so that both operate as a unit. See single-pole, double-throw switch.

double sideband A type of modulation in which both sidebands are transmitted and the carrier is not. This results in more transmission efficiency since the carrier does not contain any information and would require two-thirds of the total transmission power if it were transmitted. Not as efficient as single-sideband transmission. See single sideband.

double-tuned amplifiers Electronic circuits that have the inputs and outputs of amplifer circuits constructed with resonant (tuned) circuits. These types of amplifiers are commonly used as RF and IF amplifiers.

DPDT Abbreviation for double-pole, double-throw switch. See double-pole, double-throw switch.

DPSK Abbreviation for differential phase-shift keying. See differential phase-shift keying.

DPST Abbreviation for double-pole, single-throw switch. See double-pole, single-throw switch.

drain One of the connections of a field-effect transistor. The voltage applied to the gate controls the current between the source and the drain.

drain curves In a JFET, the graphical relationship between the channel current (I_{DS}) and the control voltage (V_{DS}, voltage from drain to source).

DRAM 1. Abbreviation for dynamic RAM. 2. Dynamic random access memory. See dynamic RAM.

drift The small change in an electrical characteristic of a component or circuit over time. Drift is usually caused by temperature changes.

driver In electronics, a circuit whose purpose it is to increase the power of a signal in order to

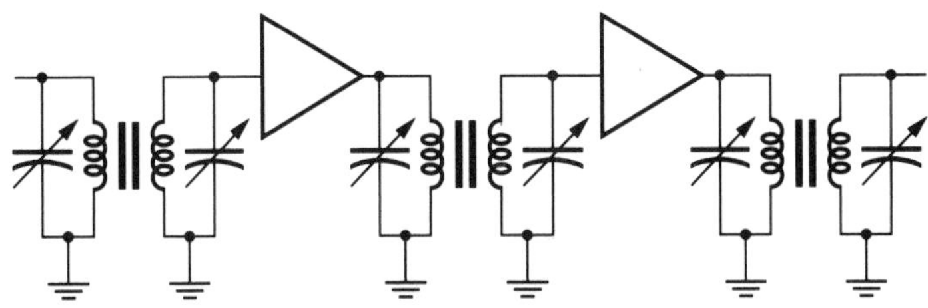

Double-tuned amplifiers

operate another circuit properly. For example, in a transmitter, the audio driver is the circuit that feeds the audio signal into the audio power amplifier.

droop In a pulse waveform, a decrease in the pulse amplitude with time.

dry cell 1. The most basic unit of a battery that does not use a liquid as an electrolyte but instead uses a type of paste. 2. A sealed source of voltage constructed from an electrolyte paste contained in a zinc container surrounding a carbon rod. The carbon rod is the positive terminal while the zinc contained is the negative terminal.

DSB Abbreviation for double sideband. See double sideband.

dual in-line package An electrical device or integrated circuit packaged in a rectangular form where its connectors are rows of pins along the long sides of the container.

dual J-K Two J-K flip-flops in the same IC package. An example

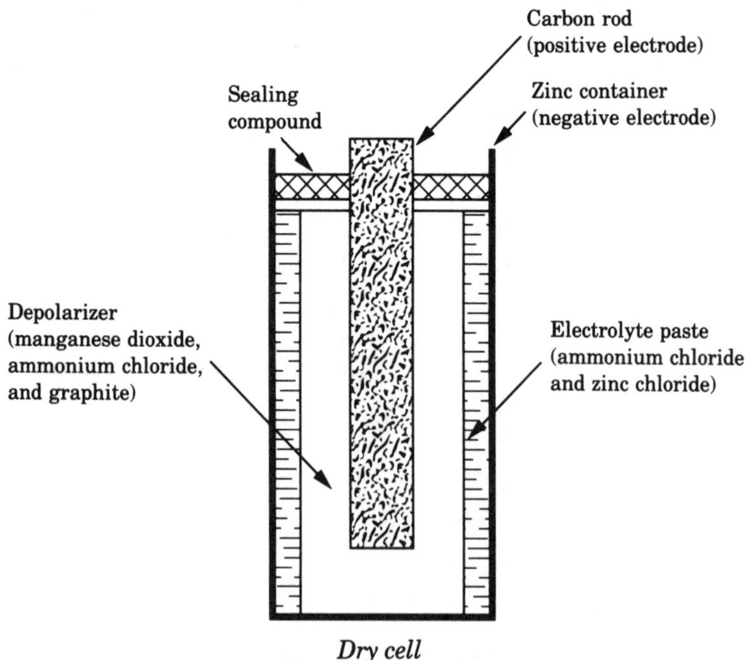

Dry cell

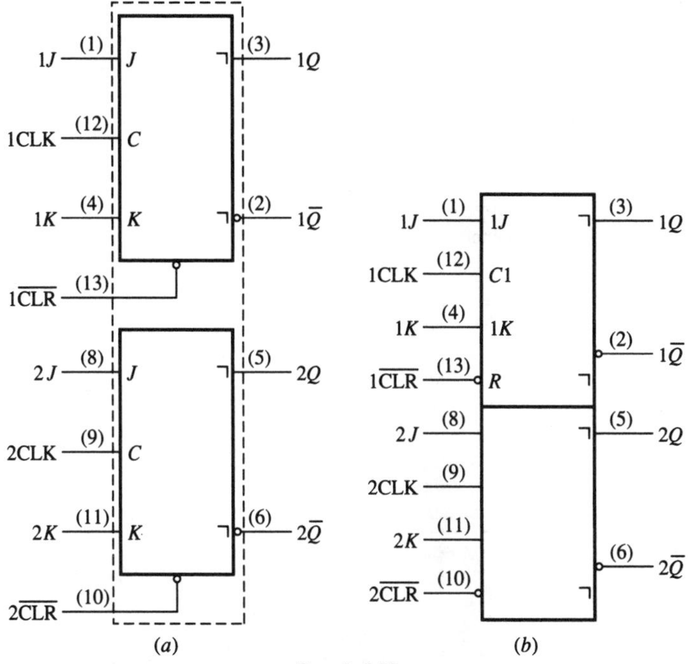

Dual J-K

of a dual J-K flip-flop is the 74107 dual master-slave flip-flop.

dual power supply A source of voltage that supplies both a positive and negative polarity. For example, a 12-volt dual power supply would supply +12 VDC and −12 VDC.

dual-slope A/D converter An analog-to-digital converter that is commonly found in digital voltmeters and similar equipment. This type of converter is similar to that of the single-slope A/D converter except that both a variable-slope ramp and a fixed-slope ramp are used.

dual-trace scope An oscilloscope capable of displaying two waveforms in such a manner that they appear to be displayed at the same time.

dummy load A device that simulates the transmitting antenna but does not allow any signal to be transmitted. Dummy loads are used when adjusting a transmitter; they prevent transmission of the signal while providing correct termination for the transmitter.

duplexer An electronic switch that allows the same antenna to be used for transmitting and receiving.

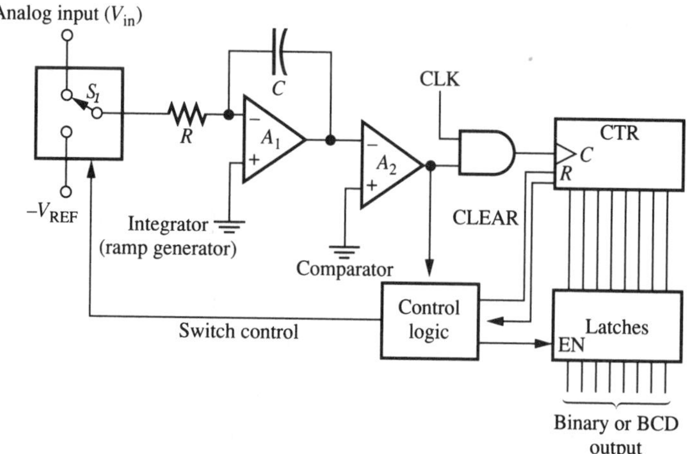

Dual-slope A/D converter

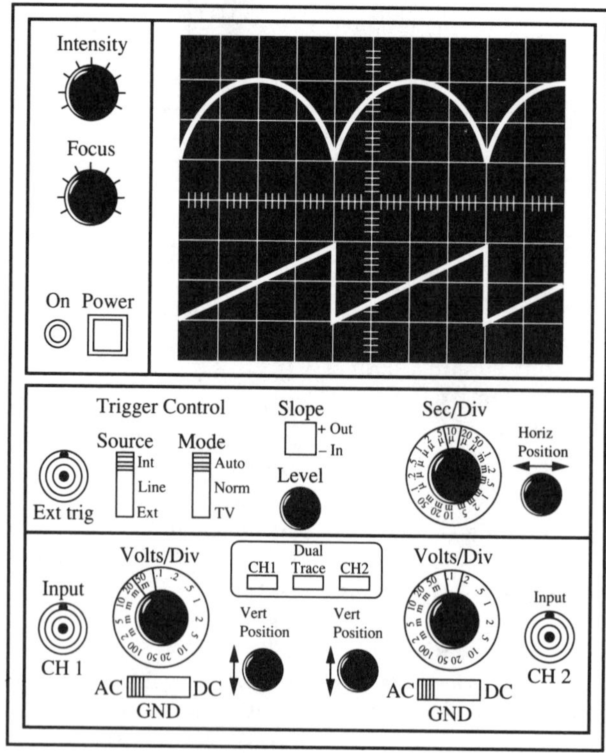

Dual-trace scope

duplex mode A method of communicating between two systems. The duplex mode allows for two-way communication between the two systems. See half-duplex and full-duplex mode.

duty cycle The ratio of the ON time to that of the OFF time of a given waveform. Usually expressed as a percentage.

dynamic input indicator The triangular symbol used on the clock input of a flip-flop to indicate that the device is synchronized by a logic transition of the clock. The edge-trigged flip-flop allows for greater timing accuracy over that of a simple latch.

dynamic RAM Volatile, randomly accessible, electrical memory that tends to lose stored information over a short period of time and requires "refreshing" in order to retain its bit pattern.

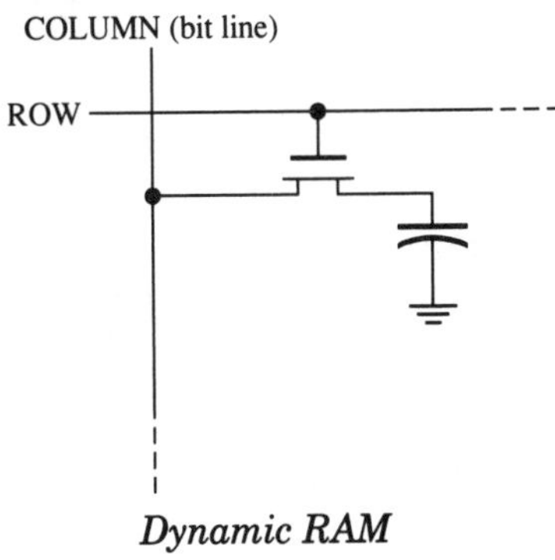

Dynamic RAM

E bend A waveguide that is bent 90 degrees along the axis of its electrostatic field (E field).

EC Abbreviation for emitter collector in a transistor.

ECL Abbreviation for emitter-coupled logic. Sometimes called current-mode logic.

eddy currents Electrical currents induced in the body of a conducting material by changing magnetic fields. Eddy currents in the core of a transformer can cause transformer losses and make the device less efficient. To reduce the effect of eddy currents, the core of the transformer is laminated (made up of thin core slices electrically insulated from each other).

edge connector A mechanical device mounted on a printed circuit board used to connect a second printed circuit board. In a microcomputer, the main circuit board contains edge connectors so that other printed circuit boards (such as the printer interface board) may be connected to it.

edge triggered Any electrical device that is activated by an electrical change. This usually takes place in the form of an electrical transition from 0 volts to +5 volts.

edge-triggered flip-flop A flip-flop that is activated by an electrical transition, usually from 0 volts to +5 volts or +5 volts to 0 volts, or 0 volts to +5 volts.

Edison effect When a material is heated in a vacuum, it will give off a cloud of electrons. If a positively charged plate is present, electrons will flow from the heated material to the positive charge. The Edison effect, named after Thomas A. Edison, ushered in the beginning of the vacuum tube era.

editor In computer programming, an editor is a program that allows you to enter and manipulate words and other keyboard symbols.

EDP Abbreviation for electronic data processing.

Edge-triggered flip – flop logic symbols (top – positive edge-triggered; bottm – negative edge-triggered).

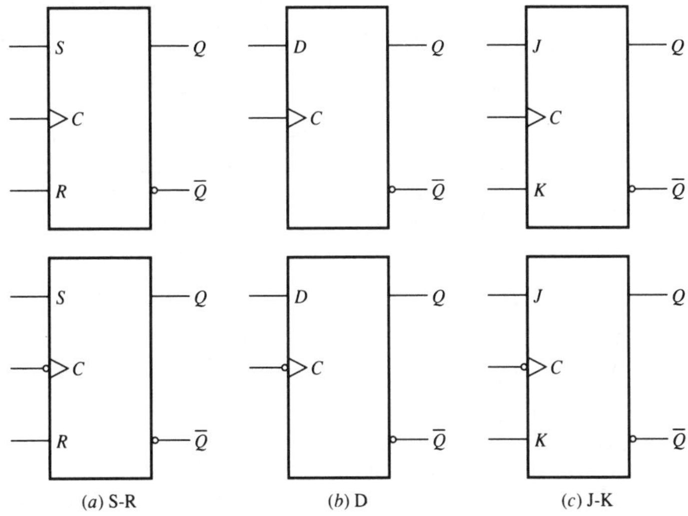

Truth table for a positive edge-triggered S-R flip-flop.

Inputs			Outputs		
S	R	C	Q	$\overline{Q}$	Comments
0	0	X	Q_0	$\overline{Q}_0$	No change
0	1	↑	0	1	RESET
1	0	↑	1	0	SET
1	1	↑	?	?	Invalid

↑ = clock transition LOW to HIGH
X = irrelevant ("don't care")
Q_0 = prior output level

Edge triggered

EEPROM Abbreviation for electrically erasable programmable read-only memory.

effective address The actual memory location that is used to store data.

effective value See root mean square.

efficiency In electronics, the ratio of output power to input power for a passive device. As an example, for a transformer, the output power is always less then the input power because of core losses and other imperfections.

EIA Abbreviation for Electronic Industries Association. See Electronic Industries Association.

8421 BCD code A binary code where the weight of each binary

8421 (BCD)	Decimal
0000	0
0001	1
0010	2
0011	3
0100	4
0101	5
0110	6
0111	7
1000	8
1001	9

8421 BCD code

digit is given by 8, 4, 2, and 1. This code uses only values from 0 through 9 (even though 16 values can be represented). The advantage of this code is its ease of conversion between itself and decimal numbers.

elastance The reciprocal of capacitance. Elastance is a measure of the difficulty of placing a charge on a capacitor. Measured in the daraf.

electrical 1. Having to do with the movement and control of electrical charges. Usually implies the use of passive devices such as resistors, inductors, and capacitors. 2. Related or pertaining to electricity.

electrical charge Electric energy stored in a device. Electrical charge is either positive or negative where like charges repel and unlike charges attract.

electrical ground The zero voltage reference for the power supply in an electronic device. Electrical ground is usually connected to the equipment chassis.

electrical noise Undesirable and unpredictable electrical signals caused by natural phenomena and person-made equipment.

electrical shielding Conducting material that is used to protect a circuit or connecting wires from electrostatic or radio frequency interference.

electric circuit An electric circuit, in a most fundamental sense, consists of a voltage source, a load, and a path for current from the source to the load.

electric field The region around a charged body that indicates the strength of the influence of the charge. For example, the electrical field of a capacitor is generally confined to the region

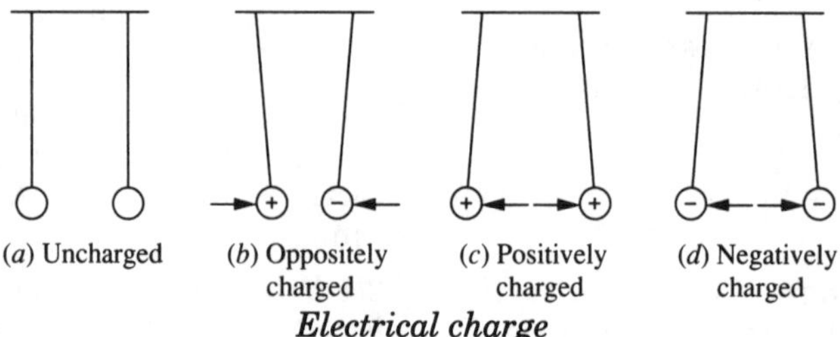

Electrical charge

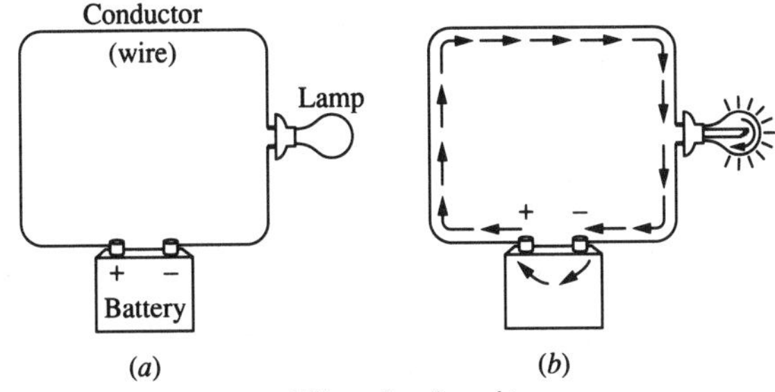

Electric circuit

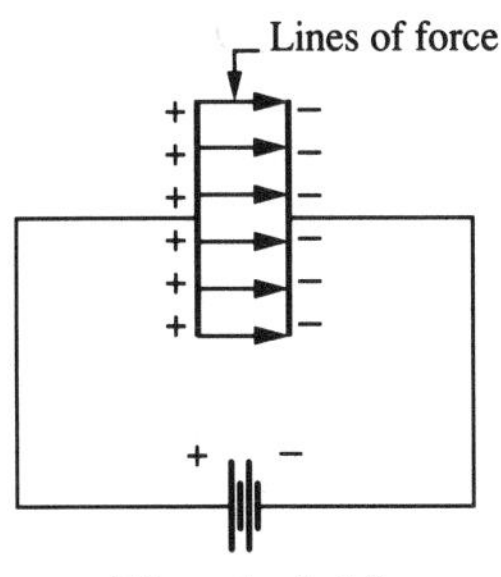

Electric field

between the charged plates of the capacitor.

electric field intensity A measure of the strength of the electric field. See electric field.

electricity The area of knowledge that deals with the control of electrical charges. Electricity deals with static (nonmoving charges) and dynamic (moving charges) charges.

electrochemical Dealing with the interaction of a chemical process that results in an electrical process or an electrical process that results in a chemical one. For example, the charging and discharging of a lead-acid battery is an electrochemical process.

electrodynamometer A type of meter movement simular to that of the d'Arsonval movement but uses an electromagnetic rather than a permanent magnetic. See d'Arsonval movement.

electroluminescence Light caused by some direct electrical action. For example, electroluminescence is caused by the current flow in a pn junction. Light-emitting diodes use electroluminescence to emit light. The process of producing light by electrical means in a solid-state device. The light-emitting diode uses electroluminescence.

electrolyte 1. A chemical substance where the conduction of electricity is accompanied by a chemical action. For example, the material used in an electrolytic capacitor is an electrolyte. 2. Contact element that is the source of ions between the electrodes of a battery.

electrolytic capacitor An electrical element constructed from two electrical isolated conducting materials separated by an electrolyte. Electrolytic capac-

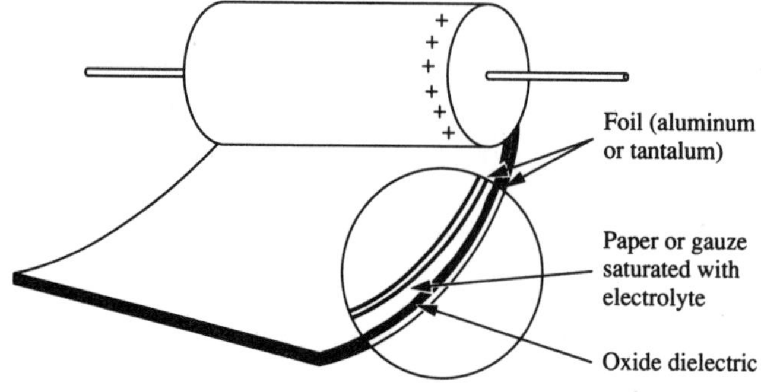

Basic construction

Electrolytic capacitor

itors are an economical way of having large capacitance values. These capacitors are usually found in low-frequency applications such as power supply filtering.

electromagnet A temporary magnet whose magnetic properties are produced from the application of an electrical current. An electromagnet consists of a coil of wire wrapped around a core (called the solenoid) of magnetic material. When current flows in the coil of wire, a magnetic field is induced in the core.

electromagnetic recording A method of placing information on a magnetic material for storage and later retrieval of that information.

electromagnetic spectrum A chart that shows the characteristics of a continuous range of frequencies from 0 Hz to cosmic rays.

electromagnetic waves Waves of energy that do not require a medium for transmission. They travel at a rate of 3×10^8 meters in free space. Can be analyzed as

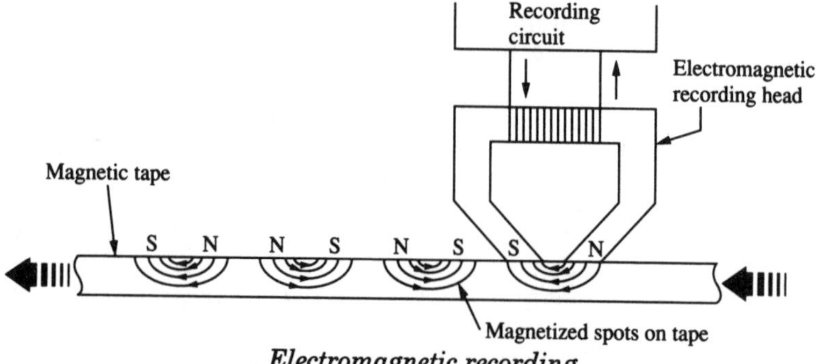

Electromagnetic recording

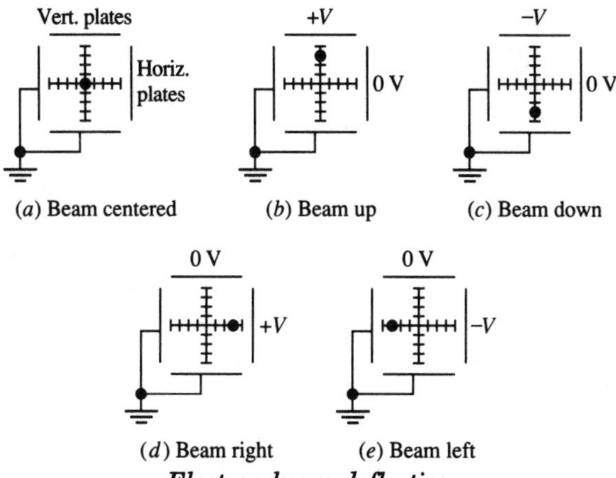

(*a*) Beam centered (*b*) Beam up (*c*) Beam down

(*d*) Beam right (*e*) Beam left

Electron beam deflection

a magnetic wave with a perpendicular electrostatic component. Radio waves and light are electromagnetic waves.

electromagnetism The relationship between electricity and magnetism. A moving charge creates its own magnetic field. A moving magnetic field can control the motion of charges.

electromotive force The force that causes electrical current flow when there is a difference of potential between two points. This is not the same as potential difference, which is the voltage developed across a circuit element as a result of the current in the element.

electron An atomic particle with a negative charge. See Bohr model.

electron beam deflection The moving of an electron beam (as in a cathode ray tube) through the use of electronstatic, magnetic, or electromagnetic means.

electron drift The movement of electrons in a definite direction in a material. Electron drift is usually caused by the application of some external difference of potential.

electron flow The direction of the flow of electricity that is in the same direction as the motion of electrons.

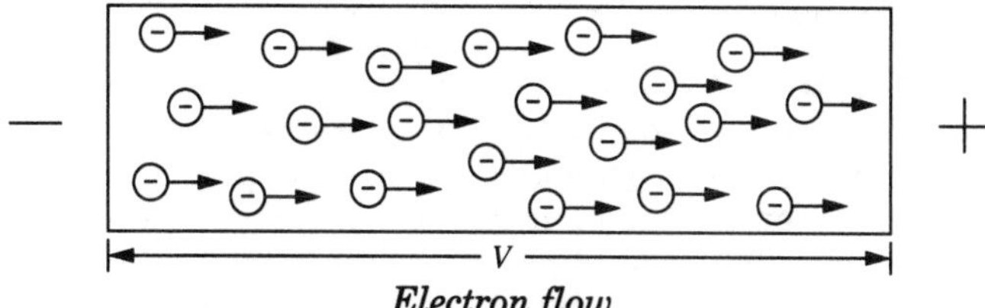

Electron flow

electron gun A mechanical device that is contained in the neck of a cathode ray tube. The electron gun is used to generate and in some cases control the flow of electrons that will strike the phosphor-coated face of the cathode ray tube to help produce a desired image.

electron optics The field of study that deals with the behavior of the electron beam under the influence of electrostatic and electromagnetic forces.

electron-pair bond A valence bond formed by two electrons, one from each of two adjacent atoms.

electron scanning The movement of an electrical beam across a plane done by electrostatic or magnetic means. Electron scanning in an oscilloscope is accomplished by electrostatic means, while a television receiver uses magnetic means.

electron volt Amount of energy gained by electron when it falls through a potential difference of 1 volt. One electron volt equals 1.602189×10^{-12} erg.

electronic Having to do with the movement and control of electrical charges. Usually implies the use of active devices such as transistors and diodes.

electronic mail Any system for delivering a hard copy using electrical means. As an example, sending text from one computer to another over telephone lines, where the receiving computer prints the text on its printer is a form of electronic mail (E mail).

electronic switch A circuit element (such as a transistor) used to make an ON/OFF condition as opposed to a mechanical device used as a switch. Thousands of electronic switches are used inside computers to help form logic and data storage networks.

Electronic Industries Association A trade association of the electronics industry. Abbreviated EIA. One of the functions of the EIA is to help set manufacturing standards in electrical equipment and components.

electronics The field of study that includes the control of electricity through the use of circuit elements (resistors, capacitors, and inductors) and circuit devices (such as transistors, FETs, etc.). Electronics includes the study of equipment for making electrical measurements as well as the design, maintenance, and repair of electrical systems.

electro-optics That field of study dealing with the effects of electric fields on optical phenomena.

electrostatic deflection The deflection of an electronic beam using an electrostatic field. A cathode ray tube found in oscilloscopes uses electrostatic deflection. As electrons leave the electron gun, they pass through a pair of vertical plates and a pair of horizontal plates. By varying the charge applied to these plates, the position of the electron beam on the face of the CRT can be precisely controlled.

element A chemical substance that cannot be reduced to a simpler substance by chemical means. For example, hydrogen is an element because it cannot be reduced to a simpler substance; water is not an element because it can be reduced to hydrogen and oxygen.

E mail Abbreviation for electronic mail. See electronic mail.

emf Abbreviation for electromotive force. See electromotive force.

emitter bias In a transistor amplifier, using two power supply connections, one above ground (positive), the other below (negative) between the emitter and collector with the base referenced to ground in order to establish the dc operating point of the device.

emitter follower See common collector.

empty term Part of a Boolean expression used in the evaluation of a larger Boolean expression where the contents of the smaller expression have not yet been evaluated.

EMS Expanded Memory Specifications. Describes a method where four contiguous physical pages of 16 k each can access up to 32 M of expanded memory space. See expanded memory.

encoder A logic network that converts information into a form of a code. Usually has several input lines, the activation of which determines the binary code generated on the output. See decoder.

energy The ability to do work. One of the most common units of measurement for work is the joule (J); another is wattseconds (Ws).

energy diagram In semiconductor physics, a pictorial representation of the energy levels for semiconductor materials.

engineering The profession of applying scientific principles to practical applications in an economical and reliable manner.

enhancement mode A method of operating a MOSFET where a positive gate voltage is applied to the device.

envelope The resulting amplitude variations of an amplitude-modulated wave. See amplitude modulation.

EPROM Abbreviation for erasable programmable read-only memory. See erasable PROM.

equipment A complete system as opposed to something being a substructure. For example an oscilloscope is called a piece of equipment, meaning that the oscilloscope is a complete system in itself.

equivalent circuit A circuit whose arrangement of common circuit elements (such as resistors, capacitors, and inductors) has the identical electrical characteristics for a given set of conditions as a different or more complex circuit using active (transistors, FETs, etc.) devices.

erasable memory Memory where information can be deleted without destroying the medium containing the digital information. An example would be magnetically erasing a floppy disk.

erasable PROM A ROM whose bit pattern can be changed by the user and later erased and then changed again. Usually contains a transparent plastic window that allows ultraviolet light to erase the memory so it may be programmed again by the user.

error detection In digital circuits, a method of detecting errors caused by the transfer of processing of digital information. One of the simplest methods of error detection is through parity checking. See parity.

ev Letter symbol for electron volt. See electron volt.

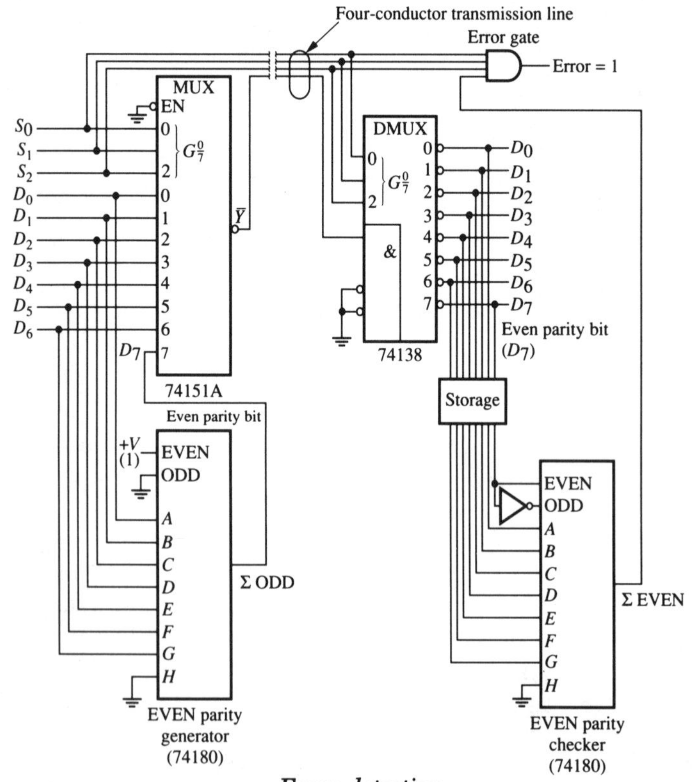

Error detection

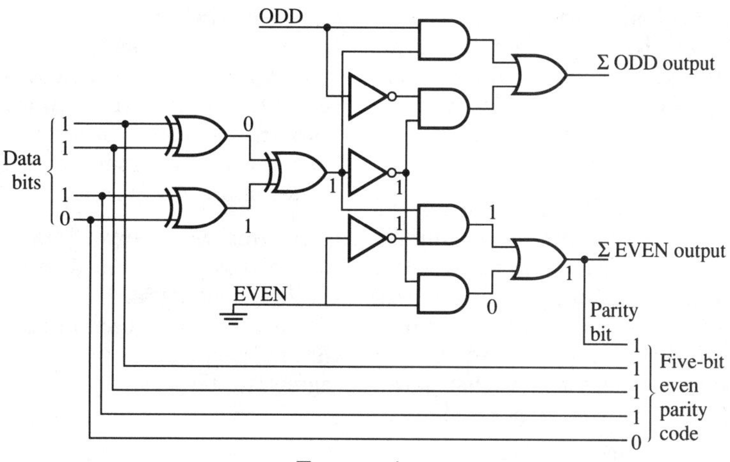

Even parity

even harmonics Even multiples of a fundamental frequency.

even parity Creating a binary word with an extra bit so that the number of 1's in the word adds up to an even number. See parity.

excess three code Similar to BCD but with a 3 added to the groups of 4 binary bits used in the BCD representation of a number. See binary-coded decimal.

exclusive NOR The Boolean function that is TRUE only when there is an even number of TRUE variables.

exclusive NOR gate An electronic digital circuit with one output and two or more inputs that follows the logic of the Boolean exclusive nor function.

exclusive OR The Boolean function that is TRUE only when there is an ODD number of TRUE variables. Represented as "+." Thus, in $A = B \oplus C$, A will be TRUE only when B is TRUE or C is TRUE, but will be FALSE when B and C are both TRUE or both FALSE. May also be expressed as: $A = \overline{B}C + B\overline{C}$.

exclusive OR gate An electronic digital circuit with one output and two or more inputs that follows the logic of the Boolean exclusive OR.

expanded memory Computer memory that is not directly accessible to the microprocessor. Ex-

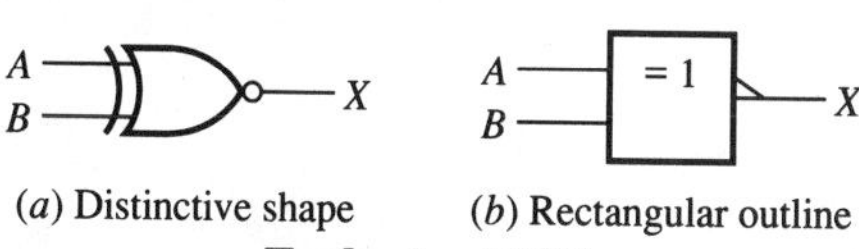

(*a*) Distinctive shape (*b*) Rectangular outline

Exclusive NOR

panded memory is accessed by *bank switching*. This technique provides small windows of memory through which blocks of expanded memory are traded with base memory.

exponential Concerning exponents or an expression using exponents. When a quantity is said to change in an exponential manner, it means that the quantity will increase by the square (or square root) or some other power rather than a simple linear change.

exponential curves Curves that follow a precise mathematical formula that are used to represent the electrical charging and discharging characteristics of a capacitor and magnetic field behavior of an inductor.

extended memory The portion of computer memory past the first megabyte. Used by the 80286 and 80386 microprocessor that support 16 megabytes and 4 gigabytes, respectively.

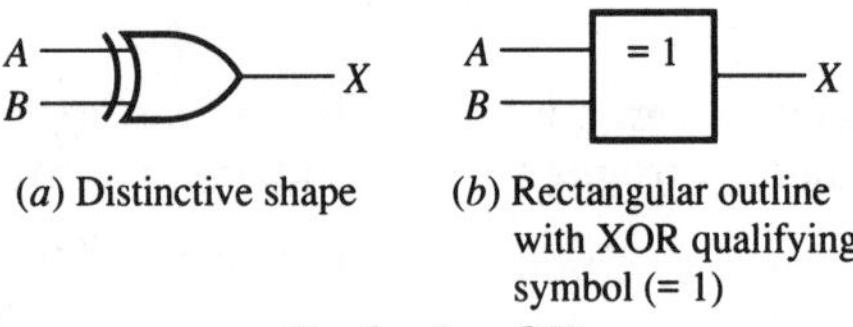

(*a*) Distinctive shape (*b*) Rectangular outline with XOR qualifying symbol (= 1)

Exclusive OR

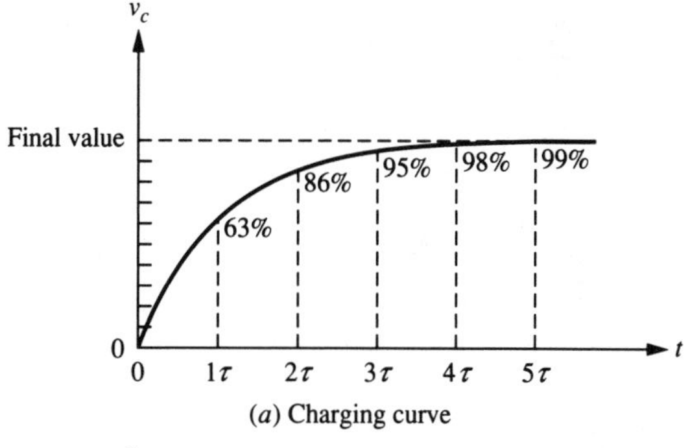

(*a*) Charging curve

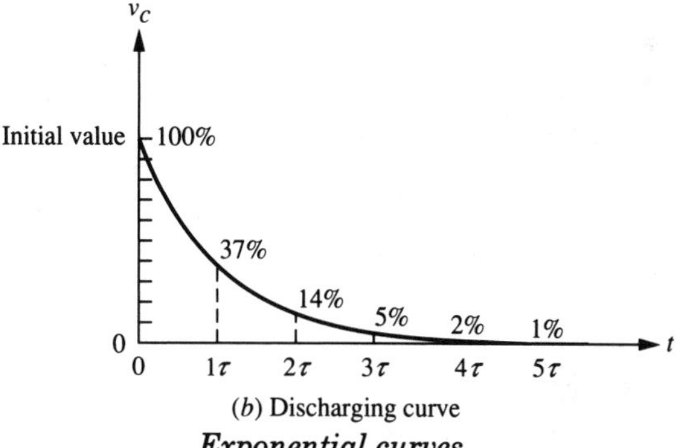

(*b*) Discharging curve

Exponential curves

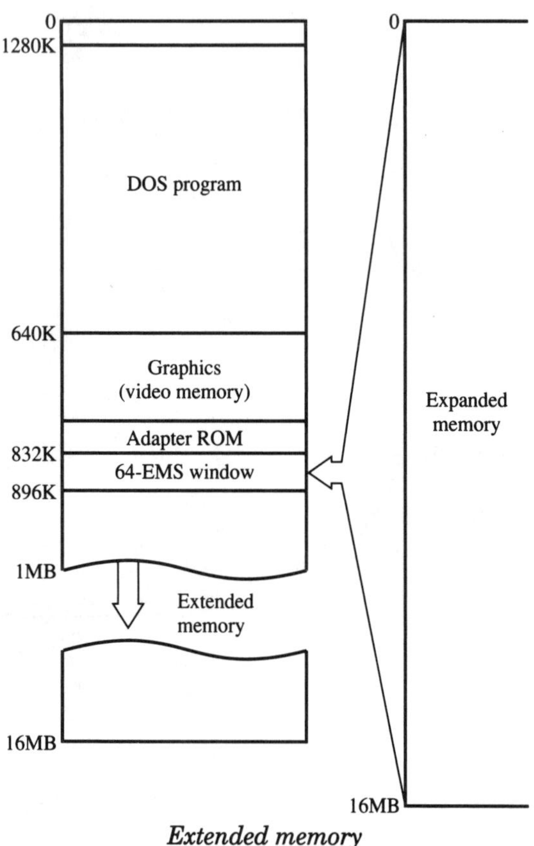

Extended memory

fade The gradual decrease in amplitude or signal strength of a signal. A radio signal is said to fade when the received station sounds gradually weaker. Fading can be caused by changes in atmospheric conditions.

fading The result of decreased signal strength in a receiver due to multiple hop paths from a reflected sky wave. In some cases when an electromagnetic wave is reflected back from the ionosphere, it will have several return paths, some of which may be out of phase and cause a reduction in signal strength. Because of this phenomenon, reception strength may actually be increased by moving farther away from the transmitter to a point where the reflected waves are not out of phase.

failure The inability of a component or system to perform in its intended manner. A component failure means that the component is no longer useful for its originally intended purpose.

fall time The amount of time it takes for a pulse waveform to go from 90% to 10% of its maximum amplitude.

false In a two-state logic system, the opposite of true. False is usually represented by a numeric zero or a low (0) voltage.

fan-out 1. In digital integrated circuits, the maximum number of logic devices that can be controlled by the output of a single logic device. For example, a standard TTL NAND gate may have a fan-out of 10, meaning that its output can be connected to up to 10 different TTL inputs without degradation in the quality of its output pulse. 2. In logic circuits, the number of parallel loads that a single output may reliably operate. For example, an AND gate will no longer have a reliable output condition if its fan-out is exceeded by connecting it to too many inputs of other logic gates. Most gates have a fan-out of 10, meaning that they may be reliably connected to at least 10 other gate inputs. Fan-out can be increased through the use of line drivers.

farad Unit of measurement for capacitance. One farad is the amount of capacitance when a charge of one coulumb produces a change of one volt in the potential difference between the plates of the capacitor.

Faraday's law The moving of a coil in a magnetic field will produce a voltage across the coil. Faraday's law is expressed mathematically as $e = n(d\Phi/dt)$, where e is the resulting voltage (in volts), $d\Phi$ an instantaneous change in flux (in webers), and dt is an instantaneous change in time (in seconds).

FCC Abbreviation for Federal Communications Commission. See Federal Communications Commission.

FDM Abbreviation for frequency division multiplexing. See frequency division multiplexing.

Federal Communications Commission The electronic communications licensing and enforcement authority of the United States.

feedback The process of causing some of the output to be brought back to the input. In an amplifier, negative feedback can be used to help improve the linearity of the amplified signal. In an oscillator, positive feedback is used to sustain oscillations.

female connector Pertaining to a mechanical device con-

A demonstration of the first part of Faraday's law: The amount of induced voltage is directly proportional to the rate of change of the magnetic field with respect to the coil.

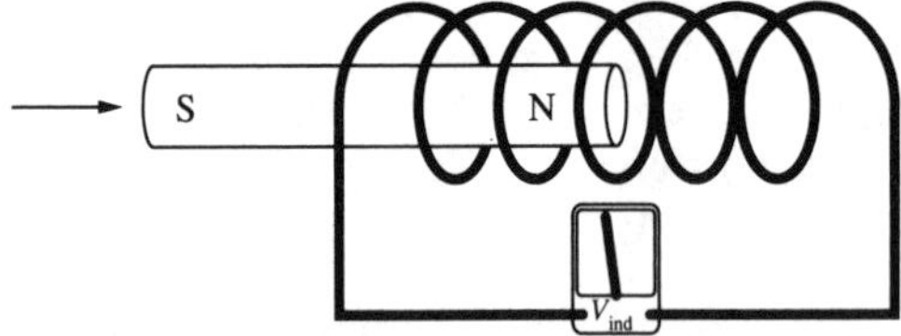

(a) As magnet moves to the right, magnetic field is changing with respect to coil, and a voltage is induced.

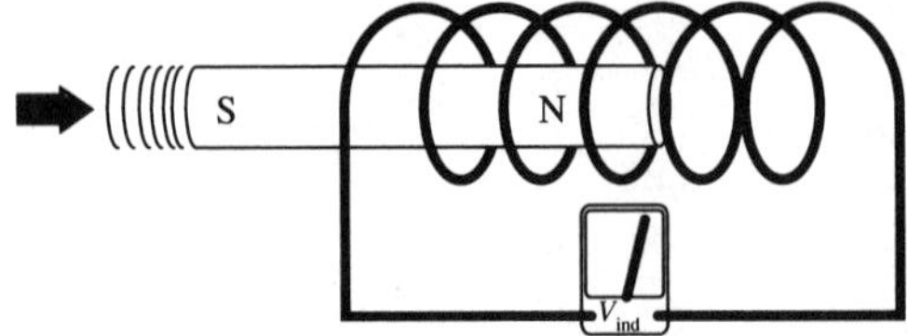

(b) As magnet moves more rapidly to the right, magnetic field is changing more rapidly with respect to coil, and a greater voltage is induced.

Faraday's law

A demonstration of the second part of Faraday's law: The amount of induced voltage is directly proportional to the number of turns in the coil.

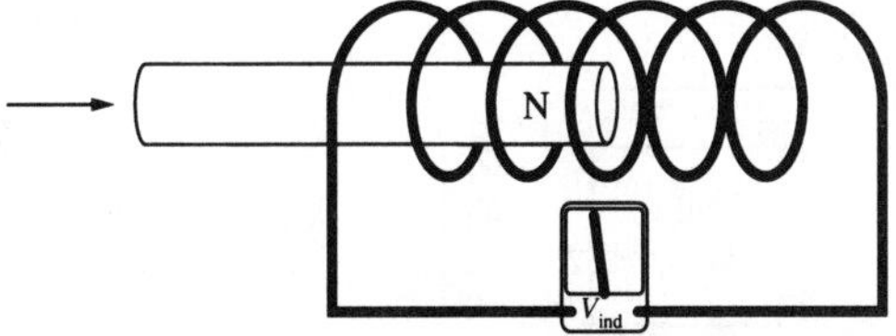

(*c*) Magnet moves through coil and induces a voltage.

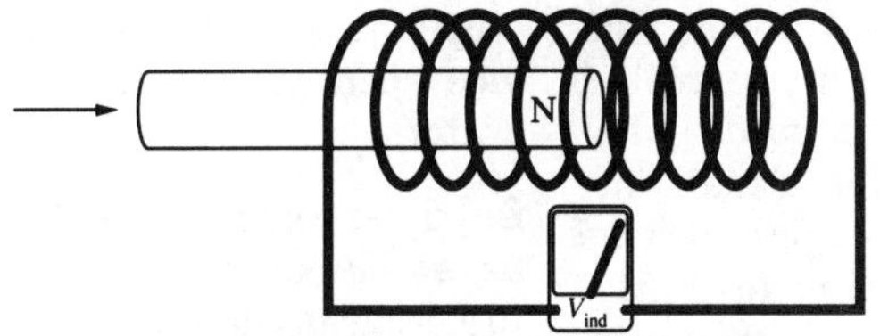

(*d*) Magnet moves at same rate through a coil with more turns (loops) and induces a greater voltage.

Faraday's law (con't)

taining recesses into which another matching device will fit.

ferromagnetic materials Materials that have permeabilities many times greater then those of free space.

FET Abbreviation for field-effect transistor. See field-effect transistor.

fetch That part of a microprocessor cycle spent getting data from an external source.

fetch/execute In microprocessor systems, the process of the microprocessor getting an instruction from memory and then performing the process required of that instruction.

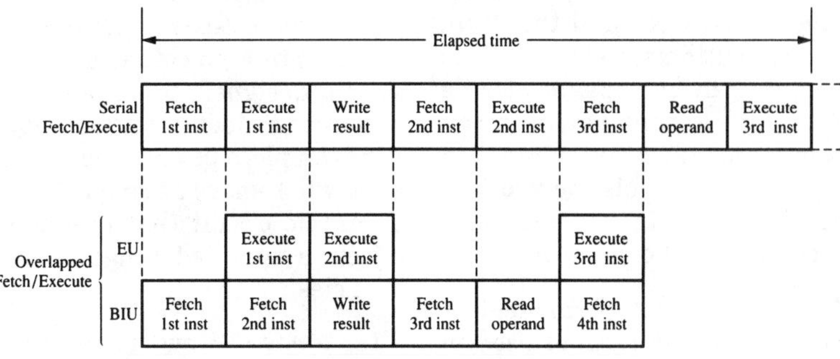

Fetch/execute

A basic FPLA block diagram.

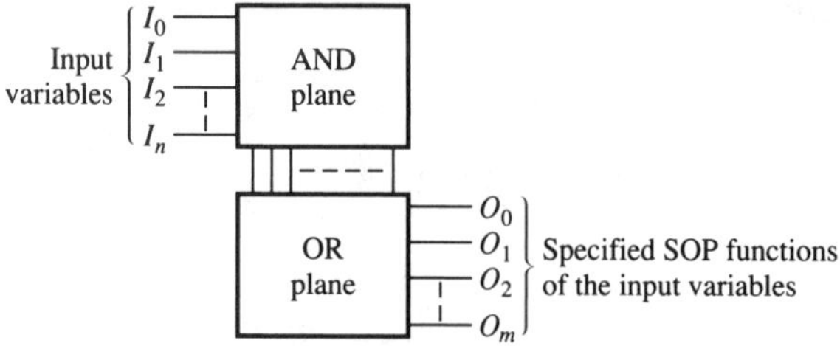

Field-programmable logic array

fiber optics The transmission of information in a controlled path where light is used as the transmission carrier. Usually accomplished by material such as glass made into tiny fiber filaments that is capable of transmitting light.

fidelity The accuracy in which a weak signal is reproduced into a stronger one. The term high fidelity is usually used to describe an audio system capable of reproducing voice and music in its original and natural form.

field-effect transistor A solid-state device consisting of a source, gate, and drain. Voltage applied to the gate controls the current between the source and the drain. The current path between the source and the drain is referred to as a channel. The channel can be thought of as a path for current between the source and drain. Voltage applied to the gate controls the width of the channel and thus the amount of current in the channel.

field magnet A permanent or electromagnetic magnet used in the construction of a speaker, microphone, or other electrical device such as an electrical generator.

field-programmable logic array A digital device similar to a programmable ROM but used to implement Boolean logic functions. An FPLA can replace many separate gating chips and greatly simplify the design of a digital system. See PLD, PLA, PAL, and PLS.

FIFO Abbreviation for first in, first out. See first in, first out.

film capacitor A capacitor constructed in such a manner as using a thin film as its dielectric.

filter 1. In electronics, the process of removing electrical variations or changes. As an example, a capacitor is used to filter pulsating dc into a smoother dc. 2. A circuit designed to remove an unwanted electrical quantity. For example, a power supply uses a filter to remove ac ripple. 3. An electronic circuit that passes or rejects a specified range of frequencies.

filter order How a filter responds to a change in frequency.

Basic concept of an FPLA matrix.

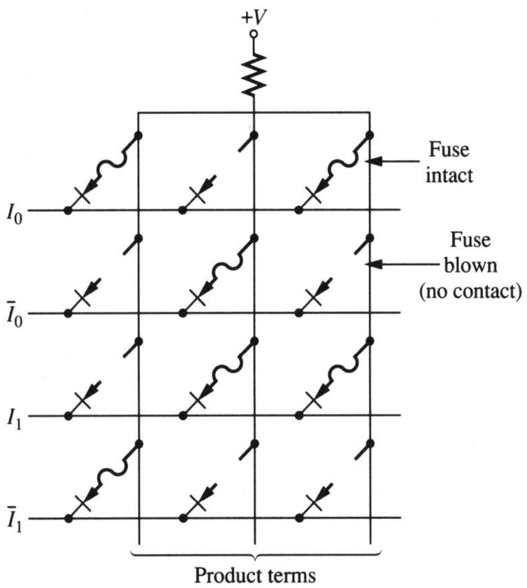

(*a*) Basic AND matrix

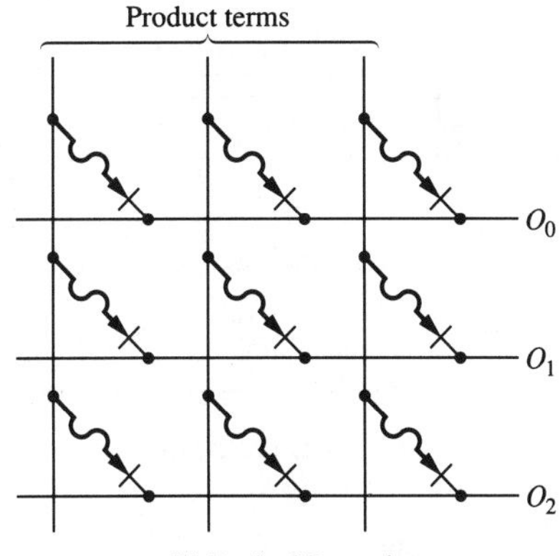

(*b*) Basic OR matrix

Field-programmable logic array (con't)

A first-order filter approaches a constant slope of 20 dB per decade; a second- order filter has 40 dB per decade.

firmware Permanent digital patterns contained in some kind of electronic medium. For example, programs and data stored in read-only memory are firmware.

first in, first out Refers to an arrangement of digital memory in which the first bit of data

Comparison of conventional and FIFO register operation.

Conventional shift register

Input	X	X	X	X	Output
0	0	X	X	X	⟶
1	1	0	X	X	⟶
1	1	1	0	X	⟶
0	0	1	1	0	⟶

X = unknown data bits.

In a conventional shift register, data stay to the left until "forced" through by additional data.

FIFO shift register

Input	—	—	—	—	Output
0	—	—	—	0	⟶
1	—	—	1	0	⟶
1	—	1	1	0	⟶
0	0	1	1	0	⟶

— = empty positions.

In a FIFO shift register, data "fall" through (go right).

Block diagram of a typical FIFO serial memory.

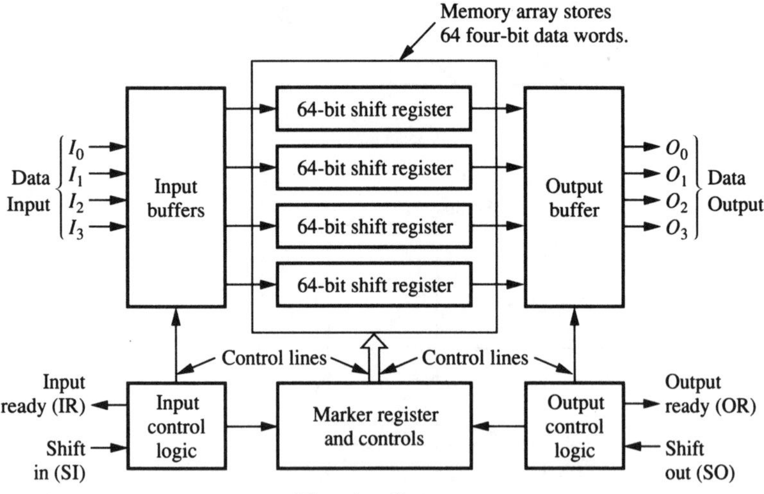

First in, first out

written into the memory is the first bit of data read from the memory.

5-bit decoder A logic circuit that will decode all possible combinations of a 5-bit binary number. This type of decoder has 5 inputs and 32 outputs. Only one output is active at a time.

555 timer An integrated circuit that can be used as an astable or monostable multivibrator. The 555 timer can also be used in other applications such as a voltage-controlled oscillator.

fixed inductor An inductor whose inductance value is not designed to be changed.

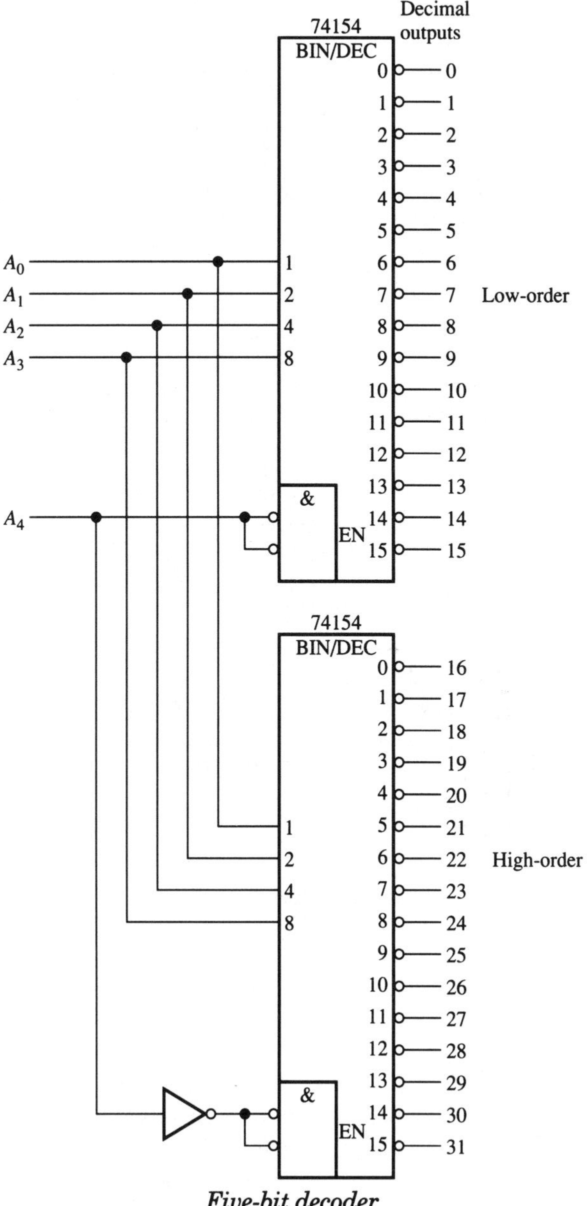

Five-bit decoder

A 555 timer.

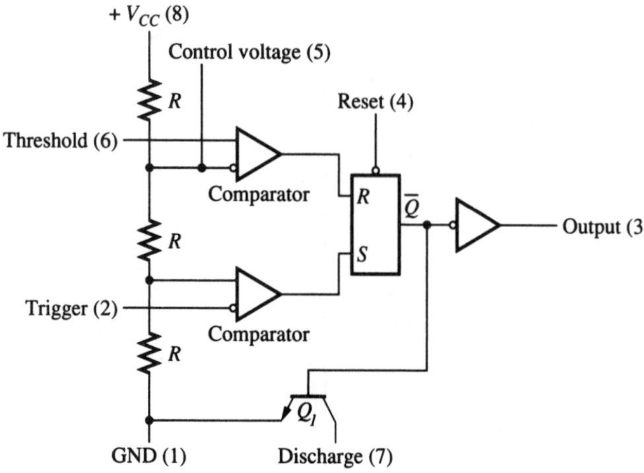

A 555 timer connected for astable operation.

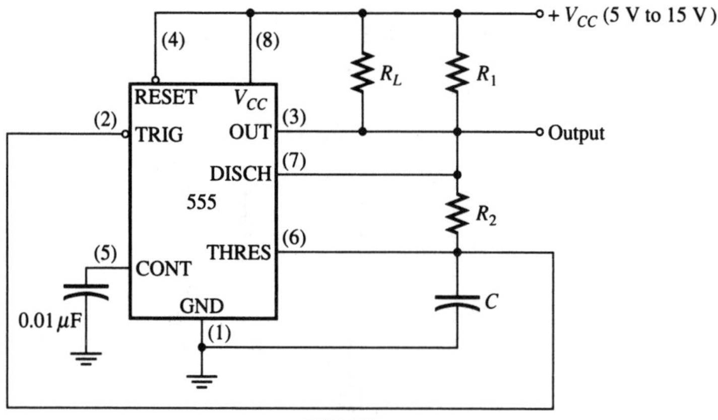

555 Timer

Fixed inductor

fixed resistor A resistor whose resistance value is not designed to be changed. The standard carbon composition resistor using color-coded bands is a fixed resistor.

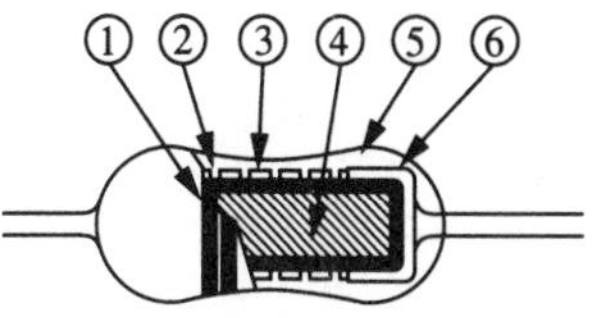

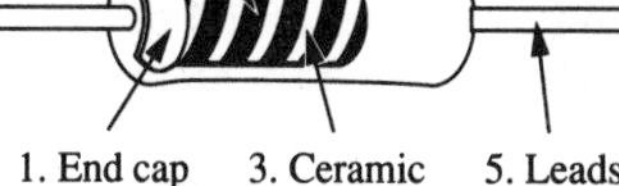

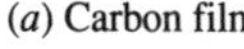

(*a*) Carbon film

(*b*) Metal film

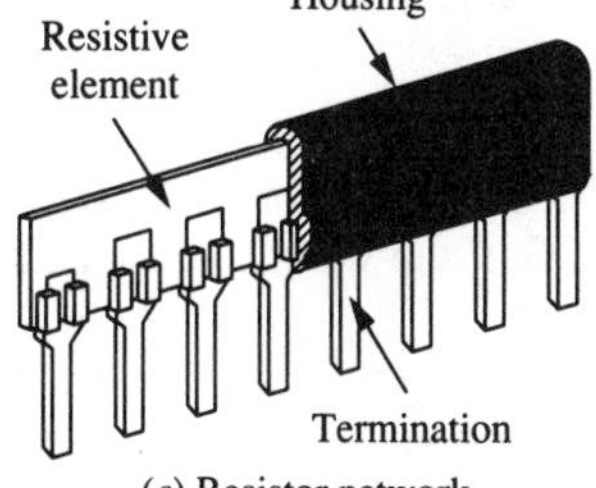

(*c*) Resistor network

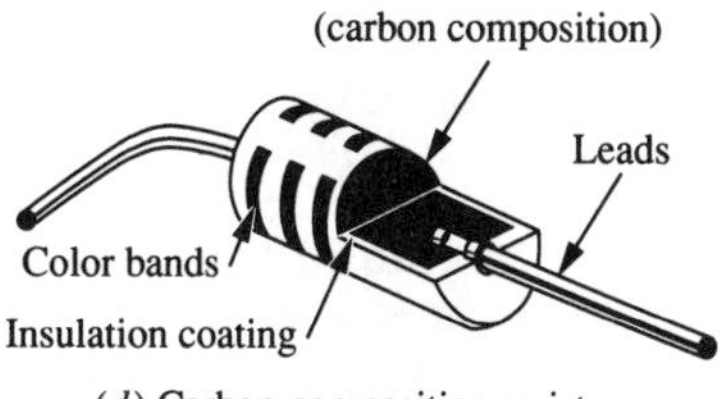

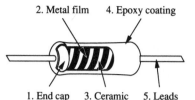

(*d*) Carbon-composition resistor

Fixed resistance

flag In programming, a single bit whose ON or OFF condition will signal another operation or provide information concerning the results of an operation. For example, most microprocessors contain a flag register, where the individual bits of the register are used as flags.

flap attenuator In a waveguide, a method of controlling the strength of the electromagnetic wave by using a flat piece of material that can be moved in or out of the guide.

flash converter An analog-to-digital converter that uses differ-

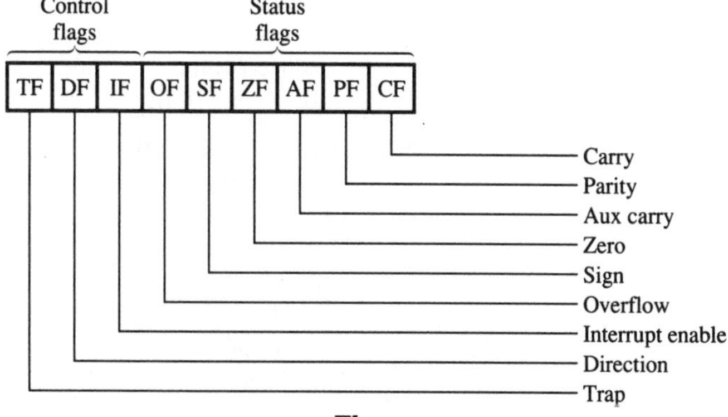

Flags

ential amplifiers and digital logic to produce a binary output that is equivalent to the magnitude of the analog input. Called a "flash" converter because it has a very fast conversion time. The main disadvantage of this type of converter is its poor accuracy.

flat pack A method of packaging an IC circuit so that it is encapsulated in a thin rectangular package with its connecting leads projecting from the edges of the unit.

F-layer Region of the ionosphere that causes most long-distance communications. The F-layer is about 130 to 160 miles above the earth (210 to 260 km).

Fleming valve The first vacuum tube diode named after its inventor.

flexible waveguide A waveguide constructed from spiral-wound ribbons of brass or copper. The outside of the waveguide is covered with a soft dielectric material such as rubber. These waveguides are used in the laboratory where continuous flexing may be necessary.

flip-flop A bistable digital device that is synchronized by an external source.

floppy disk A thin flexible plastic disk coated with a magnetic material used to store information in its magnetic material. Floppy disks are most commonly used for long-term storage of computer programs and data.

flops Floating-point operations per second. A measure of how fast a computer can perform calculations involving numbers that are not integers. These types of calculations are important in computer programs that describe the physical world.

flowchart The graphical representation of a sequence of operations. (See illustration on pg. 117.)

flowcharting A graphical means of showing a sequence of activities. Flowcharting is frequently used to show the sequence of operations in a computer program. One of the advantages of flowcharting a computer program is that it is computer language independent.

fluctuating dc Fluctuating direct current. See fluctuating direct current.

fluctuating direct current A current that changes in value but always travels in the same direction.

fluorescent Having the property of emitting light when activated by electrical energy. See fluorescent light.

fluorescent light An electrically activated source of visible radiation activated by ionized gas that in turn activates a fluorescent material that produces the visible radiation. Fluorescent lights most commonly come in the form of long tubes.

flutter Frequency deviation produced by irregular motion of

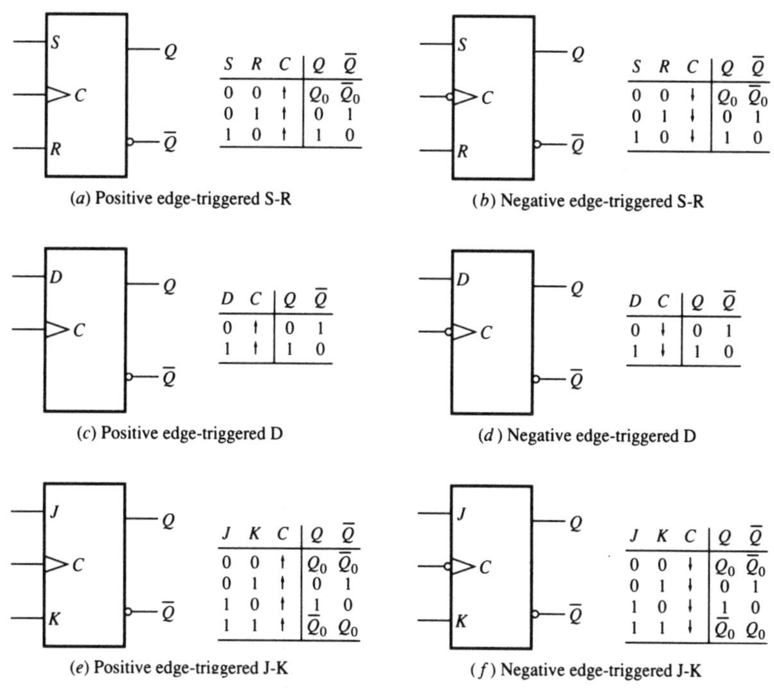

S	R	C	Q	$\overline{Q}$
0	0	↑	Q_0	$\overline{Q}_0$
0	1	↑	0	1
1	0	↑	1	0

(*a*) Positive edge-triggered S-R

S	R	C	Q	$\overline{Q}$
0	0	↓	Q_0	$\overline{Q}_0$
0	1	↓	0	1
1	0	↓	1	0

(*b*) Negative edge-triggered S-R

D	C	Q	$\overline{Q}$
0	↑	0	1
1	↑	1	0

(*c*) Positive edge-triggered D

D	C	Q	$\overline{Q}$
0	↓	0	1
1	↓	1	0

(*d*) Negative edge-triggered D

J	K	C	Q	$\overline{Q}$
0	0	↑	Q_0	$\overline{Q}_0$
0	1	↑	0	1
1	0	↑	1	0
1	1	↑	$\overline{Q}_0$	Q_0

(*e*) Positive edge-triggered J-K

J	K	C	Q	$\overline{Q}$
0	0	↓	Q_0	$\overline{Q}_0$
0	1	↓	0	1
1	0	↓	1	0
1	1	↓	$\overline{Q}_0$	Q_0

(*f*) Negative edge-triggered J-K

Flip-flops

a turntable or tape transport during recording, duplication, or reproduction. Also called wow.

flux density 1. Measure of the strength of a wave, expressed in watts per square centimeter or lumens per square foot. Visualized as the number of lines or maxwells per unit area in a section that is perpendicular to the direction of the flux. 2. A measure of the flux lines per unit area that are perpendicular to a magnetic flux path. Flux density (B) is measured in teslas (T) or in webers per square meter (Wb/m^2).

flux leakage A cause of reduced secondary voltage in a transformer. Caused by some of the magnetic flux lines of the primary winding of the transformer breaking out of the core and passing through the surrounding air back to the other end of the winding. (See illustration on pg. 118.)

flyback In a sawtooth waveform, the shorter of the two time intervals making up the wave. In a CRT, flyback is the retrace of the electron beam across the screen.

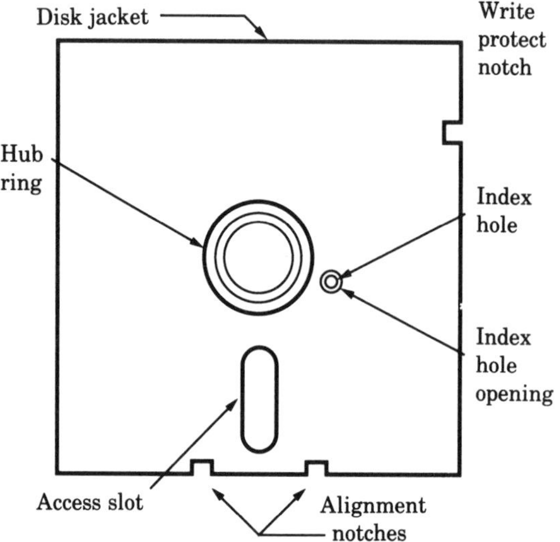

(a) The disk is contained within the jacket.

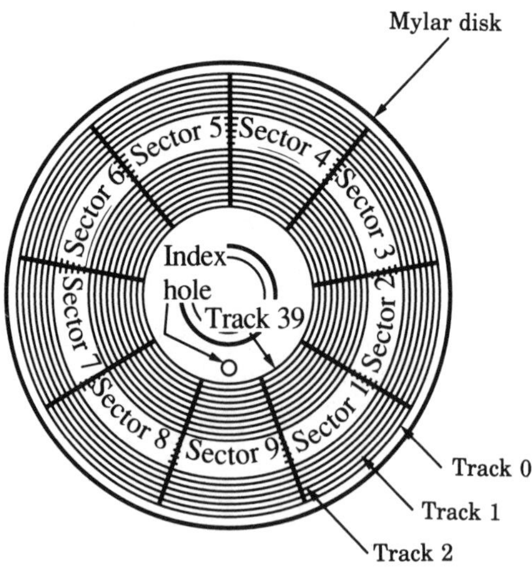

(b) Track/sector format of a typical floppy disk.

Floppy disk

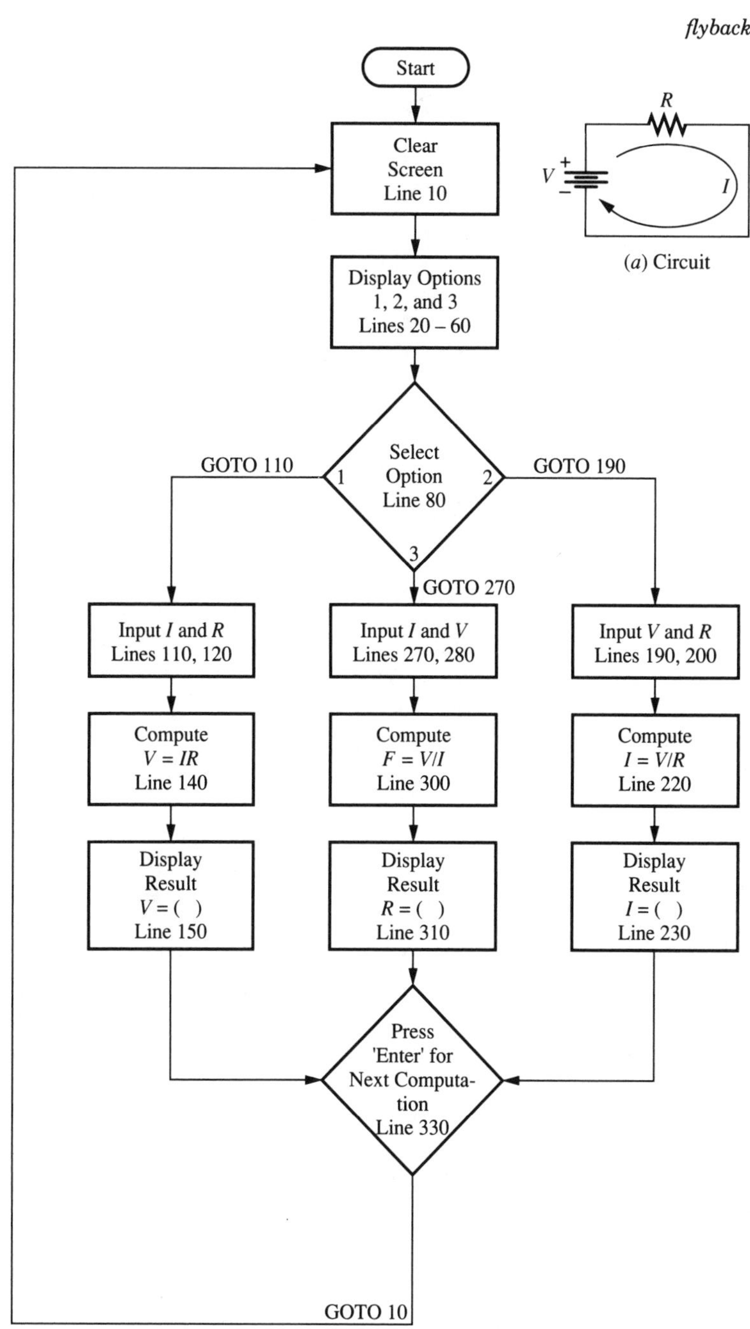

(*a*) Circuit

(*b*) Flowchart

Flowchart

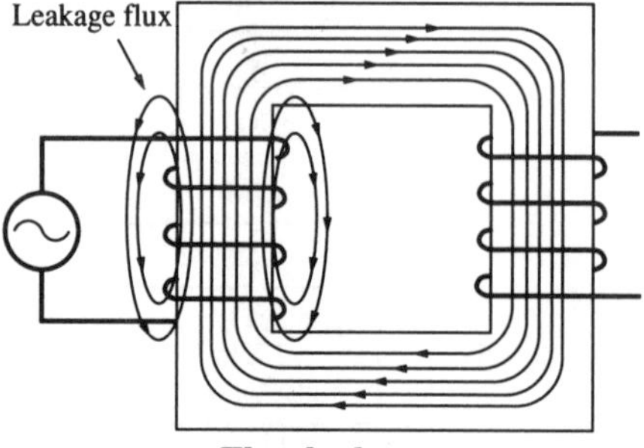

Flux leakage

FM Abbreviation for frequency modulation. See frequency modulation.

FM modulator An electrical circuit that causes a carrier's frequency to be modified by a lower-frequency signal that represents information. FM modulators are found in FM transmitters as well as other types of electronic systems.

FM receiver Abbreviation for frequency modulation receiver. A communications receiver capable of receiving an FM transmission, removing the intelligence, and reproducing it in a usable form.

focus The converging of light rays or of an electron beam on a specific point. To focus an oscilloscope means to make the spot formed on the face by the electron beam as small and sharply defined as possible.

focus control A control on a display device used to focus a display manually. See focus.

Forster-Seely discriminator A historically popular circuit consisting of diodes used to detect FM signals.

FORTRAN A high-level computer language used to perform mathematical, scientific, and engineering computations. Comes from FORmula TRANslation.

forward bias To cause a device to conduct. In semiconductors to forward bias a pn junction, a positive voltage is applied to the p-material and negative voltage to the n-material causing conduction of the junction.

forward current The current that flows in a pn junction when the junction is forward biased. See reverse current.

forward direction The direction of current flow when a pn junction is forward biased. See reverse direction.

forward voltage The voltage needed to forward bias a pn junction. The forward voltage for a silicon pn junction is approxi-

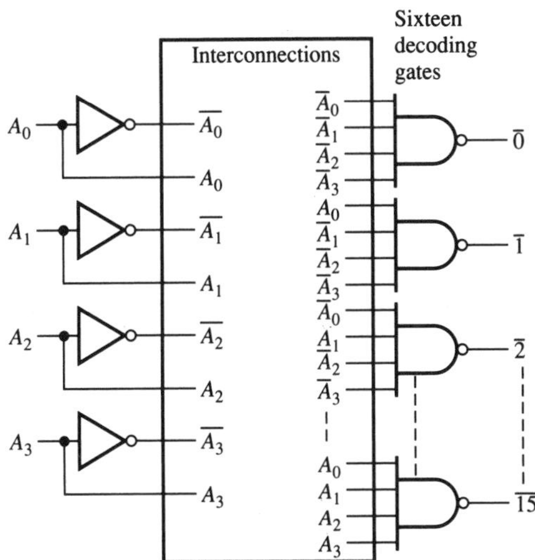

4-line to 16-line decoder

mately 0.7 volts while the forward voltage for a germanium pn junction is approximately 0.3 volts.

Fourier series A mathematical formula showing that all periodic functions are made up of pure sine waves.

4-line to 16-line decoder A logic circuit capable of decoding all possible combinations of 4 bits. There are 4 input lines and 16 output. This type of decoder will have only 1 of the 16 output lines active at one time.

FPLA Abbreviation for field-programmable logic array. See field-programmable logic array.

FPLS Abbreviation for fuse-programmable logic sequencer. See fuse-programmable logic sequencer.

fractals A term introduced by Benoit B. Manderlbrot of the IBM research center in 1975. Comes from the Latin *fractus*, which means "to break." Used to describe the generation of complex geometric shapes using mathematical techniques rather than the production of these same shapes by line and curve drawing that is done in classical Euclidean geometry.

free electron An electron that is not bound to a specific atom. A free electron has gained enough energy to break loose from the bonding force of the atom. Electron flow is accomplished by free electrons.

free-running frequency The frequency at which an oscillator normally synchronized will operate when it is not being synchronized.

free-running multivibrator See astable multivibrator.

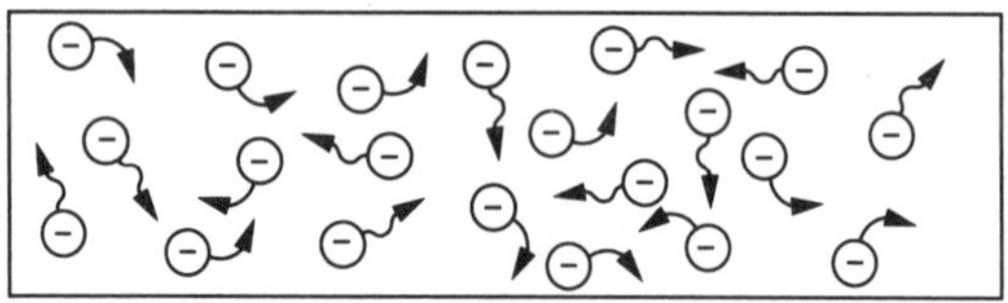

Free electrons

free space A region that is empty of any electrical charge or any other forms of electrical interference.

frequency The measure of the rate of change of a periodic function. Measured in hertz.

frequency counter An instrument used to measure the frequency of a periodic waveform.

frequency deviation The amount of frequency change of a waveform from its normal unmodulated frequency. Frequency deviation is a measure used in FM systems.

frequency distortion The unequal treatment of all frequencies of a complex waveform by an electrical circuit. For example, frequency distortion will result in a poor reproduction of a square wave by an amplifying circuit.

frequency division multiplexing A form of separating different sources of information by using different subcarrier frequencies to transmit the information.

frequency drift A slow change in the frequency of an oscillator. Frequency drift can be caused by an oscillator circuit that is sensitive to temperature changes. Usually frequency drift is an undesirable condition that can be corrected for by special circuits.

frequency modulation The process of having information change the frequency of a carrier wave. Referred to as FM.

frequency multiplier An electrical circuit that will produce a multiple of the input frequency. Frequency multipliers are used

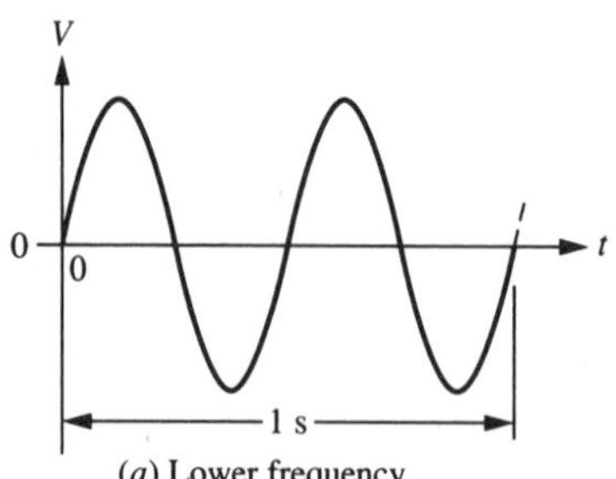

(*a*) Lower frequency

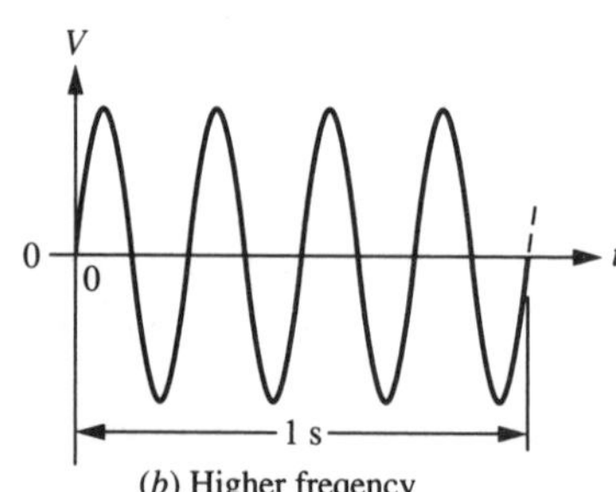

(*b*) Higher freqency

Frequency

when higher frequencies than those produced by the systems oscillator are required; usually found in FM transmitters. Frequency multipliers depend upon the harmonics of the fundamental frequency being multiplied.

frequency response 1. A measure of how a given range of frequencies is affected by a given circuit. For example, the frequency response of a low-pass filter is such that low frequencies are easily passed while higher frequencies are not. 2. The characteristics of a circuit or a device to a continuous range of frequencies. See bandwidth.

frequency spectrum The entire range of frequencies of electromagnetic radiation.

frequency standard A source that generates a given frequency used as a standard of comparison. Frequency standards are precise instruments used as references for measuring the accuracy of oscillators and similar frequency-producing circuits.

front end 1. That part of a communications system that is connected to the antenna. 2. The portion of a communications receiver that first receives the incoming signal. The front end usually consists of an RF amplifier along with frequency selective circuits. In a television receiver, the tuner is sometimes referred to as the front end.

FSK Abbreviation for frequency-shift keying. A method of transmitting digital information that uses FM. For example, a frequency of 1,270 KHz can represent a space and a frequency of 1,070-KHz a mark.

fuel cell An electrochemical generator that uses hydrocarbons for fuel and operates continuously as long as fuel and oxygen are available.

full adder A logic network consisting of three inputs. The sum of these three inputs is reflected on two outputs; one is the sum, the other is the carry.

full-adder logic Logic circuits used to construct a full adder. The Boolean expression used to express the full-adder logic.

full address decoding The process of using all the address lines to decode the binary pattern of the address.

full-duplex mode 1. A method of communicating between two systems where both systems can communicate at the same time to each other. Full-duplex mode is

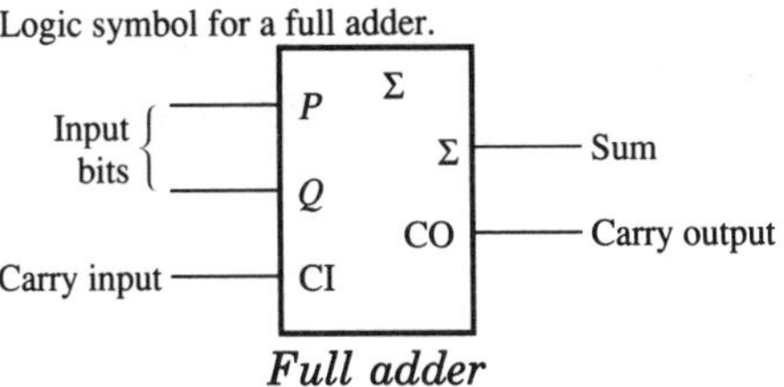

Full adder

Truth table for a full adder.

P	Q	CI	CO	Σ
0	0	0	0	0
0	0	1	0	1
0	1	0	0	1
0	1	1	1	0
1	0	0	0	1
1	0	1	1	0
1	1	0	1	0
1	1	1	1	1

CI = carry input
CO = carry output
Σ = sum
P and Q = input variables (operands)

Full adder (con't)

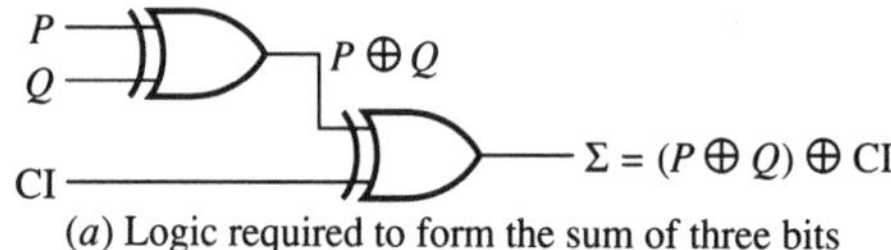

(*a*) Logic required to form the sum of three bits

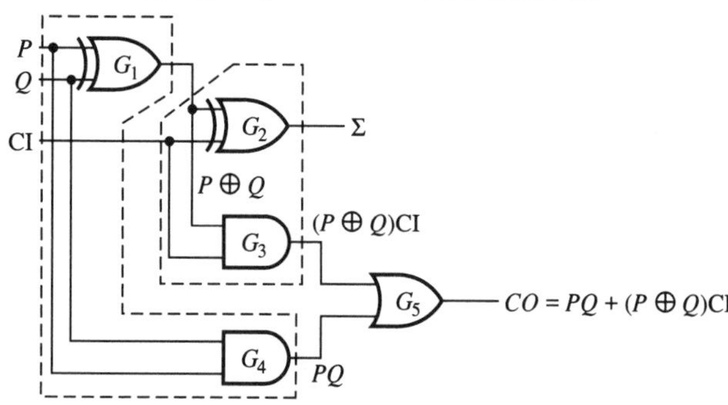

(*b*) Complete logic circuitry for a full adder (each half-adder is shown in a dashed outline)

Full-adder logic

usually accomplished by frequency division multiplexing where the transmitting and received signals are at different frequencies. 2. A mode of transmission where two systems may transmit and receive data at the same time.

full load 1. The greatest amount of power output that a given circuit is designed to deliver. See load. 2. A full load condition means that maximum power is being taken from the source. See no-load condition.

full scale The total amount at which an instrument is designed to be operated. For example, in a voltmeter, a full-scale reading is the maximum amount of voltage the meter is capable of reading on a given range.

full wave A total cycle of a waveform. A full wave of a sine wave is equal to one complete cycle of the sine wave.

full-wave dipole A two-conductor antenna cut to the wavelength of its transmission or reception frequency. Presents a high impedance to the electrical power source.

full-wave rectification The process of rectifying both parts of an input ac waveform. See full-wave rectifier.

full-wave rectifier A circuit that converts a sine wave (ac) into pulsating dc, where both the positive and negative peaks of the sine wave are felt on the output. The output frequency of a full-wave rectifier is twice the frequency of the input wave.

function generator An electrical instrument that is used to generate electrical signals of various types such as sine waves, square waves, and triangular waves. Function generators are sometimes referred to as signal generators.

fundamental In the analysis of frequencies, the most basic frequency in a given complex waveform consisting of many other frequencies. Used to determine the value of harmonic frequencies, as harmonics are multiples of the fundamental frequency.

fundamental frequency In a complex wave consisting of many frequencies, the lowest frequency making up the wave.

fuse A protective device that will cause an open circuit when the current through it exceeds a specified amount. Fuses are given current ratings. When replacing a fuse, *never* replace it with one of a higher current rating.

fuse box A container that houses fuses used to protect specific electrical circuits.

fuse-programmable logic sequencer Similar to fuse-programmable array logic except that flip-flops are included. This allows the FPLS to be programmed as sequential logic such as a binary counter.

fusible link A type of connection in a read-only memory that allows the memory to be permanently programmed (PROM). A fusible link is a tiny integrated circuit connection that represents a HIGH or TRUE condition. It may be opened by the application of a specified voltage. The open will now represent a LOW or FALSE condition.

fuzz box An electronic device used with music amplifiers to add distortion to the sound, thus producing a sought-after “fuzzy” sound.

g Letter symbol for gram. See gram.

G Letter symbol for conductance. See conductance.

gain In electronics, the ratio of the output value of an electrical quantity to the input value of that same quantity. For example, the voltage gain of an amplifier is the ratio of the output signal voltage to the input signal voltage. Gain can be less than unity as in the case of the voltage gain of an emitter follower, which is always less than one, or the power gain of a practical transformer, which, because of transformer losses, is always less than one.

gain bandwidth product A characteristic of amplifiers where the product of the gain and the bandwidth is always a constant. This means that if the gain of an amplifier is caused to increase, its bandwidth will decrease, and if the gain of an amplifier is caused to decrease, its bandwidth will increase.

game controllers See joystick.

ganged capacitor Two or more variable capacitors connected to a common shaft causing their values to change at the same time by the rotation of a common shaft. Any electrical arrangement where two or more capacitance values are being changed at the same time by a single action.

ganged tuning capacitor See ganged capacitor.

garbage In computer terminology a term for meaningless data stored in memory or produced as an output.

garbage collection A programming process of placing meaningless data in memory in the same memory location and then reducing its size or eliminating it altogether.

gas laser A laser where the lasing medium is a gas.

gate 1. The controlling lead of a field-effect transistor consisting of a source and drain. The voltage applied to the gate controls the

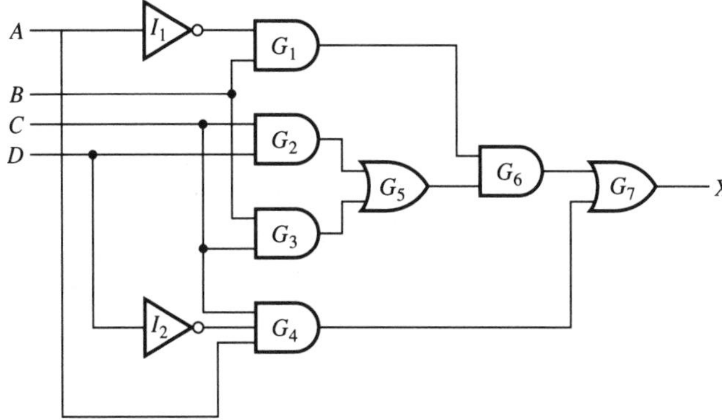

Gate minimization

current between the source and the drain. 2. An electrical circuit that emulates a logic function.

gate array An arrangement of logic gates inside one integrated circuit package. A gate array may be used to perform a complex logic task.

gated D latch A latch with two inputs called the ENABLE and the D (for data) input. The logic level at the D input will determine the state of the latch when the ENABLE input is active. Contains two complementary outputs.

gate minimization The process of reducing a combinational logic circuit to the absolute minimum number of required gates and achieving the exact same logical results. Two methods of doing this are by using Boolean algebra techniques and Karnaugh mapping.

gated S-R latch An S-R latch with a third input called the ENABLE input. The ENABLE input must be active before the SET

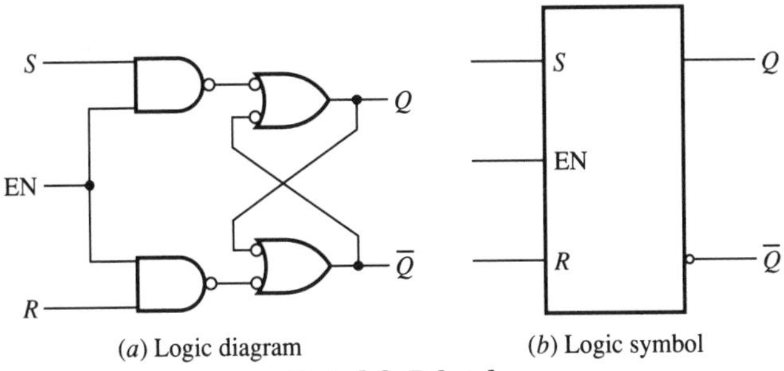

(*a*) Logic diagram (*b*) Logic symbol

Gated S-R latch

The GPIB connector.

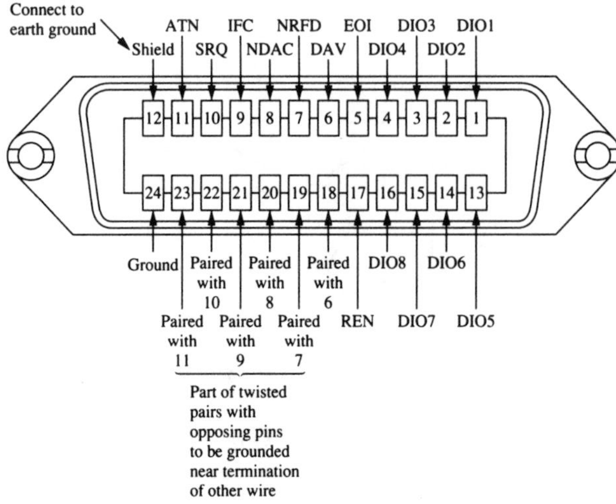

A typical GPIB interface arrangement showing the three bus groups.

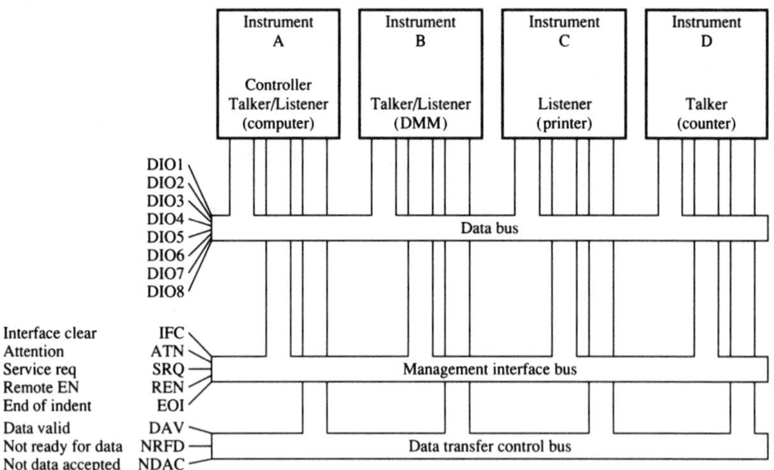

General-purpose interface bus

RESET inputs have any effect on the output of the latch.

general-purpose computer An electronic device designed for a wide variety of operations that involve word processing, mathematical operations, programming, and other activities commonly associated with computers. A general-purpose computer is not built for just one specific task, but designed for many.

general-purpose interface bus A bus system designed to allow control between different digital instruments. One of the major applications of this bus system (referred to as the IEEE-488) is the interfacing of test and measurement instruments with a computer to create an automated test system.

generator An electromechanical device for converting mechanical energy into electrical energy. In electronics, an instrument for producing a specified waveform, such as a sine wave generator.

generic computer A computer built by one manufacturer to have the characteristics of a computer built by another manufacturer. A generic computer may have some extra features and a lower price than the computer it is attempting to emulate.

geostationary orbit Orbiting the earth in such a manner so as to always be over the same point on the surface of the earth. Communications satellites usually have geostationary orbits. Sometimes called geosynchronous orbits.

geosynchronous orbit See geostationary obit.

germanium An element used in the manufacturing of semiconductor devices. Germanium has an atomic weight of thirty-two.

GHz Letter symbol for gigahertz. See gigahertz.

gigahertz A frequency of 1,000,000,000 or 1×10^9 cycles each second.

glitches An undesirable pulse or signal that affects the operation of the circuit. Certain types of digital decoders will experience glitches.

GPIB Abbreviation for general-purpose interface bus. See general-purpose interface bus.

graded-index fiber Optical fiber where the index of refraction is largest in the center and smaller toward the surface. See step-index fiber.

gram A unit of mass in the metric system.

graph A picture showing the relationship between variables. In electronics, a graph can show the relationship between an electrical variable, such as voltage, and time. A two-dimensional graph uses two mutually perpendicular lines as the axes. The vertical line is referred to as the *Y*-axis and the horizontal as the *X*-axis. (See pg. 129 for illustration.)

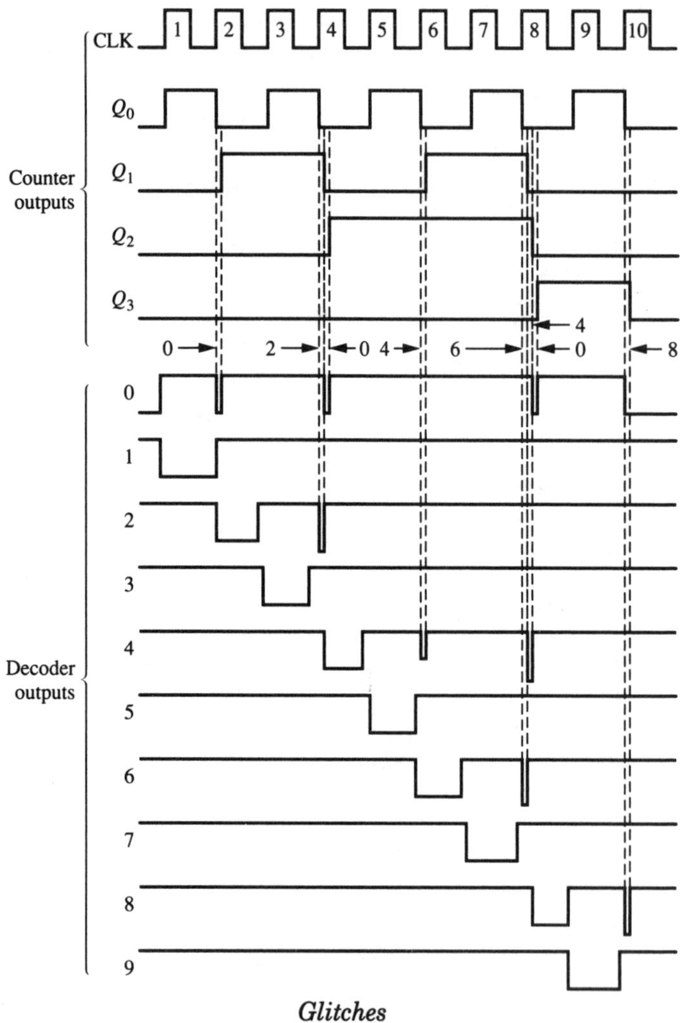

Glitches

graphical analysis A method of analyzing the characteristics of an electrical device. Graphical analysis usually consists of two coordinates that show the relationship between two quantities. For example, the graphical analysis of a transistor amplifier shows the relationship between collector current (I_C) and the voltage between the emitter and collector (V_{CE}), for different values of the base current (I_b).

graphics mode In computers, the operation of a computer so

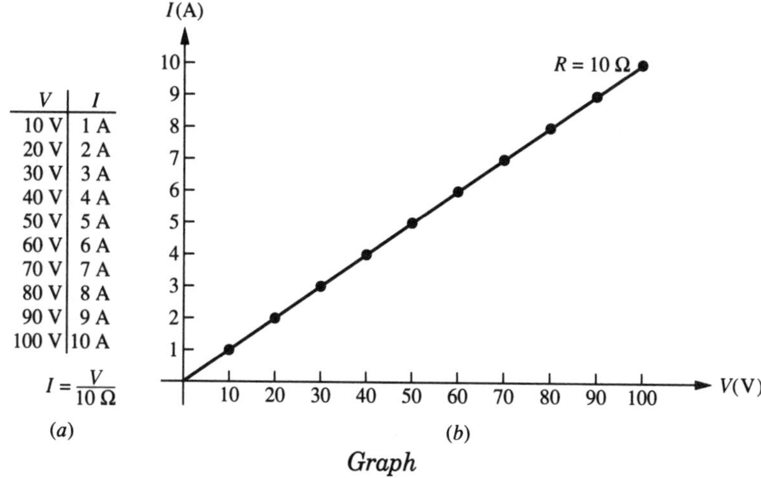

Graph

that images may be displayed on the screen that are different from the standard text display. When in graphics mode, most computer systems use a different area of memory for display purposes. See text mode.

Gray code A two-symbol numeric code (0 and 1) that allows only one bit to change at a time in its counting sequence. Useful as a digital code that reduces errors when reading data from mechanical devices such as shaft positions and converting the data to usable digital information.

Gray code counter A digital counter that produces the Gray code sequence. See Gray code.

ground Traditionally, a physical connection to the earth. In modern electronics, ground is a common voltage reference point in a circuit.

ground absorption The loss of radiated energy in a transmis-

Decimal	Binary	Gray
0	0000	0000
1	0001	0001
2	0010	0011
3	0011	0010
4	0100	0110
5	0101	0111
6	0110	0101
7	0111	0100
8	1000	1100
9	1001	1101

Gray code

Three-bit Gray code counter.

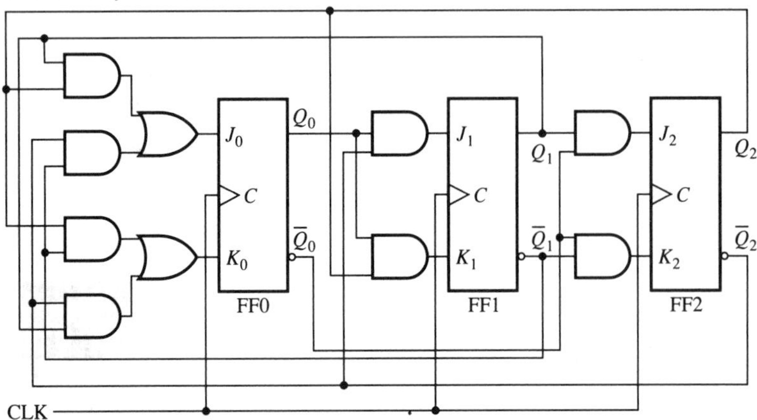

Transition table for three-bit Gray code counter.

Output State Transitions			Flip-Flop Inputs (present state)
Q_2	Q_1	Q_0	
Present → Next	Present → Next	Present → Next	J_2 K_2 J_1 K_1 J_0 K_0
0 → 0	0 → 0	0 → 1	0 X 0 X 1 X
0 → 0	0 → 1	1 → 1	0 X 1 X X 0
0 → 0	1 → 1	1 → 0	0 X X 0 X 1
0 → 1	1 → 1	0 → 0	1 X X 0 0 X
1 → 1	1 → 1	0 → 1	X 0 X 0 1 X
1 → 1	1 → 0	1 → 1	X 0 X 1 X 0
1 → 1	0 → 0	1 → 0	X 0 0 X X 1
1 → 0	0 → 0	0 → 0	X 1 0 X 0 X

Gray code counter

sion system because of the energy dissipated to the ground.

ground clutter The pattern produced on a radar screen because of objects located on the ground. An undesirable part of a radar signal due to these types of objects.

grounded Meaning a connection to the earth or to some conduction body in place of the earth or conductive body that is electrically connected to the earth.

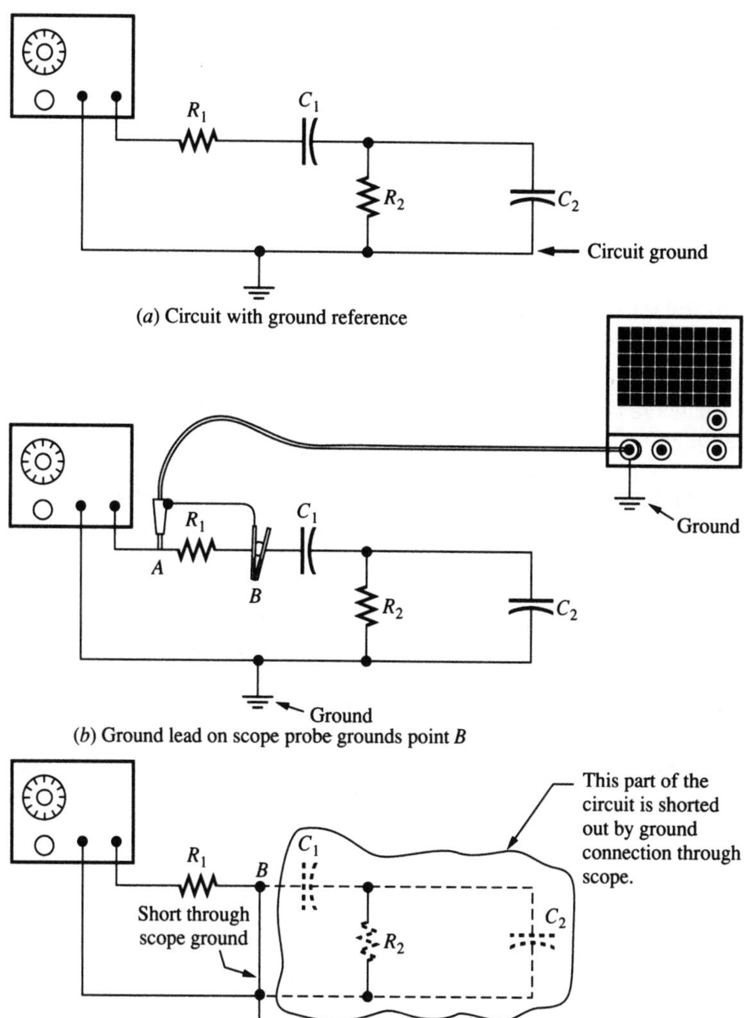

Ground measurement

grounded grid amplifier An electrical circuit consisting of a vacuum tube where the input signal is applied between the cathode and the control grid and the output signal is taken from between the control grid and the plate. In a grounded grid amplifier, the control grid is common to both the input and output signals.

ground measurement The use of a grounded piece of test

equipment when making measurements on a grounded piece of electrical equipment. The grounds of both the test equipment and electrical circuit under test should be connected in common.

ground potential A reference point for measuring voltages. Ground potential is usually a reference of 0 volts.

ground wave Electromagnetic wave that travels along the earth's surface. Confined to low radio frequencies of about 300 kHz or less. Sometimes called a surface wave.

grounding The process of making a connection to ground or to an electrical conductor connected to ground or to a conduction body that is used in place of ground.

guard band A range of frequencies used between two adjacent transmitting frequencies for the purpose of preventing interference from the two transmitting frequencies. In commercial FM transmission, the guard bands are 25 kHz wide.

guide wavelength In a waveguide, the measured wavelength along the length of the waveguide is longer than the free-space wavelength of the wave. This is because the electromagnetic wave inside the guide is reflected off the sides of the guide as it travels its length. The guide wavelength is expressed as $l_g = l/\sqrt{l - (l/l_0)^2)}$ where l_g = guide wavelength (in centimeters), l = free-space wavelength, l_0 = cutoff wavelength.

gunn diode A solid-state device constructed from a thin slice of n-type gallium arsenide sandwiched between two metal conductors and exhibits negative resistance characteristics. Used in high-frequency microwave applications.

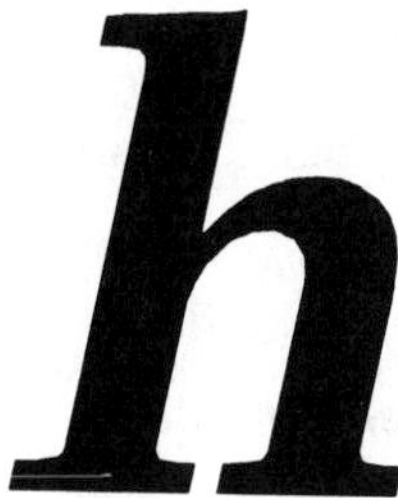

H bend A waveguide that is bent 90 degrees along the axis of its magnetic field (H field).

half adder Logic network with two inputs, the sum of which is represented by two outputs, the sum and the carry.

half-duplex mode A method of communicating between two systems that allows for two-way communication but not at the same time. See full-duplex mode.

half-power point The place, on a frequency response curve, where the power output of a circuit is one-half of its maximum power output. The half-power points are used as the measuring points from which the bandwidth of a circuit is measured.

half wave One hundred and eighty degrees of a symmetrical waveform.

Logic symbol for a half-adder.

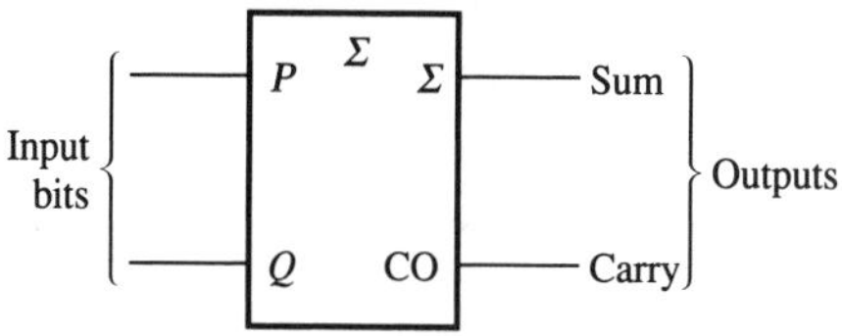

Half-adder truth table.

P	Q	CO	Σ
0	0	0	0
0	1	0	1
1	0	0	1
1	1	1	0

Σ = sum
CO = carry output
P and Q = input variables (operands)

Half adder

half-wave dipole A type of antenna sometimes referred to as a hertz antenna. Basically a quarter-wave transmission line that has been spread apart. Appears as a very low impedance to the source, usually about 73 Ω.

Hall effect The phenomenon of producing a voltage that is proportional to the strength of a magnetic field in a current-carrying conductor.

Hall generator A solid-state device that produces an output voltage that is proportional to the strength of a magnetic field. Hall generators are used as a means of measuring the strength of a magnetic field.

Hall sensor See Hall generator.

ham Slang for amateur radio operator.

Hamming code An error correction code system used in data transmission named after its inventor.

hand capacitance The capacitance caused by the human hand brought next to an electrical circuit or device. The effect of hand capacitance is to change the capacitance value sensed by the circuit. Hand capacitance is used to turn on and off touch control circuits such as table lamps.

handler A section of a computer program used to control or communicate with an external device.

handset That portion of a telephone that has both the mouthpiece and earphone contained in the same unit.

handshaking In computer communications, the matching of the computers that will communicate with each other so that accurate communications can take place.

hard copy Computer data, such as a program that is in a permanent, nonelectrical form. A hard copy of a computer program is a printout or hand-written copy of the program.

hard disk A storage system of one or more nonremovable disks protected in a permanently sealed case.

hard sector A method of keeping track of how data are stored on a computer disk by the use of a physical indicator rather than

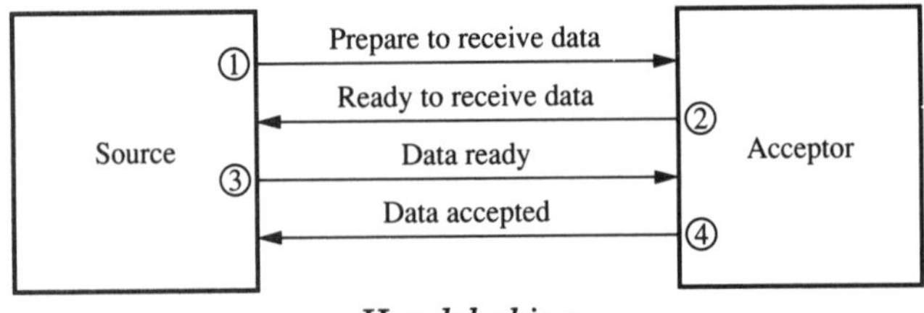

Handshaking

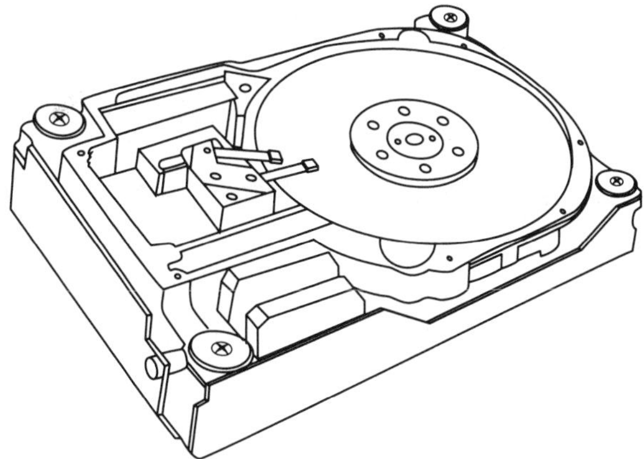

Hard disk

by electrical means. A hard-sector disk usually uses a physical hole on the disk as an aid for storing and retrieving data. See soft sector.

hardware The tangible part of a processing system such as a computer.

hardware buffer A set of registers acting as a temporary storage of data in a computer system. A hardware buffer is usually used when transmitting data from a faster to a slower device, such as between a computer and a printer.

hardwired Electrical devices interconnected though physical wiring. The implementation of logic and other computer functions through the use of physical devices as opposed to software.

harmonic A sine wave that is an integral multiple of a fundamental or basic frequency. For example, the second harmonic is twice the frequency of the fundamental. Thus, 2 kHz is the second harmonic of 1 kHz.

harmonic analysis The identification and analysis of the various frequencies that make up a complex waveform. See harmonic.

harmonic distortion The production of harmonic frequencies from circuit or device when supplied with a periodic waveform such as a sine wave.

Hartley oscillator A circuit that generates its own signal. Uses an inductive voltage divider in parallel with a capacitor. Resonant frequency is determined by the equivalent value of the inductor and value of the capacitor.

Hay bridge A bridge circuit constructed in such a manner that it is possible to measure very small resistance and inductance values of coils when the resistance is a small fraction of the total coil reactance.

head An electromechanical device that senses or causes to be stored electrical patterns on a magnetic storage medium. For example, a floppy disk has a read-write head that is capable of reading or writing computer data from and to the disk. A VCR has a read-write head capable of reproducing of storing sound and video images on a magnetic tape.

head gap A small physical space placed on a magnetic recording head to force the magnetic field pattern out from the head as to direct the flux toward the recording medium.

headphone An electromechanical device that converts electrical signals into sound waves and is constructed to be worn next to the ear. Headphones are used for private listing or to help drown out surrounding environmental sounds.

heap In computer programming, memory that is used for data when the program is running as opposed to memory used by the static structure of the program. For example, in C programming, the heap is used by dynamically allocated variables using pointers, where statically assigned arrayed variables do not use the heap.

heat sink A mechanical device that is a good conductor of heat used to protect an electrical device from damage due to overheating. For example, most power devices, such as power diodes and transistors, must be mounted on a heat sink consisting of a metal plate with fins to help increase the surface area of the device for heat dissipation purposes. A heat sink is also used in soldering where the sink consists of a mechanical device placed between the soldering iron and the device itself.

hermetic seal An airtight fitting. For example, most solid-state devices are hermetically sealed.

hertz The unit of measurement for frequency.

hertz antenna See half-wave dipole.

heterodyne The process of combining (mixing) frequencies to produce resultant sum and difference frequencies. Heterodyning is used in modern communication receivers in order to produce a constant bandwidth while tuning to different stations. Receivers employing the heterodyning principle are commonly referred to as superheterodyne receivers.

heuristic program A group of computer instructions that simulate the behavior of human operators in their approach to a similar problem.

hex The hexadecimal numbering system. See hexadecimal.

hexadecimal The number system to the base 16. Characters

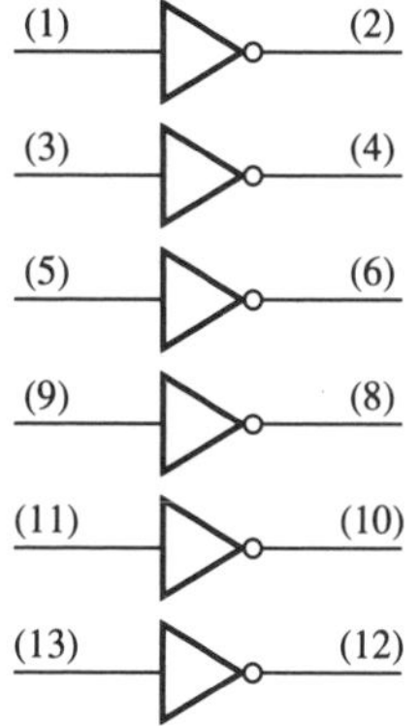

(*a*) Distinctive shape logic diagram.

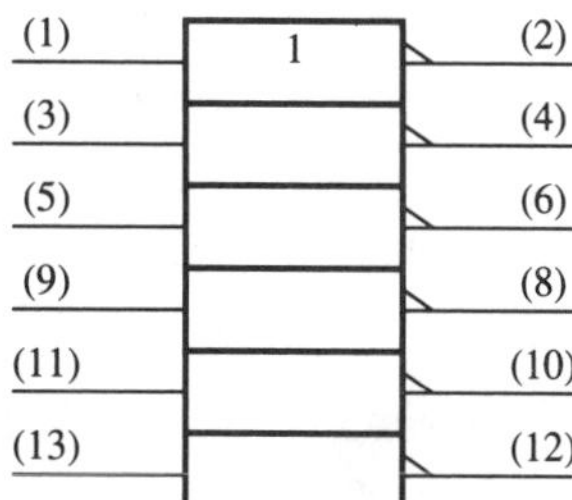

(*b*) Rectangular outline logic symbol with polarity indicators. The inverter qualifying symbol (1) appears in the top block and applies to all blocks below.

Hex inverter

are 0, 1, 2, 3, 4, 5, 6, 7, 8, 9, A, B, C, D, E, and F. Useful in converting directly from binary numbers. For example, the binary number 10010111_2 may be represented as 97_{16}.

hexadecimal display A device capable of displaying the Arabic numbers 0 through 9 and the first six characters of the English alphabet (A through F). A hexadecimal display is used to display the values of hexadecimal numbers.

hexadecimal loader A device for entering hexadecimal numbers into a digital system.

hexadecimal number system A number system to the base 16. Consists of the symbols 0, 1, 2, 3, 4, 5, 6, 7, 8, 9, A, B, C, D, E, F.

hex inverter A digital integrated circuit consisting of six logically independent inverters.

hi fi See high fidelity.

high fidelity Characteristics of a system that causes it to reproduce sound as nearly as possible to the original source of the sound. See fidelity.

high-level language A computer program whose code is similar to our everyday language. For example, the programming language Pascal is a high-level language because it uses everyday words such as BEGIN, END, FOR, WHILE, and so forth to express computer commands. See low-level language.

high-level transmission In an AM transmitter, the carrier wave is modulated at the power amplifier stage. The advantage of high-level transmission is greater sys-

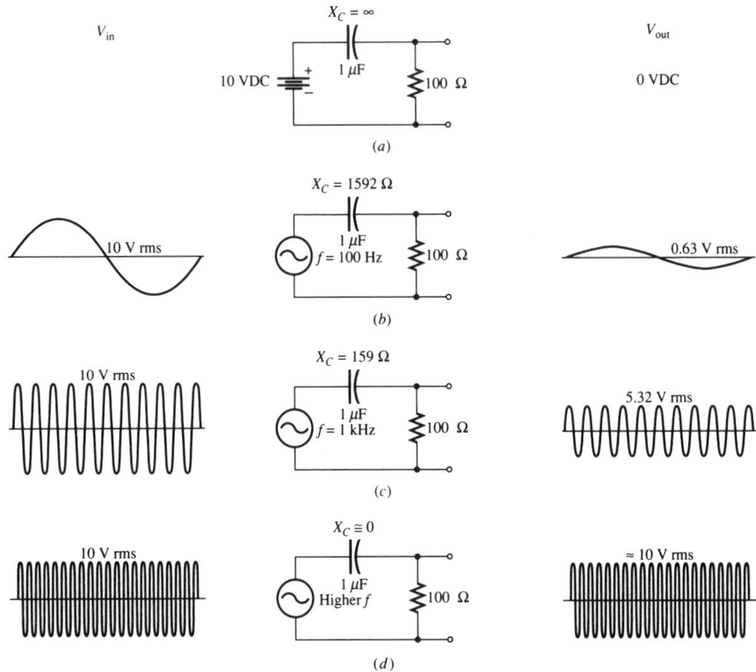

High pass RC filter

tem efficiency with an increase in system cost.

high-order address The most significant part of the first half of the total number of address bits. For example, the high-order address of an 8-bit address bus is the 4 most significant bits.

high-order nibble First 4 bits, including the most significant bit of an 8-bit word.

high-pass filter A circuit that blocks to passages of low frequencies but allows the passage of high frequencies. A high-pass filter has a critical or cutoff frequency that is the frequency that separates the high, passed frequencies from the low, unpassed (blocked) frequencies.

high-pass *RC* filter A circuit consisting of a resistor in series with a capacitor where the output is taken from across the resistor.

high-pass *RC* filter response The frequency characteristics of a high-pass *RC* filter.

high-pass response The electrical action of a circuit designed to pass high frequencies and block low frequencies.

high *Q* The property of having a large reactance to the effective resistance. An inductor is said to have a high *Q* when its inductive

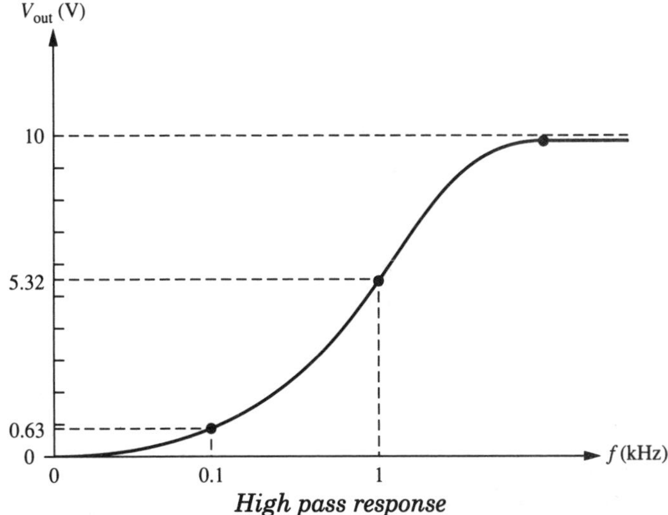

High pass response

reactance is large compared to its effective resistance.

high voltage A general term usually implying that the voltage is of such a value that it is dangerous to humans. Extreme caution should always be used when troubleshooting equipment marked "high voltage."

hipot A device used to test large resistance or insulating properties. Uses a high voltage to accomplish these measurements.

H network A circuit consisting of five impedance branches. An H network has two input and two output terminals and looks like the capital letter "H" lying on its side; hence the name H network.

hold time The minimum amount of time required for the logic levels of a digital device to remain on the inputs after the triggering edge of the clock pulse (as with a flip-flop), in order for the levels to be reliably clocked into the flip-flop.

hole In solid-state electronics, a positive charge created by the absence of an electron. To help explain the phenomena of solid-state electronics, holes are thought of as positively charged particles.

hole conduction Current flow that occurs in a semiconductor material when electrons move to fill holes (the electrical absence of an electron) under the influence of an externally applied voltage. The electrons create new holes as they begin their journey and fill holes when they end their journey. Hole conduction is thus

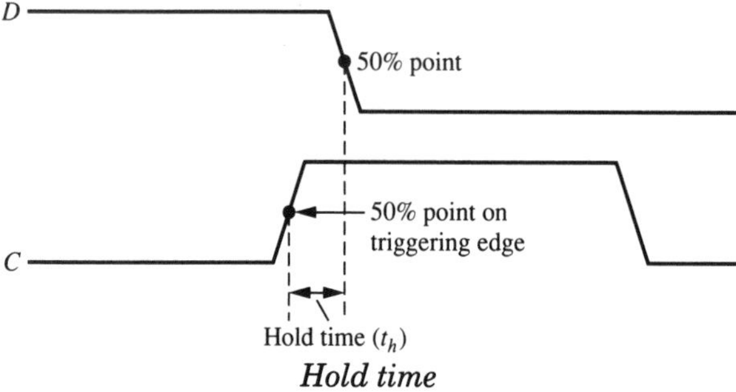

Hold time

visualized as current flow opposite to that of electrons.

hookup wire Wire used to interconnect electrical components or subassemblies. Hookup wire is usually 18- to 20-gauge soft-drawn insulated copper wire, which may be solid or stranded.

horizontal Level with the horizon or at right angles to the direction of gravity. A horizontal line is drawn on a paper from left to right. See vertical.

horizontal blanking Preventing the electron beam from hitting the face of a CRT. Used during horizontal and vertical retrace so that the retrace does not interfere with information already on the screen.

horizontal blanking pulse An electrical pulse applied to the control grid of a CRT to prevent the electron beam from hitting the face of the CRT during horizontal retrace. See horizontal retrace.

horizontal hold Control on a television receiver that helps control the frequency of the horizontal oscillator, which prevents the picture from tearing horizontally when adjusted properly. Any such control for a CRT visual display system.

horizontal oscillator In a television system the circuit that produces the required waveform for the horizontal sweep. Any such circuit that causes the electron beam of the CRT to move in a horizontal direction.

horizontal polarization Regarding the radiation pattern of an antenna. When the electric field is horizontal, the antenna is said to be horizontally polarized. In practice a horizontally polarized antenna will receive only signals from another horizontally polarized antenna. See vertical polarization.

horizontal retrace In a CRT the return of the electron beam from one side of the CRT face to the other. During this time, the electron beam is blanked out (prevented from being displayed

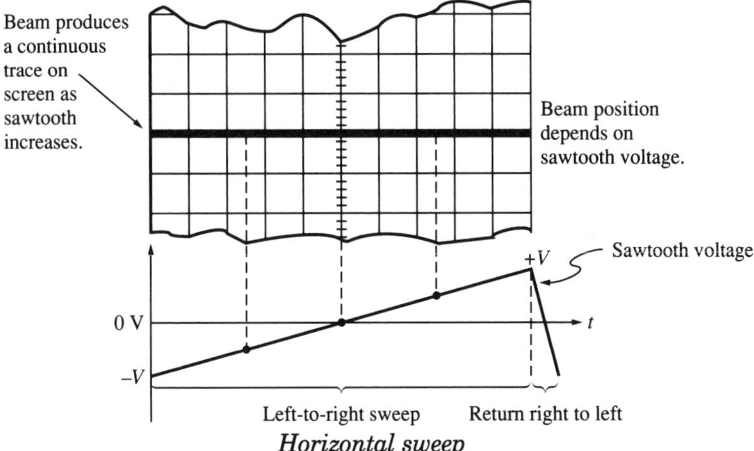

Horizontal sweep

on the CRT). See horizontal blanking.

horizontal scan In a television system, a complete movement of the electron beam from the left to the right of the CRT. Any complete movement of the electron beam horizontally across the CRT. This type of movement is caused by a sawtooth waveform to produce a linear display.

horizontal sweep See horizontal scan.

horizontal sync In a television system, a pulse that controls how often the CRT completes a horizontal scan. Any waveform that controls the horizontal scan of a CRT.

horizontal width control Control on a television receiver that adjusts the horizontal size of the picture by adjusting the amplitude of the horizontal oscillator sawtooth. Any such control on a CRT visual system.

horn antenna A tubular or rectangular antenna that is an extension of a waveguide and used in microwave communications; usually flared at the end. A horn antenna is wider at the open end from which the electromagnetic energy is transmitted and received. It is highly directional in transmission and reception of microwave signals.

horsepower In electrical circuits, equivalent to 746 watts.

hot carrier diode See Schottky diode.

hot wire In electronics wires that carry an electrical potential, usually a dc power supply voltage.

human being An analog processing and storage device. Has a bandwidth of about 50 bits per

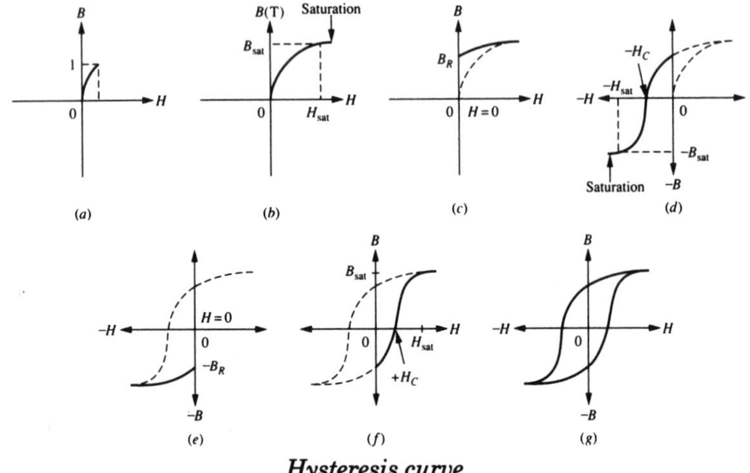

Hysteresis curve

second; excels at pattern recognition but is very slow at sequential calculations.

humanoid An electromechanical and electrochemical reproduction of a human being that attempts to simulate human characteristics and functions.

hypermedia Information arranged in a nonlinear fashion or stratified to represent varying levels of detail. In a hypermedia system, broad ideas are first presented from which the user may then select increasingly specific details.

hypertext Information presented on a computer screen in text form with certain key words in the text highlighted to make them stand out from the rest of the text. These keywords may then be accessed for more detailed information about these words or other words with similar definitions may also easily be accessed. A method of presenting information in textual form that allows the user to access easily key points and concepts.

hysteresis The difference between activating and deactivating values in a device or circuit. For example, in magnetism, hysteresis is the amount the magnetization of a magnetic substance lags the magnetizing force because of molecular friction.

hysteresis curve A graph showing the effect of hysteresis. See hysteresis.

hysteresis loss Energy expanded in a device due to the effects of hysteresis. An electromagnetic dissipates power because of magnetic hysteresis.

Hz Letter symbol for hertz. See hertz.

I Letter symbol for current. See current.

i_b Designation for transistor signal base current.

I_B Designation for the dc base current of a transistor.

i_c Designation for transistor signal collector current.

I_C Designation for the dc collector current of a transistor.

IC Abbreviation for integrated circuit. See integrated circuit.

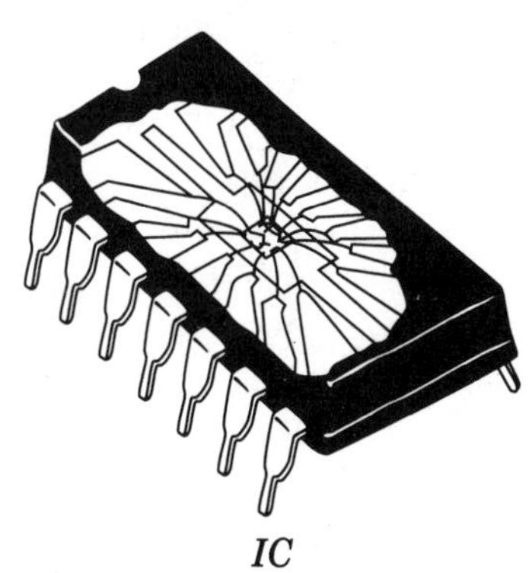

IC

IC packages A method of enclosing an integrated circuit so that it may be handled and inserted into a circuit. There are many different types of IC packages.

$I_{C(sat)}$ Designation for transistor saturation current. See saturation.

I_{CEO} Designation for transistor leakage current between the collector and emitter when the base current is zero.

i_d Designation for FET signal drain current.

I_D Designation for the dc drain current of an FET.

i_e Designation for transistor signal emitter current.

I_E Designation for transistor dc emitter current.

IEEE Abbreviation for Institute of Electronic and Electrical Engineers.

IEEE-488 bus An interface standard that defines connections for up to 15 different instruments for the purpose of exchanging data in an automated testing system.

IF Abbreviation for intermediate frequency.

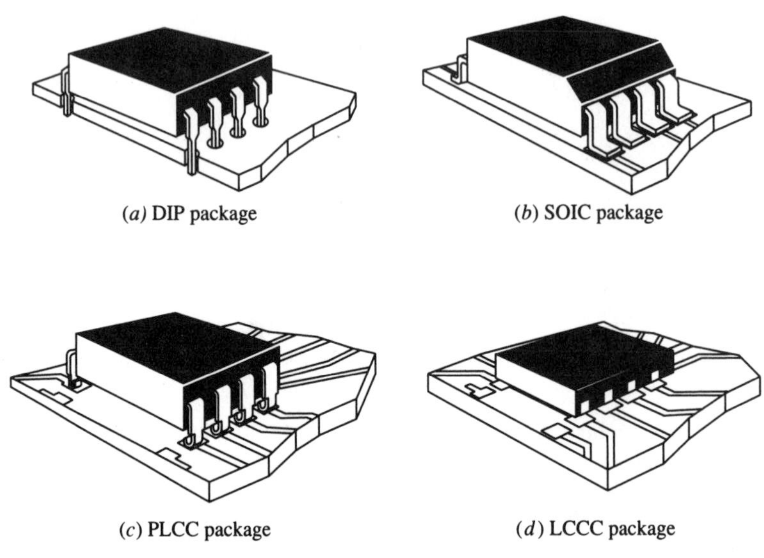

IC packages

i_g Designation for FET signal gate current.

I_G Designation for FET gate current.

i_k Designation for vacuum tube signal cathode current.

I_K Designation for the dc cathode current of a vacuum tube.

imaginary number Representation of the square root of −1. In mathematics, an imaginary number is represented by the letter "i". However, in electronics, since "i" is used to represent instantaneous current, the letter symbol "j" is used.

impact printing Printing that is caused by the action of a solid surface containing the image to be printed striking an inked ribbon onto the medium (usually paper) to contain the image.

IMPATT diode Abbreviation for impact avalanche and transit time diode. A solid-state device with very-high-frequency (10–100 GHz) characteristics.

impedance The total opposition to current flow caused by resistance, inductive and capacitive reactance. Because inductive and capacitive reactances are sensitive to frequency changes, impedance is also sensitive to frequency changes.

impedance match The condition of when the impedance of the

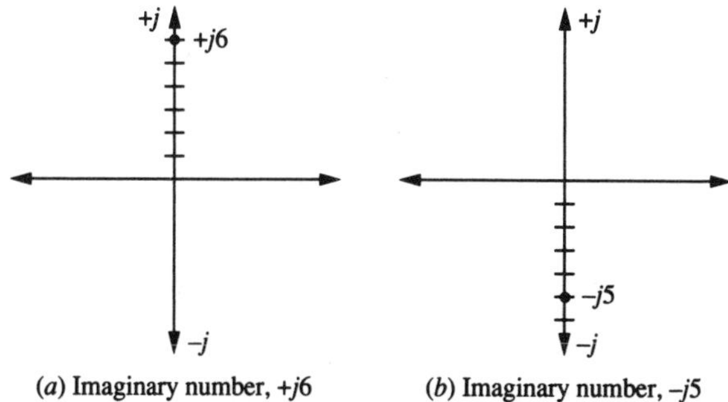

(a) Imaginary number, +j6 (b) Imaginary number, −j5

Imaginary number

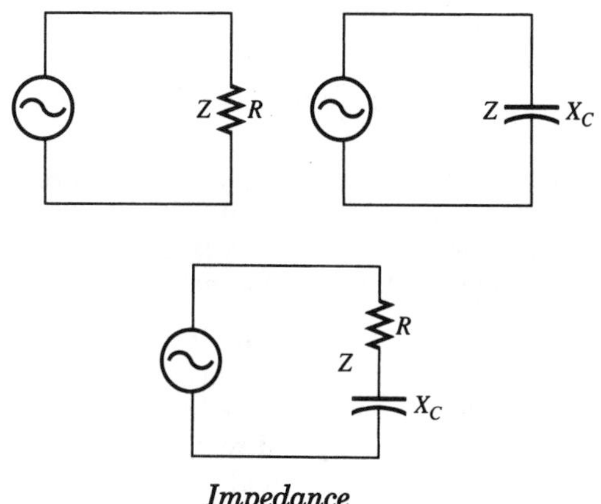

Impedance

source is equal to the impedance of the load. See impedance matching.

impedance matching Process of adjusting the impedances of the source and the load so they are equal. When these impedances are matched, there is then a maximum transfer of power from the source to the load.

impedance triangle A right triangle used to illustrate the phase and magnitude relationships of the reactive and resistive components of a circuit. On an impedance triangle, the hypotenuse represents the total impedance, and of the other two sides, one represents the circuit reactance while the other, the resistance.

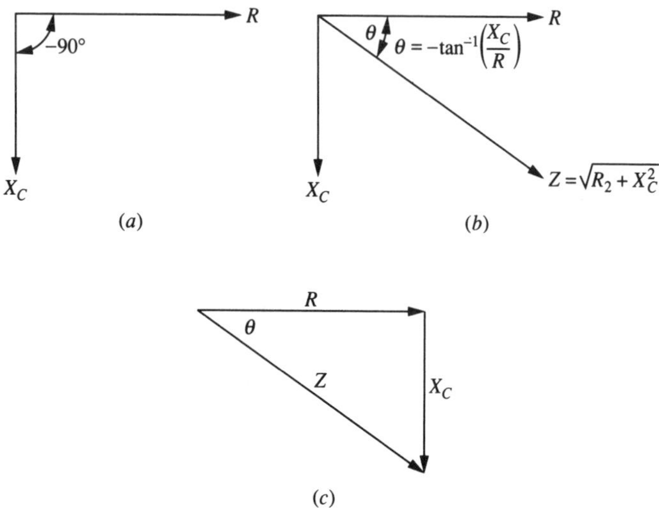

Impedance triangle

impurity In solid-state electronics, an element that is added to semiconductor material in order to make it either an n-type or a p-type material.

in-circuit tester An electronic device that is capable of testing certain components while they are still in the circuit. Using an in-circuit tester, a technician does not have to remove the component being tested.

in phase Two or more waveforms of the same frequency that pass through their maximum and minimum values at the same time. Two sine waves are said to be in phase if they are of the same frequency and their maximum and minimum values happen at the same time.

incandescence Light emitted from a high-temperature source. An incandescence light bulb generates its light from the heating of a thin piece of wire called the filament.

incident ray A ray of light that strikes the surface of an object. The incident ray illustrates the direction of travel of a beam of light.

index address In computers, the memory location to be activated is contained in a separate register inside the microprocessor that is easily incremented or decremented. This register is called the index register. See index register.

indexed addressing In computers, a method of activating memory locations by using a separate register inside the microprocessor called the index register. See index address.

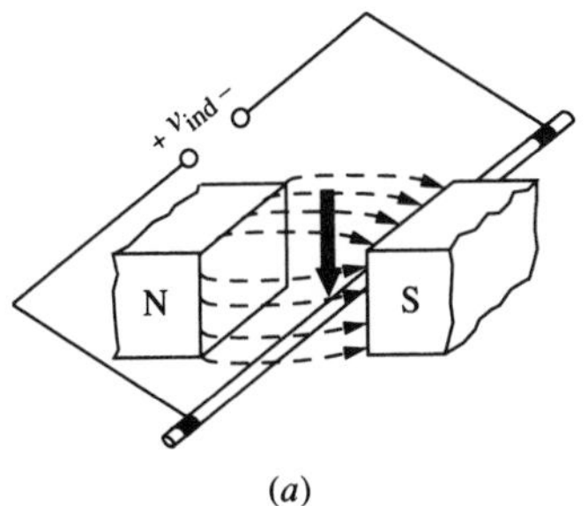

(a)

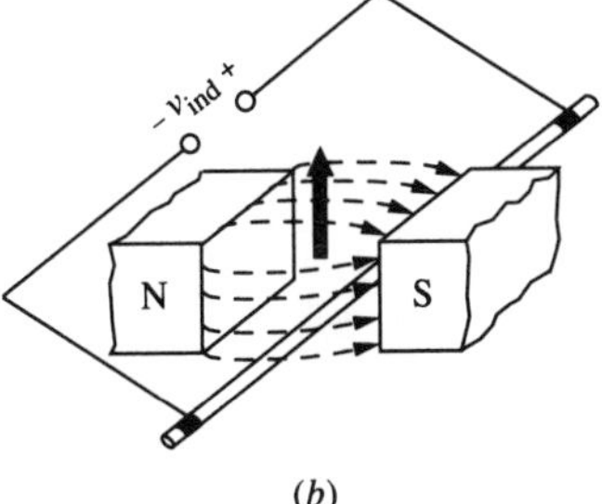

(b)

Induced voltage

indirect address In computers, the memory location of a specified operation or data is contained in memory itself. An indirect address for a specific piece of data would have the address contained in a memory location different from the data.

indirect addressing In computers, a method of activating memory locations by using memory locations to contain the locations of the memory to be activated. See indirect address.

induced current Current that flows in a conductor that is caused by the action of an external magnetic field.

induced voltage Voltage produced from an external source as from a moving magnetic field or electromagnetic field.

induction The establishment of a magnetic field or an electrical charge in a material by the proximity of a magnetic field or an electrical charge.

inductive coupling The influence of one circuit upon another through an inductance common to both.

inductive kick An induced voltage, usually many times higher than the applied voltage, caused by the rapid change of current in a coil. An inductive kick takes place when a circuit containing a coil is turned on or off.

inductive reactance Letter symbol X_L. The opposition to current flow in an inductor caused by a changing current. Inductive reactance is measured in ohms and is expressed as $X_L = 2\pi fL$, where X_L is the inductive reactance (in ohms), f the frequency (in hertz), and L is the value of the inductor (in henrys).

inductive susceptance Letter symbol B_L. The reciprocal of inductive reactance (X_L), measured in siemens (S). Mathematically, inductive susceptance is $B_L = 1/X_L$, where X_L is the inductive reactance (in ohms).

inductor An electrical device made from turns of wire. An inductor opposes a change in cur-

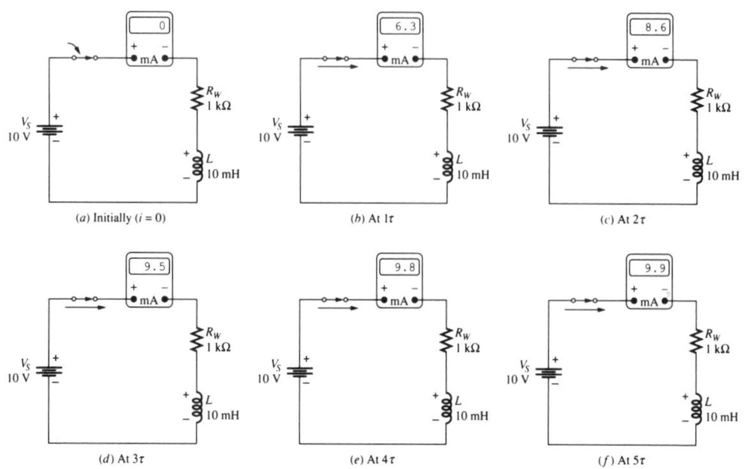

Inductor current

rent and is measured in henrys (H). The voltage produced across an inductor is directly proportional to the rate of change of current in the inductor. An inductor has the capacity to store electrical energy in its magnetic field.

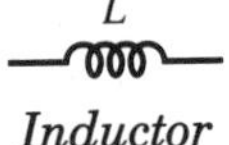

Inductor

inductor current Current in an inductor, in a series RL circuit will follow the universal time constant curve until it builds up to its maximum value.

inductor energy Stored in the form of a magnetic field in an inductor.

inductor filter A circuit that uses an inductor to help smooth out the variations in a changing voltage.

inductor parameter The inductance of an inductor is directly

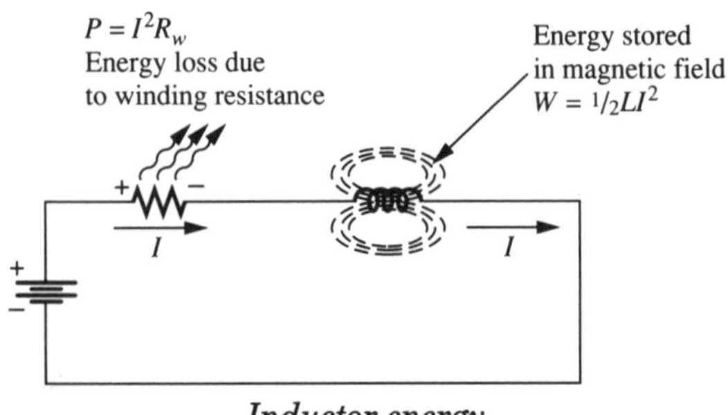

Inductor energy

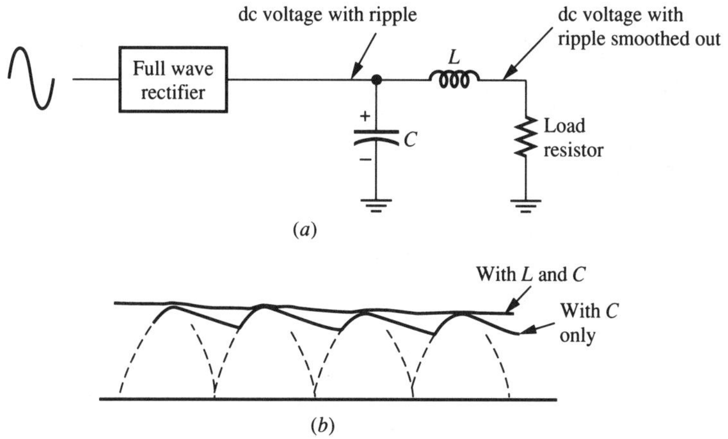

Inductor filter

proportional to the square of the number of turns of wire, permeability of the core material, and cross-section area of the wire turns and inversely proportional to the length of the coil.

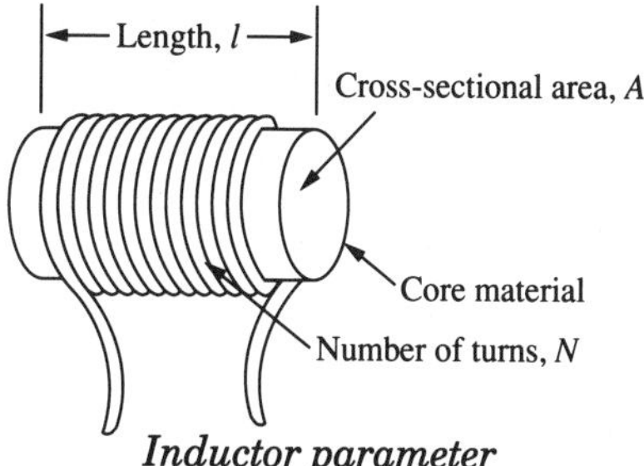

Inductor parameter

inductor phase The current in an inductor always lags the voltage by 90 degrees.

inductor testing A method of checking an inductor using an ohmmeter.

inductor time constant The rate at which the current changes in an inductor. In a series RL circuit, the time constant is proportional to the value of the inductor and inversely proportional to the value of the circuit resistance.

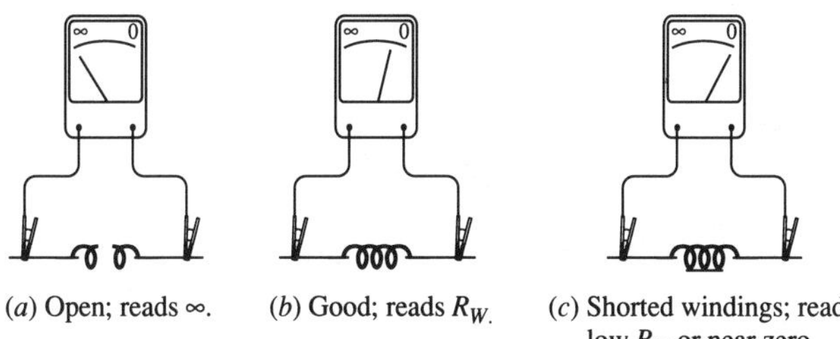

(a) Open; reads ∞.
(b) Good; reads R_W.
(c) Shorted windings; reads low R_W or near zero.

Inductor testing

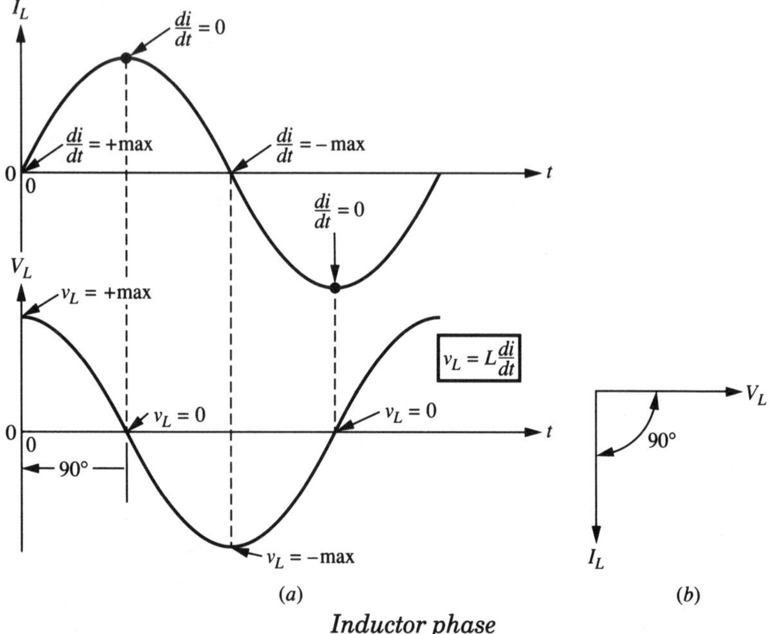

Inductor phase

Percentage of final current after each time-constant interval during current build-up.

Number of time constants	%Final Value
1	63
2	86
3	95
4	98
5	99 (considered 100%)

Energizing current in an inductor.

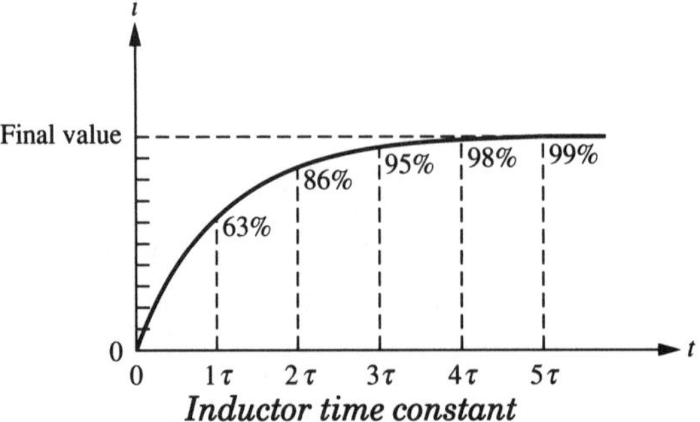

Inductor time constant

inertia switch A mechanical device that will open or close upon an abrupt change in motion.

infrared The section of the electromagnetic spectrum that produces radiation between the wavelengths of 750 nm and about 1 mm.

inhibit To prevent an electrical or mechanical action from taking place. An inhibit line would prevent a particular output from occurring when the line is active.

initialize To set an electrical circuit or mechanical device to some predetermined condition.

input That part of a circuit or an electrical device to which a signal is first applied.

input capacitance The capacitance present at the terminals of a device or circuit where the signal or control is applied. For example, an FET exhibits input capacitance between the gate and source terminals.

input profile For an electrical digital circuit, it is the specific voltage ranges of logic HIGH and logic LOW for that circuit's input. A typical input profile for TTL would be HIGH = +2 volts to +5 volts, LOW = +0.8 volts to 0 volts. This means that any voltage range on the input between +2 volts and +5 volts will be considered a HIGH, while any voltage range from +0.4 volts to 0 volts will be considered a LOW. See output profile.

instantaneous Happening at a given point, usually in time. For example, point in time along the sine wave is called the instantaneous value of the sine wave at that point.

instantaneous value The magnitude at a point in time of a time varying quantity.

instruction set In a microprocessor, the set of bit patterns understood by the microprocessor. An instruction set usually consists of the assembly language mnemonic and the corresponding process produced by the mnemonic as well as other information such as which flags are affected and how long it takes to execute the instruction.

instrumentation The use of equipment for the testing and troubleshooting electrical or mechanical systems.

insulated The electrical separation of two or more conducting materials. A conductive wire is insulated by placing it in a jacket of insulating material—it is then called insulated wire.

insulator Any material that helps prevent the flow of current. Having a large resistance. See conductor.

integer Any number that does not contain a fraction. The integers are the whole numbers from 0 through 1, 2, 3, and so on. As an example, the number 245 is an integer because it does not contain a fraction; 1,590 is also an integer for the same reason.

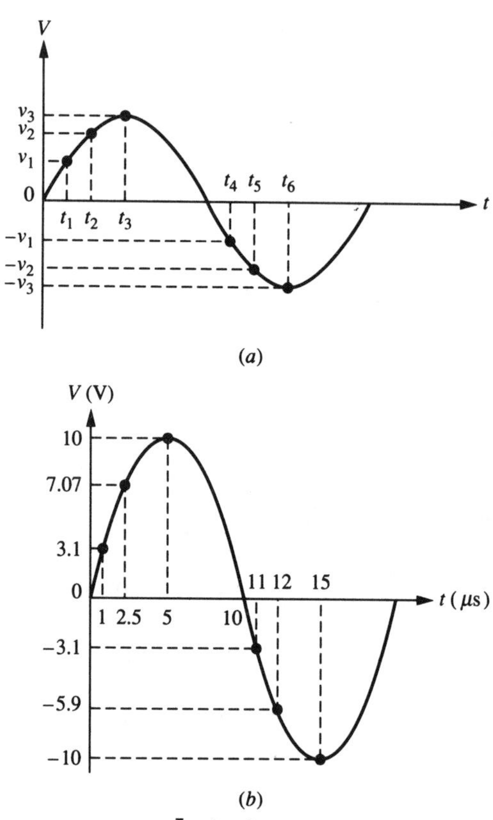

Instantaneous

However, the number 12.5 is not an integer because it contains the fraction 0.5.

integrated When two or more electrical components or functions are physically and electrically combined. For example, an integrated circuit physically and electrically combines several electrical components in one physical housing.

integrated circuit An electronic circuit where all the components are contained on a single semiconductor substrate. The advantage of an integrated circuit is its compact size, preassembled and tested condition, and low power consumption.

integrator An electrical circuit where the resultant output waveform approaches the mathematical integral of the input waveform.

interactive An application where each activity requires a response from the user. For ex-

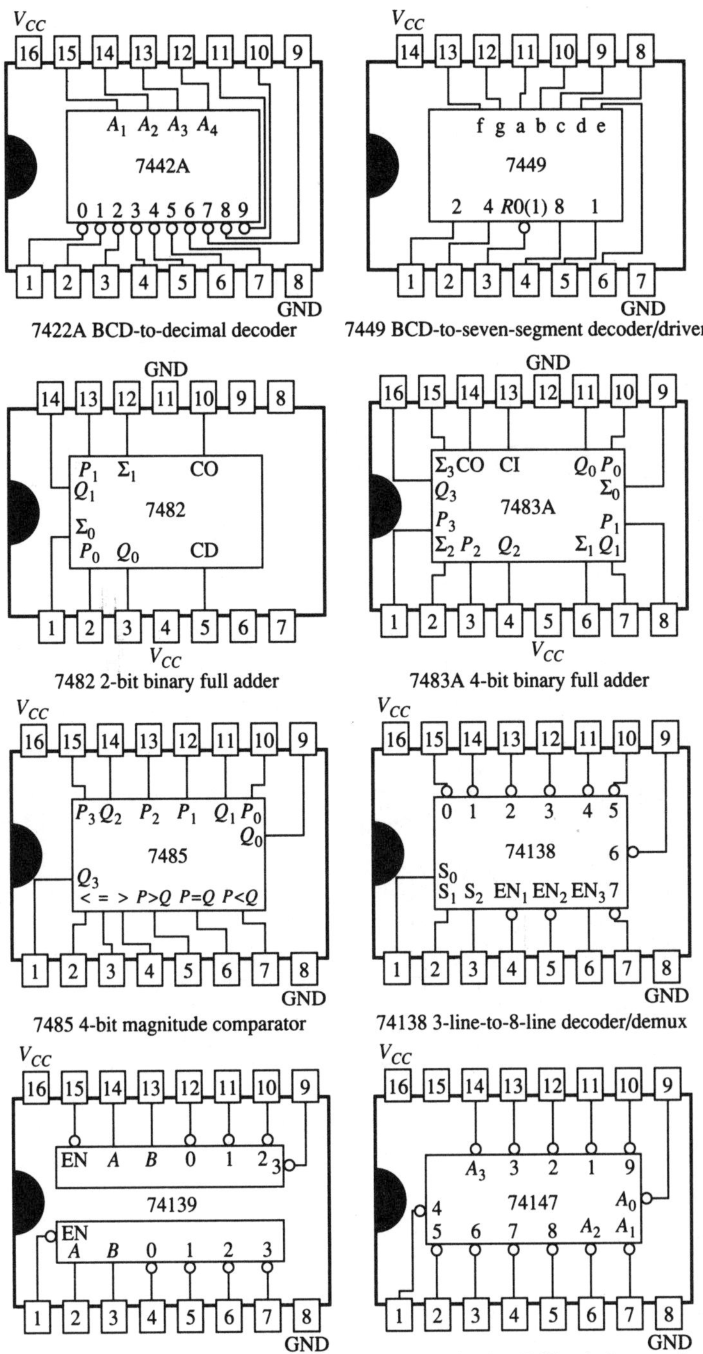

7422A BCD-to-decimal decoder

7449 BCD-to-seven-segment decoder/driver

7482 2-bit binary full adder

7483A 4-bit binary full adder

7485 4-bit magnitude comparator

74138 3-line-to-8-line decoder/demux

74139 Dual 2-line-to-4-line decoder/demux

74147 Decimal-to-BCD priority encoder

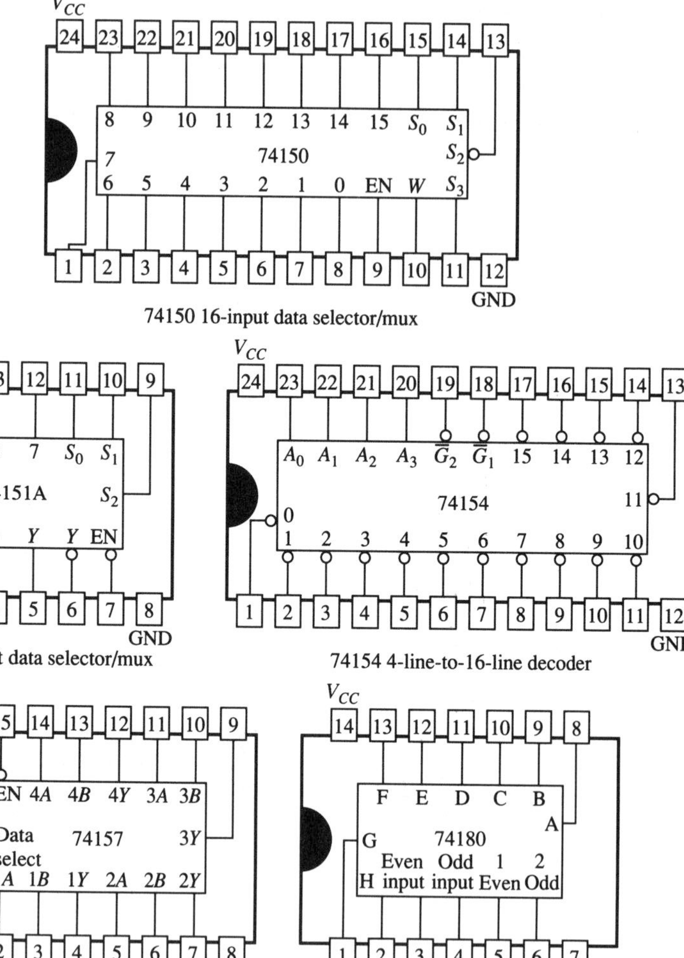

74150 16-input data selector/mux

74151A 8-input data selector/mux

74154 4-line-to-16-line decoder

74157 Quad 2-input data selector/mux

74180 9-bit parity generator/checker

Integrated circuits (con't)

ample, a computer game is an interactive computer program.

intercom An electrical system used for voice communications and music usually within the same building.

interconnection An electrical connection between two or more electrical systems that allows the transfer of electrical energy between the systems.

interelectrode capacitance The stray capacitance that exists between the electrodes of a vacuum tube.

interface card An electrical circuit that converts a computer I/O bus to some other standard configuration. An example is a parallel-to-serial interface card.

interfacing Causing the output of one type of system to be compatible with the input of another type of system. For example, between a personal computer and a printer there is an interfacing circuit.

interference An undesirable electric or electromagnetic energy that causes an unwanted response in an electrical circuit.

intermediate frequency amplifier An amplifier tuned to a single fixed frequency for the purpose of selecting and amplifying one of the frequency components of a mixer.

internal resistance The amount of resistance that is contained within a source of electrical power. As an example, all practical batteries have an internal resistance.

International System From the French (Systéme International). These are symbols used in electronics to represent both quantities and their units. In this case, one symbol is used to represent the name of the quantity, and the other is used to represent

Quantity	Symbol	Unit	Symbol
capacitance	*C*	farad	F
charge	*Q*	coulomb	C
conductance	*G*	siemen	S
current	*I*	ampere	A
energy	*W*	joule	J
frequency	*f*	hertz	Hz
impedance	*Z*	ohm	Ω
inductance	*L*	henry	H
power	*P*	watt	W
reactance	*X*	ohm	Ω
resistance	*R*	ohm	Ω
time	*t*	second	s
voltage	*V*	volt	V

International system

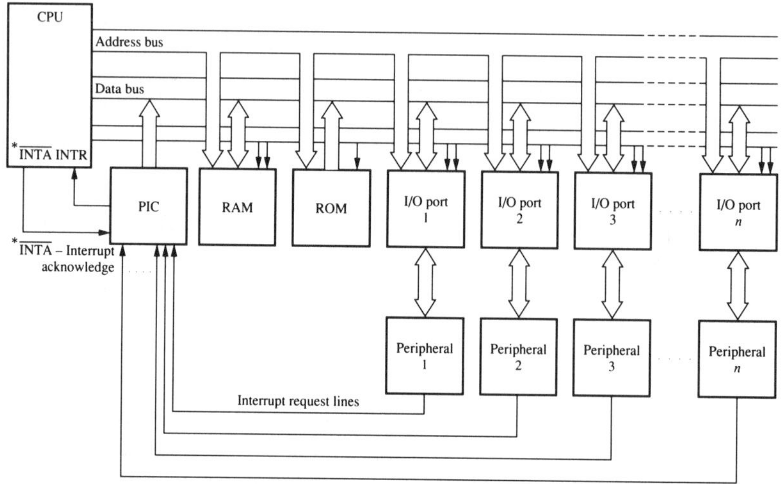

Interrupt driven

the unit of measurement of that quantity.

interpreter In computer programming a program that translates a high-level language into code understood by the microprocessor each time the high-level language is executed. See compiler.

interrupt In microprocessor-based systems, a signal that causes the microprocessor to stop its normal processing and perform some other predictable task. Interrupts are used to help control the overall operation of microcomputers.

interrupt driven In microprocessor-based systems, the microprocessor will respond to a need for service only when there is an interrupt signal given to it. This has an advantage over the polling system (where various checks are always made for special servicing) in that it does not waste processing time.

interstage Between the stages. For example, a coupling capacitor that feeds the output signal of one stage of amplification to another stage of amplification could be referred to as an interstage component.

interval A period of time from one event to another event.

inverse feedback See negative feedback.

inverter An electrical circuit that causes the phase of an input signal to change by 180 degrees. An inverter will produce a negative signal output for a positive signal input and vice versa.

inverting amplifier An electrical circuit that amplifies the

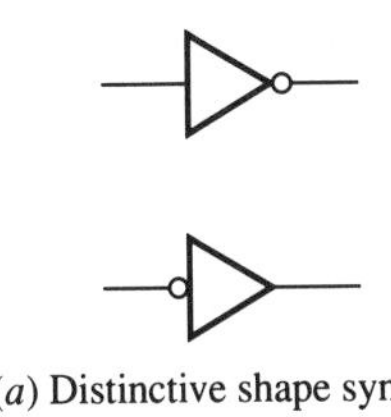

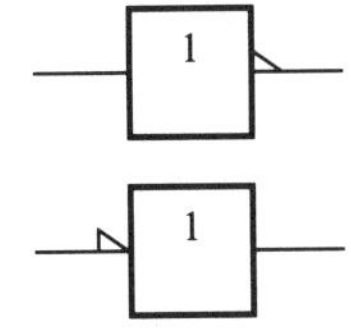

(*a*) Distinctive shape symbols with negation indicators

(*b*) Rectangular outline symbols with polarity indicators

Inverter

signal and causes the output to be 180 degrees out of phase with the input signal.

I/O Abbreviation for input and output.

ion An atom that has a net positive or negative charge. Ions are caused by the removal or addition of electrons from an otherwise electrically neutral atom.

ionization The formation of ions. Processes that lead to the formation of ions are gases exposed to an electric field, electrically charged particles produced by radiation, and the process of giving a net charge to a neutral atom.

ionize The process of producing a net charge to an otherwise neutral atom.

ionosphere That part of the earth's outer atmosphere where ions and electrons are present to affect radio wave propagation.

i_p Designation for vacuum tube signal plate current.

I_P Designation for the dc plate current of a vacuum tube.

IR Abbreviation for infrared radiation. See infrared.

iron-vane meter A type of meter movement consisting of two metal bars (called vanes) that are located next to each other within a coil, where one is stationary and the other moves. An electromagnetic field produced by a current-carrying coil causes the two vanes to repel by an amount proportional to the current in the coil.

irradiance The amount of radiated power per unit area. Expressed in watts per square centimeter.

i_s Designation for FET signal source current.

I_S Designation for the dc source current of an FET.

isolated Absolutely no connection or influence. If one circuit is isolated from another, it means that the two circuits cannot have any kind of influence on each other.

isolated

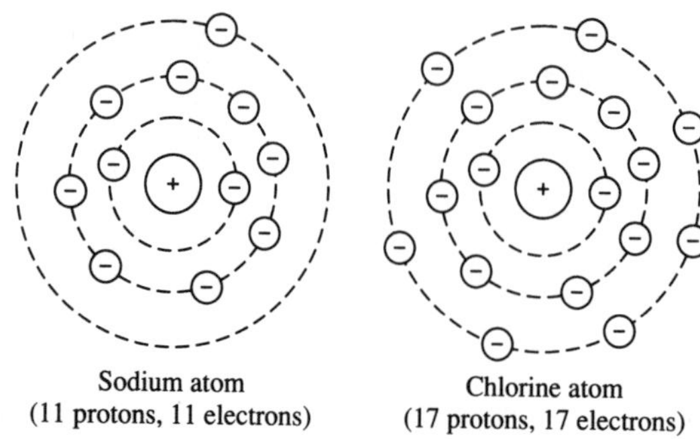

Sodium atom
(11 protons, 11 electrons)

Chlorine atom
(17 protons, 17 electrons)

(*a*) The sodium atom has a single valence electron.

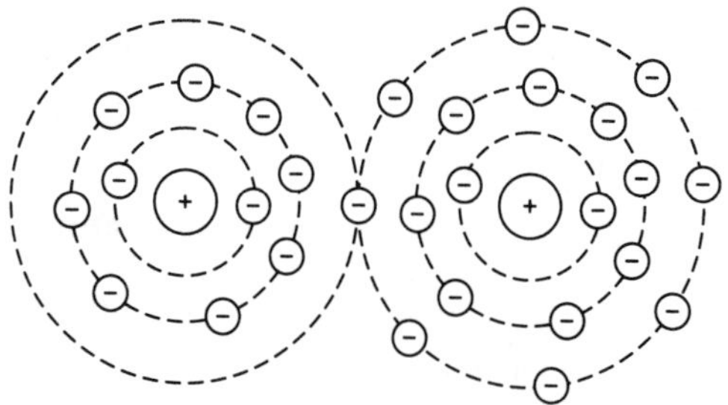

(*b*) The atoms combine by sharing the valence electron to form sodium chloride (table salt).

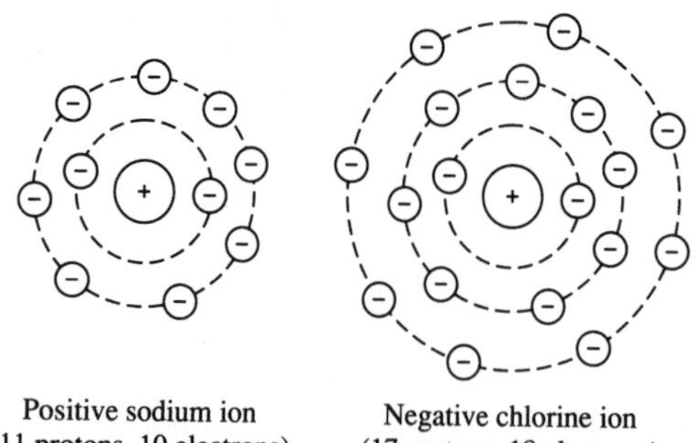

Positive sodium ion
(11 protons, 10 electrons)

Negative chlorine ion
(17 protons, 18 electrons)

(*c*) When dissolved, the sodium atom gives up the valence electron to become a positive ion, and the chlorine atom retains the extra valence electron to become a negative ion.

Ions

J Letter symbol for joule. See joule.

jack A mechanical socket that allows an electrical connection to two or more circuits and/or electrical devices.

jacket In electronics, pertaining to the outer sheath of a wire or cable that provides insulation and protects the contents.

jam An electronic method used to interfere with other signals. The process usually used by governments and the military in order to prevent the successful reception of radio and television transmissions or other forms of electromagnetic radiation, such as radar.

jamming The intentional transmission of electromagnetic energy or other means used to interfere with the reception of radio and television or other information such as radar.

JFET Abbreviation for junction field-effect transistor. See junction field-effect transistor.

J/K flip-flop A flip-flop with four useful input conditions and two complementary outputs. The inputs are referred to as J and K. When J and K are inactive, there is no change on the output; J active and K inactive will cause the *Q* output to be active; J inactive and K active will cause the NOT *Q* output to be active. Both J and K active will cause the output to toggle. All transitions are synchronized by an external clock pulse.

Johnson counter A digital counter where only one flip-flop changes state at a time. A Johnson counter consists of a shift register where the complement of the output is fed back to the input (See pg. 161 for illustration).

Johnson noise See thermal noise.

***j* operator** In electronics, the prefix *j* is used to represent an imaginary number; in mathematics, the letter *i* is used. However, in electronics, this could be confused with instantaneous current, so *j* is used instead. The

Logic diagram for a master-slave J-K flip-flop.

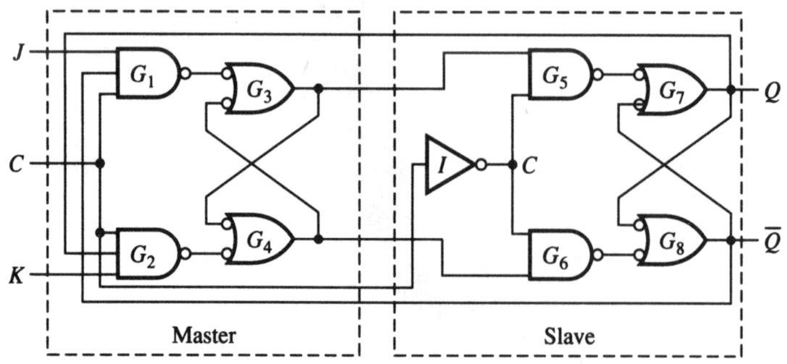

Truth table for the master-slave J-K flip-flop.

Inputs			Outputs		Comments
J	K	C	Q	$\overline{Q}$	
0	0	⎍	Q_0	$\overline{Q}_0$	No change
0	1	⎍	0	1	RESET
1	0	⎍	1	0	SET
1	1	⎍	$\overline{Q}_0$	Q_0	Toggle

⎍ = clock pulse
Q_0 = output level before clock pulse

J-K flip-flop

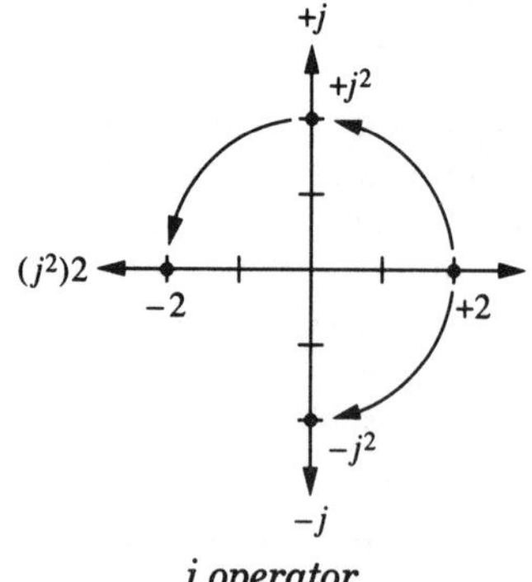

j operator

j operator is the square root of -1, where $+j$ is represented by 90 degrees on the complex plane and $-j$ by -90 degrees on the same plane.

joule 1. The unit of measurement for energy in the SI or MKS system. One joule is equal to 0.7378 foot-pounds in the English system or 10^7 ergs in the MKS system. 2. Unit of measurement for the work done when a force of one newton acts through a distance of one meter.

joystick An analog device consisting of a handle and pushbutton switches that allows the user to input information into the computer. A joystick may be used for moving the cursor to specific locations on the monitor and the push buttons to select specific points of information. Joysticks are frequently used to play computer games.

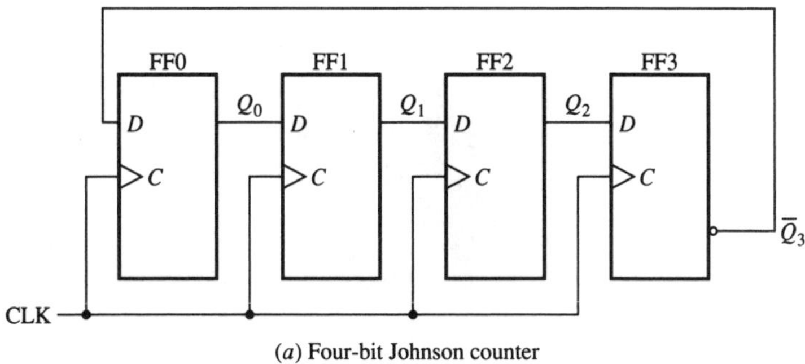

(*a*) Four-bit Johnson counter

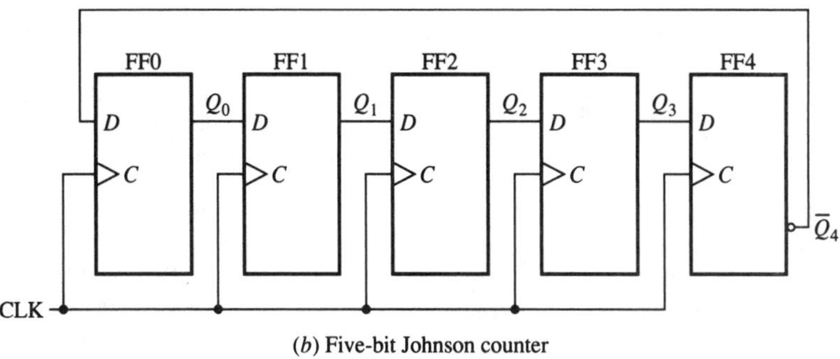

(*b*) Five-bit Johnson counter

Johnson counter

jumper Usually a short piece of wire or other conducting material used to connect one part of a circuit to another electrically.

jumper cable A short cable used to connect one circuit to another electrically.

junction diode An electrical device that conducts current more easily on one direction than the other. It is made from a single crystal that contains p-material at one end and n-material.

junction field-effect transistor 1. A solid-state device consisting of a gate region diffused into a channel region. The current in the channel region is controlled by the voltage applied to the gate. 2. A device that consists of two semiconductor materials (p-type and n-type) constructed as a sandwich where the inner part is called a channel. An external voltage applied to the device controls the current in the channel. The three external leads of the device are called the source, gate, and drain.

junction transistor A transistor consisting of three alternate sections of p-type or n-type semiconductor material. See npn or pnp transistor.

K Letter symbol for kelvin. See kelvin.

Karnaugh map A graphical means of displaying a truth table for the purpose of logic simplification. A Karnaugh map consists of a rectangular array where each rectangle represents a unique logic state of the Boolean expression to be simplified.

kelvin A unit of measurement for temperature in the SI system. One kelvin is equal to 273.15 + C in both the CGS and MKS systems.

Kerr effect Electro-optical phenomenon where specific transparent substances become double refracting when subjected to an electric field perpendicular to a beam of light.

keyboard An arrangement of push-button switches with related functions. A keyboard may be found on a calculator where the push-button switches are used for entering numbers and mathematical operations. Keyboards on computers or programmable calculators may be used for entering alphanumeric characters as well as commands for specific functions.

	$\overline{C}$	C
$\overline{A}\overline{B}$	0	0
$\overline{A}B$	0	1
AB	1	1
$A\overline{B}$	0	0

(*a*) $\overline{A}BC + AB\overline{C} + ABC$

	$\overline{C}\overline{D}$	$\overline{C}D$	CD	$C\overline{D}$
$\overline{A}\overline{B}$	0	0	0	0
$\overline{A}B$	0	1	1	0
AB	0	0	0	0
$A\overline{B}$	0	1	1	0

(*b*) $\overline{A}B\overline{C}D + \overline{A}BCD + A\overline{B}\overline{C}D + A\overline{B}CD$

Karnaugh map

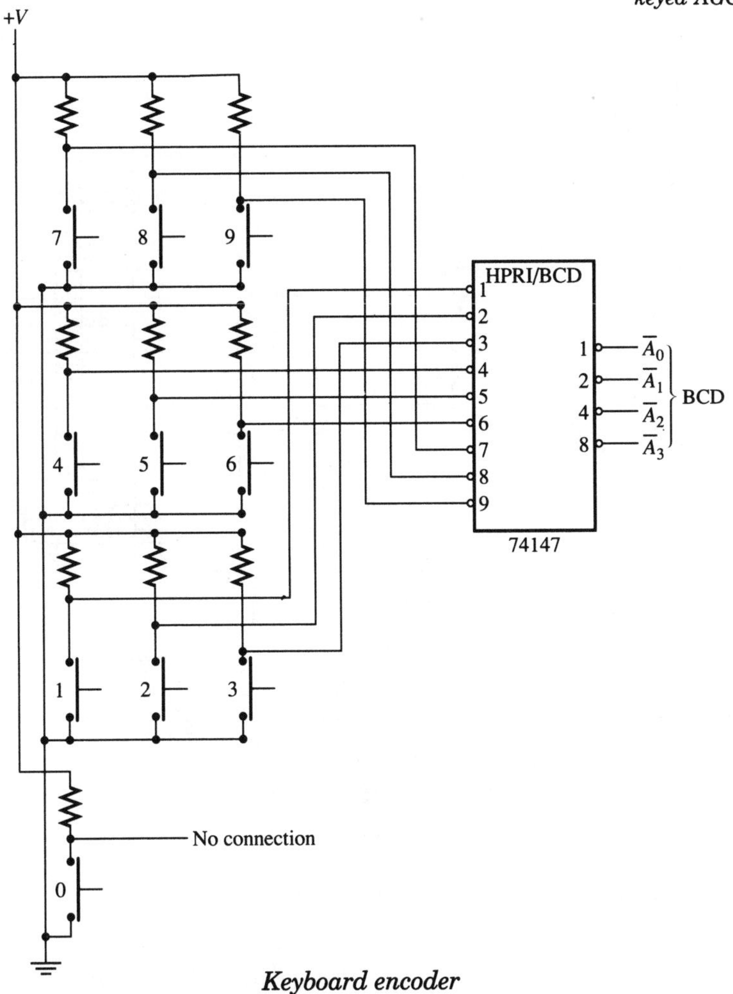

Keyboard encoder

keyboard encoder A type of encoder that allows easily depressed keys to be converted to a binary code that represents the symbol on the key depressed. For example, the 10 decimal digits on the keyboard of a calculator need to have a circuit that will convert the value marked on the depressed key to be converted into some sort of binary code.

keyed AGC A type of automatic gain control where the signal for the amount of gain is sampled only at specified times. Keyed AGC is used in television receiver circuits where AGC is developed from the strength of the horizontal sync pulse. The reason for this is because of the drastic changes that can occur in the picture signal as scenes

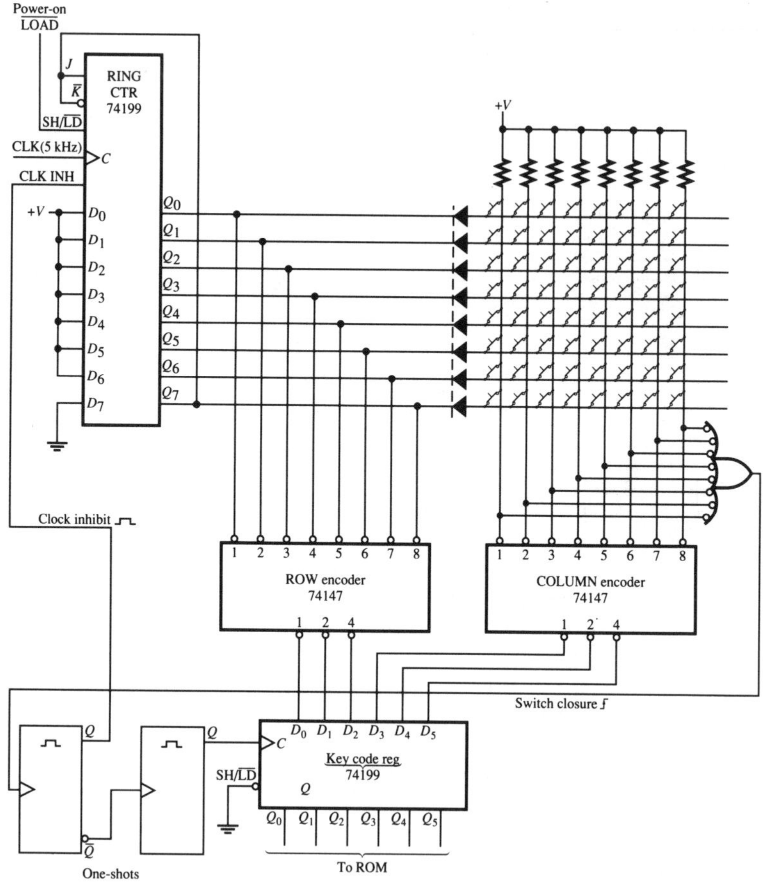

Keyboard encoder (con't)

change from dark to light. Thus, the AGC is not developed from the picture signal and is instead *keyed* to the strength of the horizontal sync pulse.

kg Letter symbol for kilogram. See kilogram.

kHz Letter symbol for kilohertz. See kilohertz.

kilo A prefix designating 1000: kilo $= 1 \times 10^3$.

kilogram The unit of measurement (kg) for mass in both the MKS and SI systems. Equal to 1000 grams in the CGS system.

kilohertz A unit of frequency that represents 1000 cycles per second.

kilovolt One kilovolt equals 1000 volts.

kilovoltampere One kilovoltampere is equal to 1000 watts.

kilowatthour A measure of the amount of power being used. A kilowatthour means that the product of the power (in watts) and the time (in hours) will equal 1000.

Kirchhoff's current law The statement that the algebraic sum of the currents entering and leaving a node is zero.

Kirchhoff's laws The two major laws are that the sum of all voltage drops around a closed loop are zero and that the sum of all currents at a node are zero. Kirchhoff's laws are useful in the analysis of series and parallel circuits.

Kirchhoff's voltage law The statement that the algebraic sum of the potential rises and drops around a closed path is zero.

klystron An electron tube used as an oscillator or amplifier at ultrahigh frequencies.

knife switch A mechanical device used to make or break electrical connections. A knife switch consists of a movable metal ribbon (like a knife) that makes or breaks contact between two metal contact clips.

knob A solid device attached to the end of a control shaft in order to make manual adjustment of the control easier. A knob may also be used for decorative purposes.

knockout A removable portion of a cabinet or container that may be easily removed to accommodate an electrical system installed in the cabinet or container.

knowbot A software servant that acts, without requiring specific information, to prioritize or scout for information. For example, research on any subject can be automatically found and tabulated in a form requested by the user. An extended system can also protect confidential information and disable computer viruses.

kV Letter symbol for kilovolt. See kilovolt.

kVA Letter symbol for kilovoltampere. See kilovoltampere.

L Letter symbol for inductance.

label In assembly language programming, the first field of the program that can be used to represent an address.

ladder D/A converter A digital-to-analog converter that uses an arrangement of resistors that take the form of a ladder network. See ladder network.

ladder network An arrangement of electrical components that consists of a cascaded set of series-parallel connections that appears as a ladder.

lag The displacement of two or more signals in time usually expressed in degrees.

lagging An event of a waveform after some other event or time.

laminate The building up of an object through the process of using many layers of the object.

LAN Abbreviation for local area network. See local area network.

language In regards to computers, the method of communicating between the computer and the user for the purpose of defining procedures to be performed by the computer.

language level This is how close a programming language is to the actual bit pattern used by the microprocessor. The closer the programming code is to that

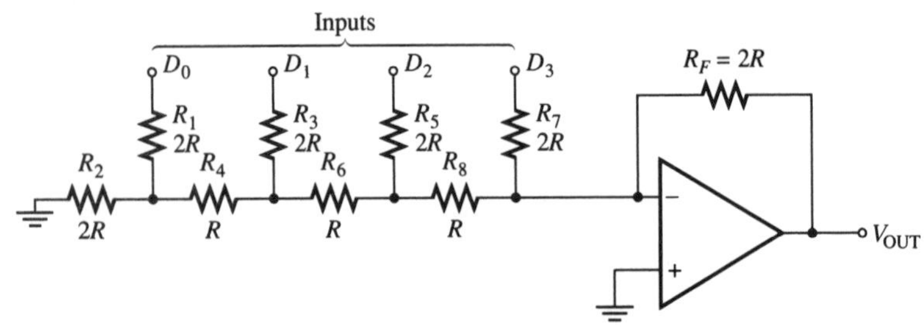

Ladder D/A converter

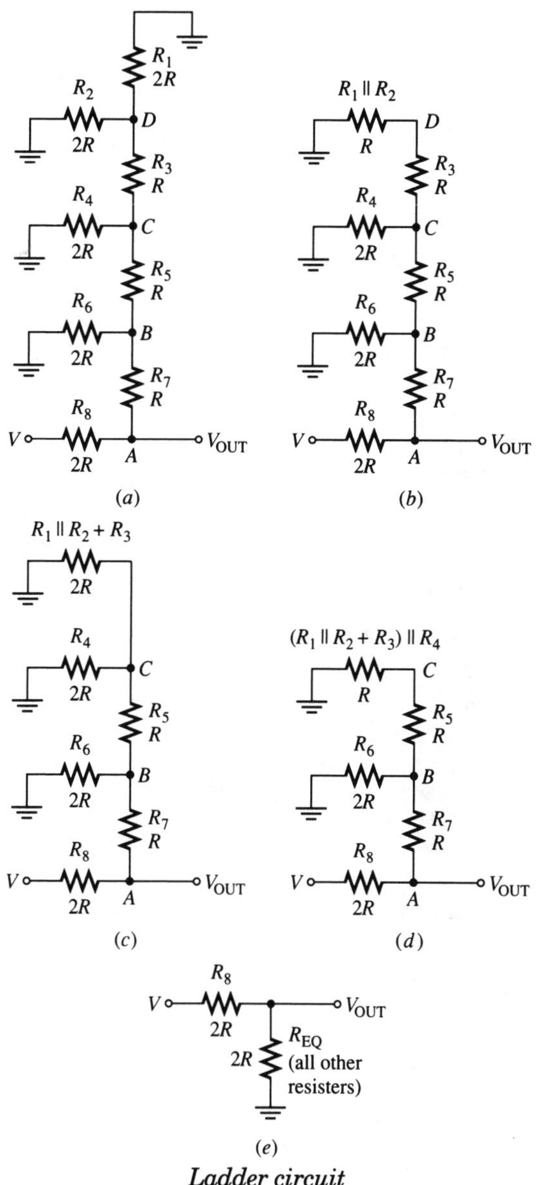

Ladder circuit

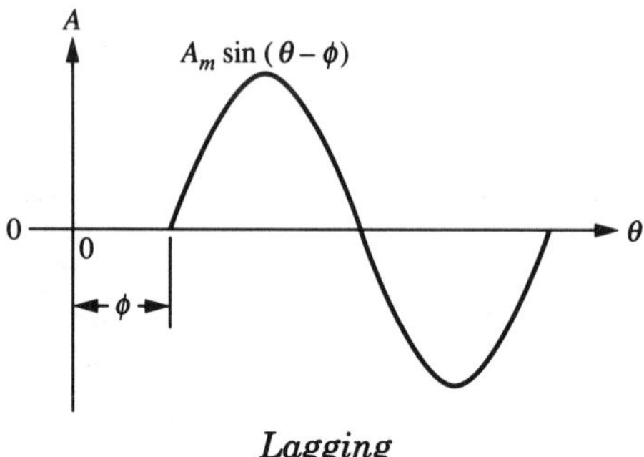

Lagging

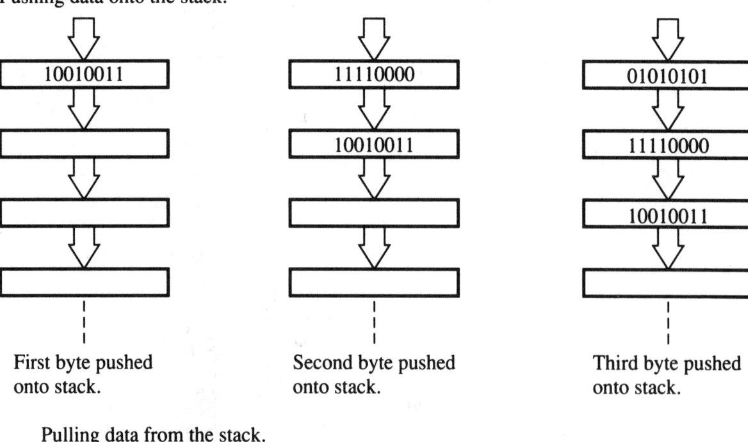

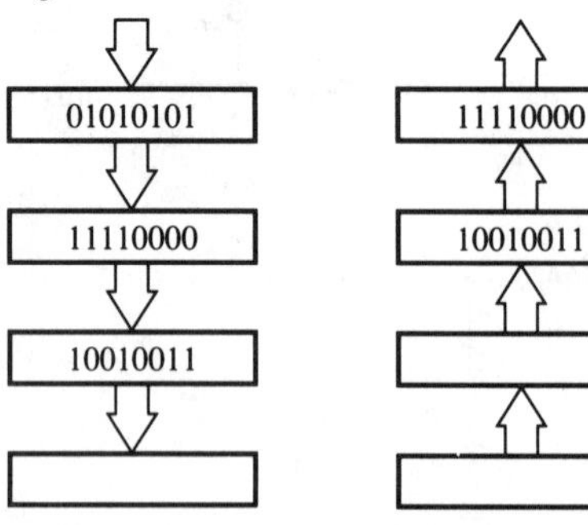

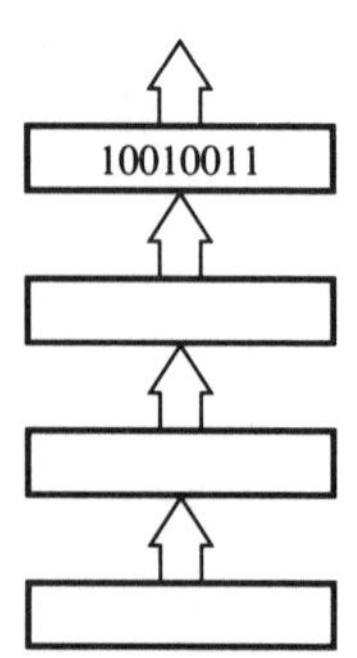

(a) Initially storing 3 bytes (b) After third byte is pulled from stack (c) After second byte is pulled from stack

Last in, first out

of the microprocessor, the lower the level of the programming language. The closer the programming language is to the way people communicate (with words and numbers), then the higher the level of the language.

Laplace transform A process of performing symbolic changes to a differential equation so it may be solved by algebraic means.

laptop computer A computer small enough to fit on your lap. Consists of a flat LCD screen and compact keyboard. Usually battery as well as a.c. powered.

large-scale integration Classification of integrated circuits by size. Large scale integration indicates integrated circuits that have more than 100 gates or circuits of equal complexity.

LASCR Abbreviation for light-activated silicon-controlled rectifier. See light-activated silicon-controlled rectifier.

laser Abbreviation for light amplification by stimulated emission of radiation.

laser diode A solid-state device capable of emitting laser light when electrical energy is applied to it.

lasing The process of producing laser light. Requires a light source and a lasing medium with opposing mirrors, one of which allows some light to pass through it.

last in, first out A type of memory arrangement where the last bit of data to be placed in the memory is the first bit of data to be removed from the memory when it is read. You can think of this arrangement of data like a stack of dishes where the last dish to be placed on the stack is the first dish to be removed from the stack.

latch A logic circuit that maintains a given logic condition until changed by an external source. Consists of two inputs and two complementary outputs. One input sets the latch the other clears the latch.

latching relay An electromechanical device that is electrically controlled that makes or breaks mechanical contact and will stay in its make or break position until reset by mechanical or electrical means.

LC Abbreviation for inductor-capacitor. See *LC* network.

LCD Abbreviation for liquid crystal display. See liquid crystal display.

***LC* filter** An electrical circuit consisting of an inductor and a capacitor designed to remove a group of frequencies or a single frequency. In a power supply, an *LC* filter consists of a series inductor connected to a parallel capacitor and designed to reduce any supply line variations on the output.

***LC* network** An electrical circuit consisting of an inductor and

LC *network*

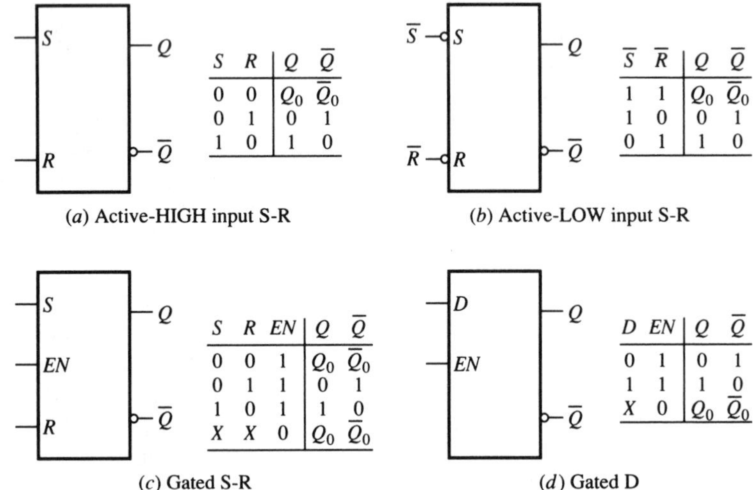

S	R	Q	$\overline{Q}$
0	0	Q_0	$\overline{Q}_0$
0	1	0	1
1	0	1	0

(*a*) Active-HIGH input S-R

$\overline{S}$	$\overline{R}$	Q	$\overline{Q}$
1	1	Q_0	$\overline{Q}_0$
1	0	0	1
0	1	1	0

(*b*) Active-LOW input S-R

S	R	EN	Q	$\overline{Q}$
0	0	1	Q_0	$\overline{Q}_0$
0	1	1	0	1
1	0	1	1	0
X	X	0	Q_0	$\overline{Q}_0$

(*c*) Gated S-R

D	EN	Q	$\overline{Q}$
0	1	0	1
1	1	1	0
X	0	Q_0	$\overline{Q}_0$

(*d*) Gated D

Note: Q_0 is the initial state

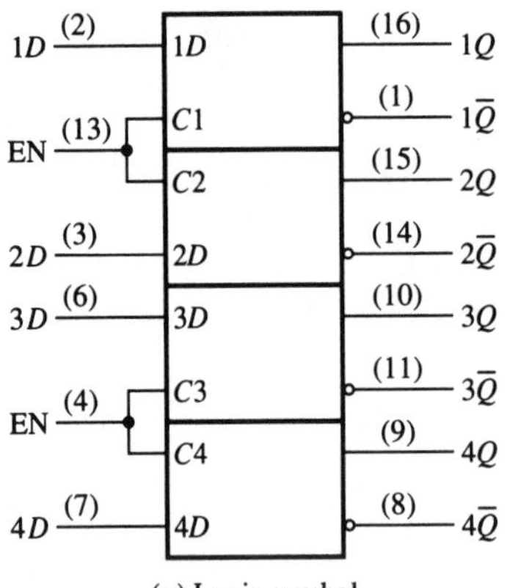

(*a*) Logic symbol

Inputs		Outputs		Comment
D	EN	Q	$\overline{Q}$	
0	1	0	1	RESET
1	1	1	0	SET
X	0	Q_0	$\overline{Q}_0$	No change

Q_0 is the prior output level.

(*b*) Truth table (each latch)

Latches

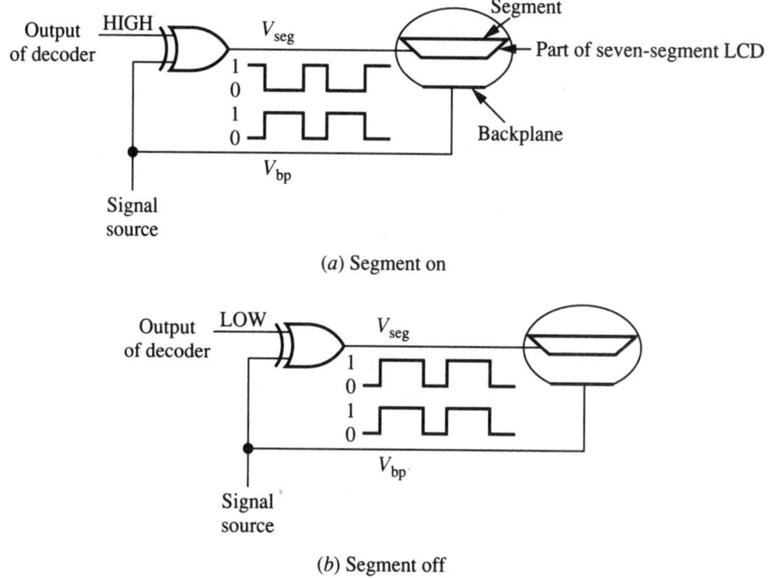

(*a*) Segment on

(*b*) Segment off

LCD

a capacitor. By nature, this type of network is frequency selective. Reference to an *LC* network usually means a frequency selective network.

***LC* oscillator** A circuit capable of generating its own signal where capacitors and inductors are used to determine the resultant signal frequency.

lead-acid cell The smallest section of a battery consisting of lead oxide plates immersed in a solution of dilute sulfuric acid. Energy is stored in the cell chemically that is converted to electrical energy when connected to a load. The cell may be electrically charged thus renewing its chemical energy.

leading Where a waveform takes place before some other event or time.

leading edge The beginning part of a pulse. For a positive-going pulse, the leading edge is the transition from a low voltage (usually 0 volts) to a higher voltage (usually +5 volts). For a negative-going pulse, the leading edge is the transition from a higher voltage (usually +5 volts) to a lower voltage (usually 0 volts).

leading-edge PDM Pulse duration modulation where the leading edge changes in accord-

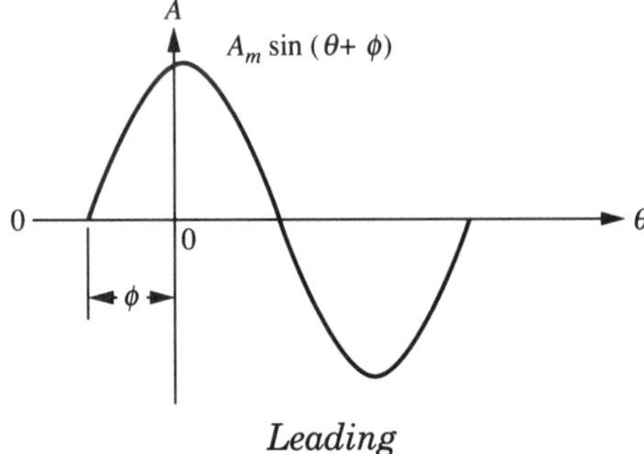

Leading

ance with the amplitude of the modulating signal.

lead-in wire Conductors used to connect an antenna to electrical equipment such as a receiver or transmitter.

leading-zero suppression A feature found in some seven-segment decoders. This function is used to blank out extra zeros in the display. In a leading-zero suppression, all zeros to the left of the most significant digit are blanked out.

leakage 1. An undesirable current flow usually of a low value in a device where the ideal condition would be zero current. For example, leakage currents exist in pn junctions due to minority carriers. 2. An undesirable flow of current across an insulator.

least significant bit In a binary number, the bit contributing the smallest quantity of the value of the number. For example, in 101_2, the rightmost 1 is the LSB. See most significant bit (MSB).

least significant digit The digit in a number that has the lowest place value.

LED Light-emitting diode. See light-emitting diode.

Lenz's law A statement that an induced effect in an inductor is such as to oppose the cause that produced it. (See pg. 174 for illustration.)

level-triggered flip-flop A two-state device that responds to the amount of an applied voltage and not to the rate at which the voltage is changing.

library In computer programming, a collection of programs or subroutines that may be used by other programs without having to rewrite anything in the library. A library is useful when developing software since it can save valuable development time.

LIFO Abbreviation for last in, first out. See last in, first out.

light See visible spectrum.

light-activated silicon-controlled rectifier A solid-state

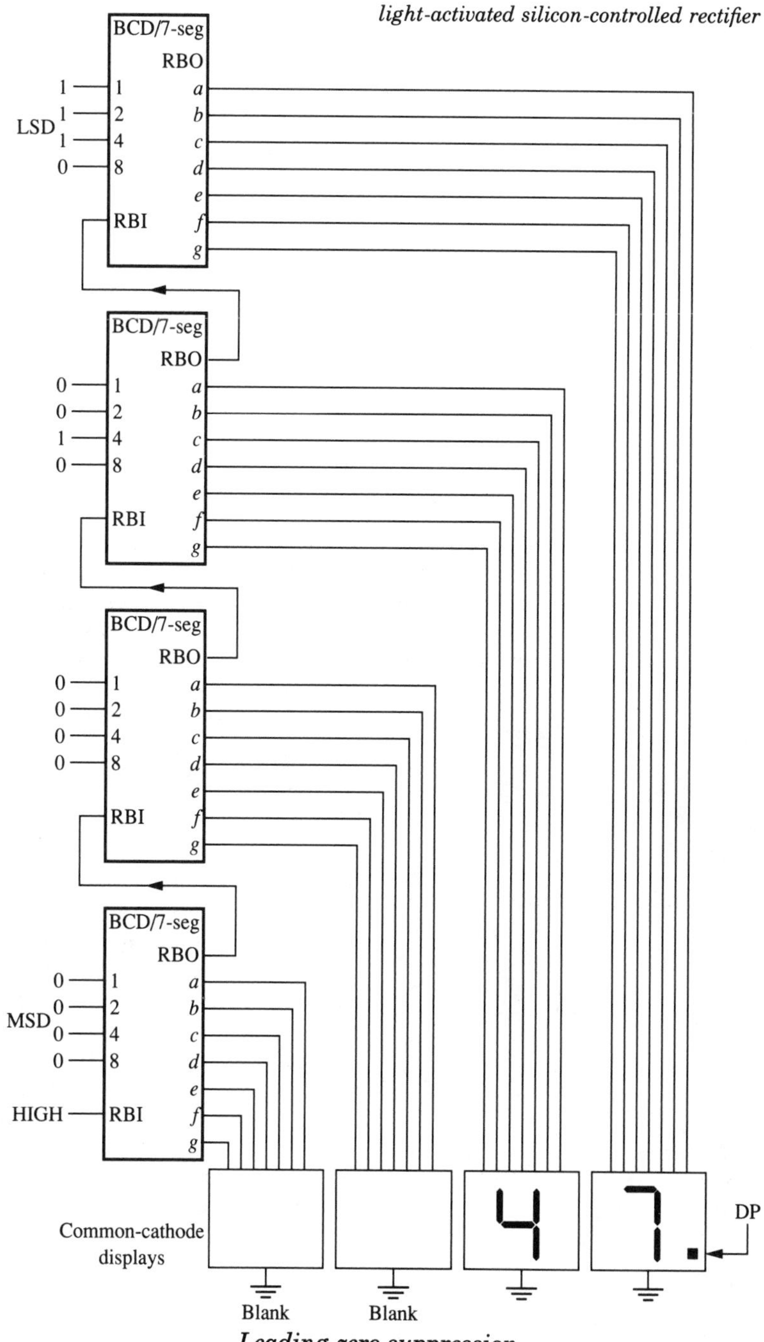

Leading-zero suppression

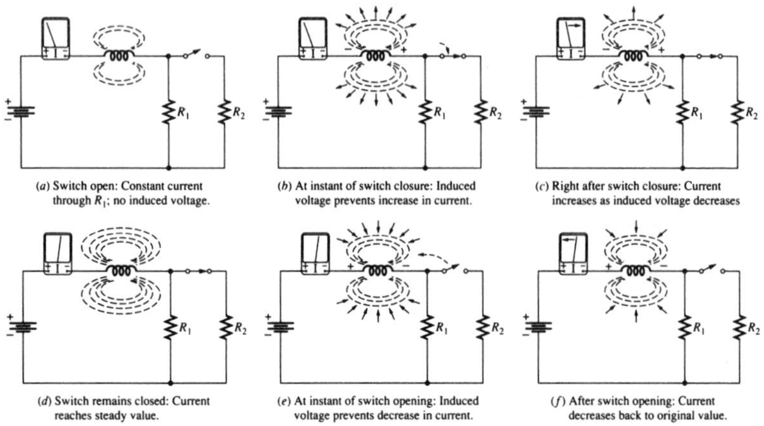

(*a*) Switch open: Constant current through R_1; no induced voltage.

(*b*) At instant of switch closure: Induced voltage prevents increase in current.

(*c*) Right after switch closure: Current increases as induced voltage decreases

(*d*) Switch remains closed: Current reaches steady value.

(*e*) At instant of switch opening: Induced voltage prevents decrease in current.

(*f*) After switch opening: Current decreases back to original value.

Lenz's law

device consisting of a pn-pn construction where light performs the function of gate current. Used in applications where light is the controlling source for electrical switching.

light-activated switch Any device that will do a make-or-break connection due to the presence, absence, or intensity of incident light.

light-emitting diode A solid-state device that emits light when the correct amount and polarity of voltage is applied across the device. Usually requires the addition of a current-limiting resistor in series with the LED and voltage source.

lighthouse tube An electrical device used to control ultrahigh frequencies and is shaped in the form of a lighthouse.

lightning rod A pointed metal shaft mounted at the highest point of an object it is to protect from a lightning strike. The lightning rod is connected to a large low-resistance cable that in turn is connected to the earth. The idea is to encourage a lightning strike to hit the rod and conduct its energy through the cable to earth, thus protecting the object (such as a house or barn).

limiter 1. An electrical circuit that prevents changes in the amplitude of a signal. Limiters are found in FM communications, where changes in amplitude of the carrier represent unwanted noise rather than information. 2. A circuit that is capable of removing part of a waveform above and/or below a specified level. A circuit that causes no more than a specified maximum amount to be produced such as a current limiter.

line cord The power cord used to supply ac power to an electrical system.

line driver A circuit that is used to prepare a signal so that its original characteristics are preserved after transmission along a controlled path. Line drivers are used between remote systems connected by cables where the characteristics of the signal must not be influenced by the electrical characteristics of the cable (line). In digital circuits, a line driver is used to increase the fan-out of a given circuit. See fan-out.

line drop A voltage drop resulting from the resistance in a power transmission line.

line equalizer A capacitor or inductor inserted to improve the frequency-response characteristics of a transmission line.

line filter An electrical circuit inserted in a power or transmission line to reduce the effects of undesirable electrical signals. A line filter usually consists of capacitors and or inductors arranged in such a manner so as to block out unwanted signals.

line impedance The total opposition to current flow as measured across the terminals of a transmission line at a given frequency or dc.

line loss The dissipation of power due to effects of a transmission line.

line noise Undesirable and unpredictable electrical signals resulting from the effects of a transmission line.

line of sight An unobstructed visual path between two points.

line regulation How well a certain voltage source stays at its rated value. It is expressed as the percentage change in output voltage for a given change in the input (line) voltage.

line spectra Energies that originate from excited atoms at distinct and predictable frequencies that in turn are capable of producing visible lines from which the elements of the excited atoms may be determined.

line transformer A transformer inserted into a circuit in order to provide impedance matching, isolation, or other desirable electrical conditions.

line voltage Voltage supplied by a source of electrical power. Line voltage may be measured at the source or at the terminal point of utilization.

linear Characterized by a straight-line relationship. A linear amplifier has an output that is an amplitude magnification of its input.

linear adder An electrical circuit that produces an undistorted output of two or more input signals.

linear amplifier Any electrical device that increases the power of an input signal where the output is proportional to the input. A transistor amplifier is said to be a linear amplifier if

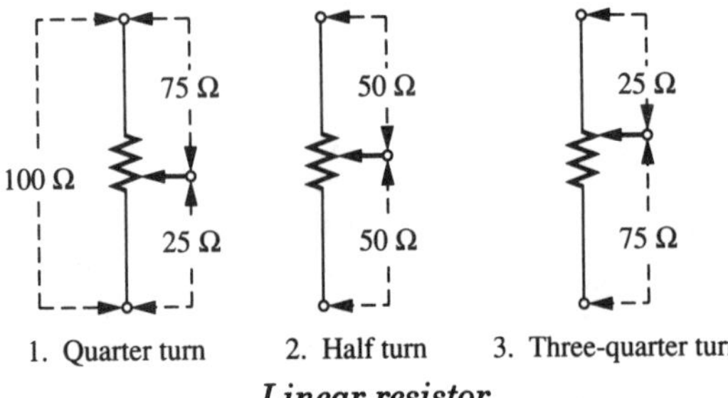

Linear resistor

it is operated within the range where the output signal is an amplified reproduction of the input signal.

linear control Any variable electrical device that produces an electrical change that is directly proportional to a change in its mechanical control. For example, a variable resistor is said to be a linear control if each incremental rotation in the shaft produces the same amount of resistance change in the variable resistor.

linear detector Any electrical device that produces an amplitude change that is directly proportional to the modulation of a radio frequency carrier.

linearity Showing the relationship between two quantities when a change in one causes a corresponding and predictable change in the other.

linear operation For an amplifier, the operational characteristics that are between saturation and cutoff. That portion of an amplifier's characteristics that result in a linear amplifier. See linear.

linear programming A mathematical method of programming so that a group of limited resources among a number of competing demands may be shared.

linear resistor A variable resistor whose value changes the same amount for each mechanical change.

lines of force Imaginary lines used to help visualize effects at a distance. Lines of force are used to help describe the effects of electrical charge and magnetism.

liquid cooling Using a circulating liquid to help equipment dissipate heat during operation.

liquid crystal display An electrostatic display medium that requires almost no power to operate because it does not generate any light. Small voltages are used to cause a polarization of reflected light from previously etched designs (such as numbers and letters) that causes incident light to be reflected from the surface at a different angle. The polarization process causes the selected symbols to become visible to the human eye.

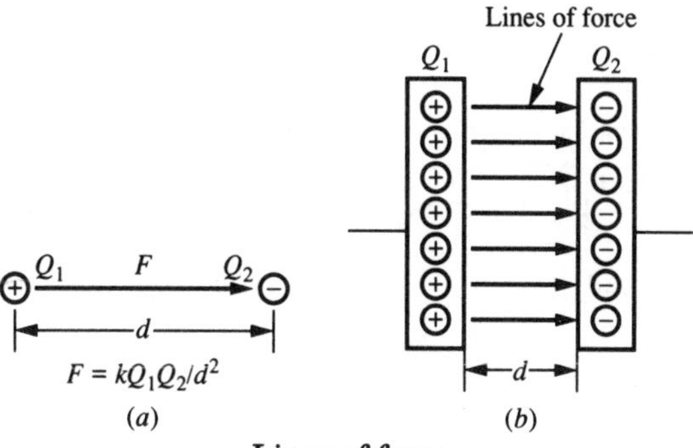

Lines of force

LISP Abbreviation for list processing. See list processing.

Lissajous figures Oscilloscope patterns formed by the applications of two signals, one to the vertical scope input the other to the horizontal. The resulting Lissajous figure is a function of the amplitude, phase, and frequency relationships of the two signals.

list processing A high-level programming language that treats program lines as data where the lines may be modified (thus modifying the program) while the program is executed.

live A term applied to any circuit that is electrically active.

LO Abbreviation for local oscillator. See local oscillator.

load Any device to which electrical power is delivered. If the load is resistive, it will also consume power. Purely reactive loads, such as an ideal inductor (zero internal resistance) or ideal capacitor (infinite internal resistance), will not consume electrical power.

loaded antenna An antenna that has had an inductor or capacitor added to it to change its electrical length. An antenna is loaded when it is not practical to change its physical length.

loading 1. Causing a circuit to deliver power. Placing a load on a power source causes the power source to deliver power to the load. 2. The process of causing power to be drawn from a source. For example, in digital logic circuits, the more circuits connected to the output of one, the greater the loading.

load line Used as part of the graphical analysis of an amplifier. A load line represents a line drawn on the voltage-current characteristics of the amplifier and shows the relationship between the controlling voltage or current and the resulting voltage or current for the amplifier and external components used with it.

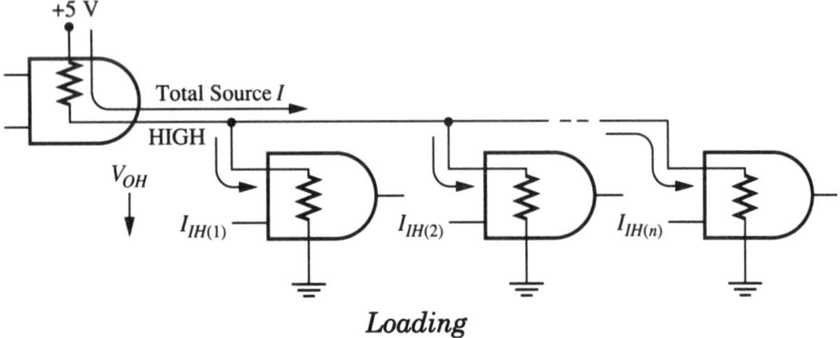

Loading

load matching The process of matching the load to the source so as to produce a maximum transfer of power between them.

load-matching network An electrical circuit inserted between the source and the load to help provide maximum transfer of power.

load regulation 1. A measure of how well the output voltage of a power source stays constant under various power requirements. Measured as the percentage change in output voltage for a given change in load current. 2. Maintaining a constant voltage across a load when the value of the load changes. See voltage regulation.

lobe An area of space emanating from an antenna showing the greatest transmission or reception strength.

local area network An interfacing of digital systems, usually by a single organization that allows communications between the systems.

local oscillator In a communications receiver a circuit that produces a sine wave to be mixed with the carrier of the incoming received signal. The frequency of

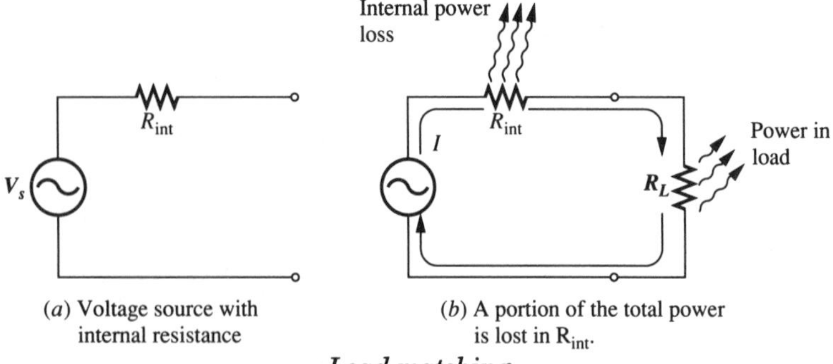

(*a*) Voltage source with internal resistance

(*b*) A portion of the total power is lost in R_{int}.

Load matching

the local oscillator is usually the IF frequency plus the frequency of the received signal. Used in superheterodyne receivers. See intermediate frequency (IF) amplifier.

local oscillator radiation The emission of electromagnetic energy from the local oscillator of a superheterodyne receiver.

log on A method of having a computer user gain access to a computer or computer system. Logging on usually requires that the user knows a code or other authorizing identification.

logarithmic amplifier An electrical circuit where the output is the logarithmic function of the input—as opposed to a linear amplifier.

logic As applied to electronics technology deals with the principles, applications, and relationships of electrical signals in solid-state switching networks.

logical 1 One of two states in a binary system, the other state being 0.

logical 0 One of two states in a binary system, the other state being 1.

logic analyzer A digital analyzing instrument that is basically a multichannel oscilloscope with the ability to detect and display logic levels in many different forms. These forms usually consist of a series of pulses, binary values, or hex values.

logic clip A digital testing device that consists of two rows of indicator lights each corresponding to one of the pins of a digital integrated circuit. The logic clip is "clipped" over the IC, and the indicator lights give the logic condition of each pin of the digital IC.

logic diagram The graphical representation of one or more logic functions and their interconnections as they may exist in an actual digital circuit. Used as an aid in the design and troubleshooting of digital circuits.

logic gate An electrical circuit that simulates a logic function. As an example, an AND gate is a logic gate.

logic level The range of voltages for a digital circuit that will result in a given logic response such as a HIGH or LOW. The level between these HIGH and LOW ranges is referred to as the "uncertain" range.

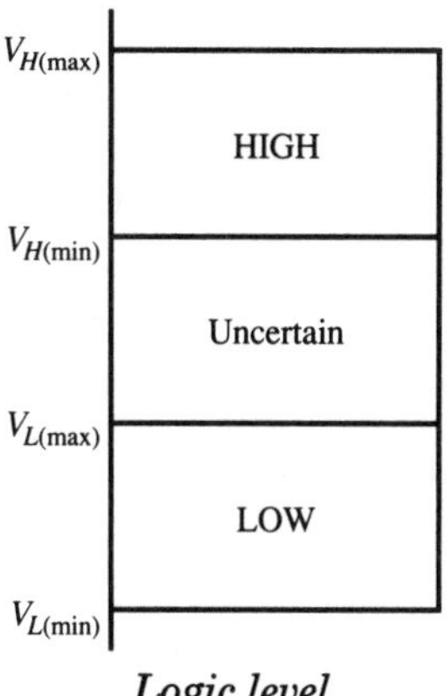

Logic level

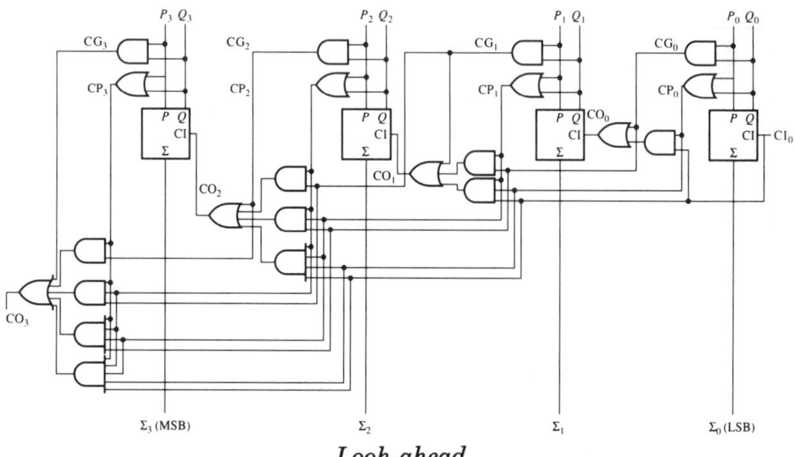

Look-ahead

logic probe A digital instrument that provides a means of detecting a HIGH or a LOW level or a group of pulses at a given point in a digital circuit.

logic pulser A digital instrument that generates a single or a series of pulses for injection into a digital circuit. A logic pulser is used to help troubleshoot a digital circuit and is usually used in conjunction with a logic probe or a current tracer.

logic symbol A drawing that represents a logic function, such as the logic symbol for a NAND gate.

long-nose pliers A hand tool consisting of narrow holding jaws used for mounting and unmounting electrical wires and components.

look ahead A method of generating a carry bit so as to minimize propagation delays. A parallel adder will usually use carry generate circuits that will "look ahead" to avoid the propagation delays inherent in the "ripple" method of generating a carry bit.

lookup table A stored program consisting of bit patterns where the address represents a value and the output represents a function of that value. For example, the sine function can be represented as a lookup table.

loop A complete current path within a circuit.

loop antenna Wires arranged in a spiral for the purpose of receiving electromagnetic radiation. Loop antennas are used for radio direction finding equipment since they are very sensitive to the direction of radio waves.

loose coupling An electrical connection between two separate circuits that transfers a mini-

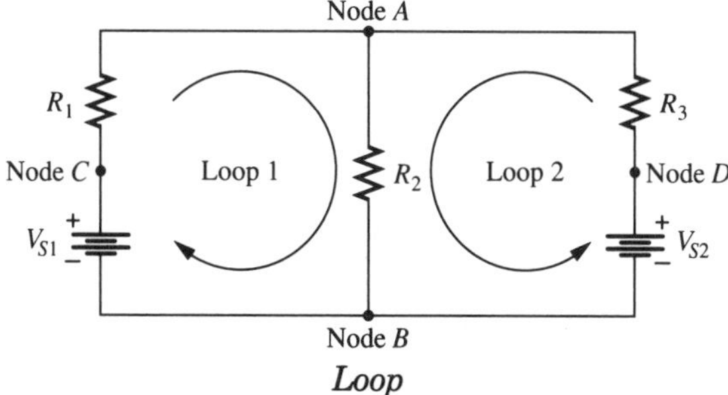

Loop

mum of energy. The advantage of loose coupling is that an impedance change in one circuit has only a very small effect on the other circuit.

loss In electronics, electrical energy dissipated without any useful effect.

loudness A measure of the intensity or volume of sound by human hearing.

loudspeaker A device used for converting electrical energy into sound waves. Usually consists of a paper cone that is set into motion by electrical currents in a coil of wire attached to the base of the cone and set inside the magnetic field of a permanent magnet. Intensity changes in the current cause changes in the resulting magnetic field that opposes or attracts the stationary magnetic field resulting in a corresponding movement of the paper cone creating sound waves.

lower threshold point The voltage at which a device or circuit is activated. For example, in a Schmitt trigger, the lower threshold point is the voltage at which the Schmitt trigger is deactivated.

low-level language A computer programming language that uses symbols and codes that are close to what the computer understands directly. As an example the lowest-level language is a binary code of HIGHs and LOWs whose bit patterns are consistent with the instruction set of the microprocessor within the computer. See high-level language.

low-level transmission In an AM transmitter, having the carrier wave modulated at stages preceding the RF power amplifier. A more economical method than high-level transmission, but not as efficient.

low-order address The least significant half of the bits of an address. For example, the low-order address of a 16-bit address bus are the first 8 least significant bits.

low-order nibble The last 4 bits, including the LSB, of an 8-bit word.

low-pass filter An electrical circuit that blocks the passage of

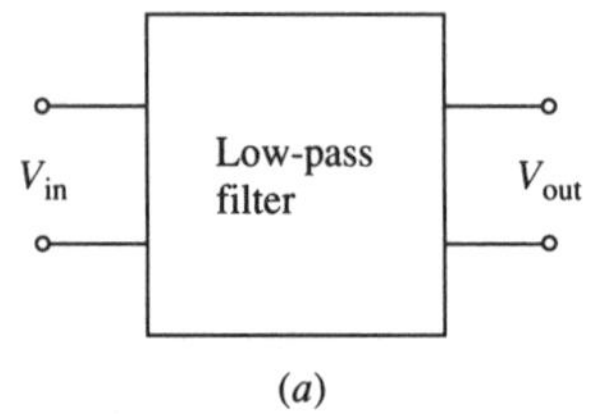

(*a*)

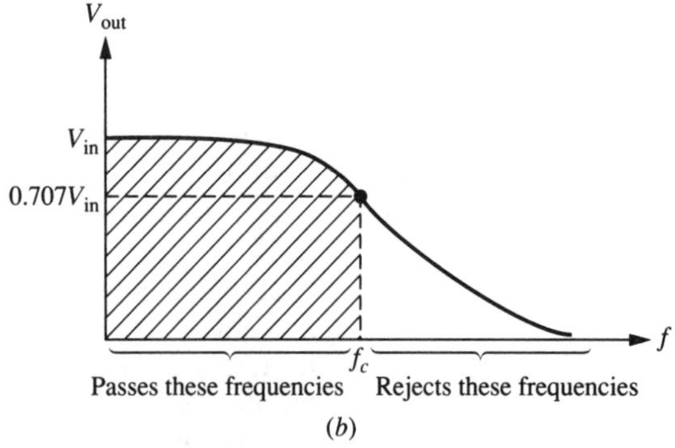

(*b*)

Low pass filter

all frequencies above a given frequency, yet allows frequencies below to be transmitted on.

low-pass *RC* filter An electrical circuit consisting of a capacitor and a resistor where the output is taken across the capacitor.

low-pass *RC* response The frequency characteristics of a low-pass *RC* filter usually presented in graphical form.

low-pass *RL* filter A series circuit consisting of a resistor and an inductor where the output signal is taken from across the resistor. Action of such a circuit is to block a range of high frequencies and pass all frequencies below that range. See high-pass *RL* filter.

LR Abbreviation for inductor-resistor. See *LR* network.

***LR* network** An electrical circuit consisting of inductors and resistors. An *LR* circuit will respond to a range of frequencies and may also be found in timing circuits.

LSB Abbreviation for least significant bit. See least significant bit.

LSI Abbreviation for large-scale integration. See large-scale integration.

LTP Abbreviation for lower threshold point. See lower threshold point.

luminescent A material that will give off light, but not heat,

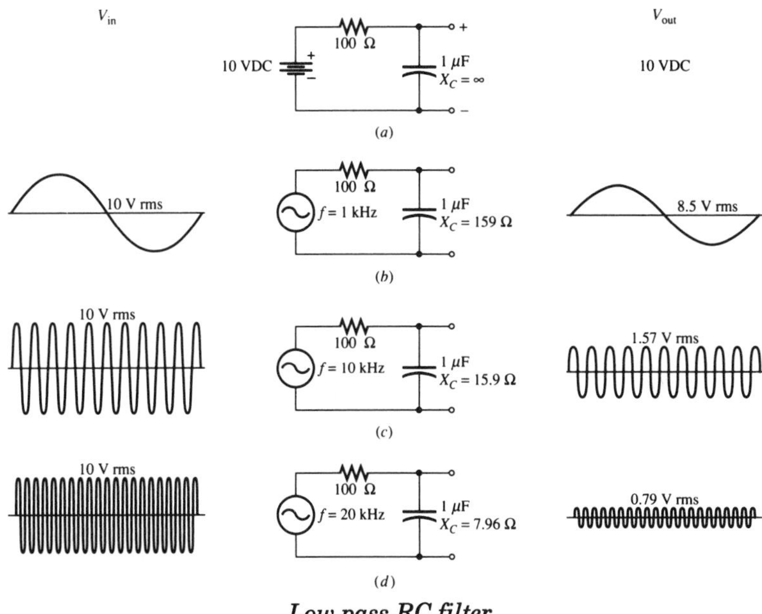

Low pass RC filter

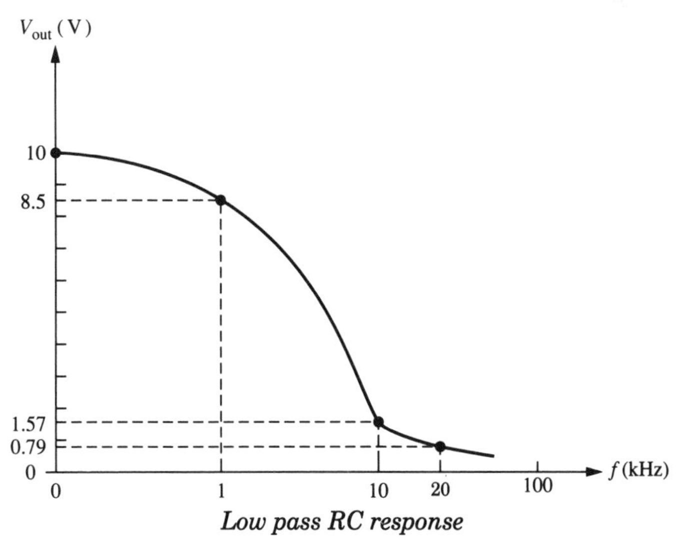

Low pass RC response

when properly energized by an external source.

lumped Viewing a circuit as being concentrated in a single discrete component rather than being distributed over space.

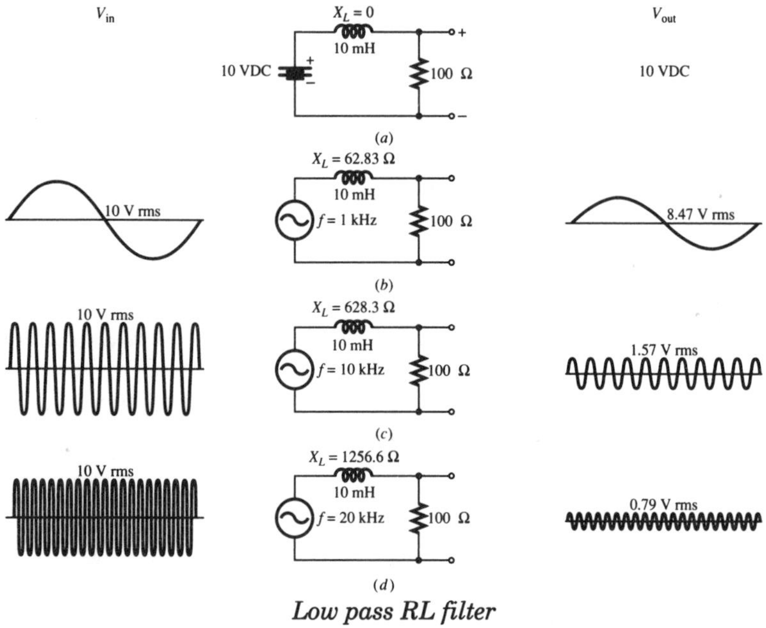

Low pass RL filter

M Letter symbol for mega. See mega.

mA Abbreviation for milliamp. See milliamp.

μA Symbol for microamp. See microamp.

μH Symbol for microhenry. See microhenry.

μm Symbol for micron. See micron.

μs Symbol for microsecond. See microsecond.

μV Symbol for microvolt. See microvolt.

μW Symbol for microwatt. See microwatt.

machine Jargon for computer.

machine code See machine language.

machine cycle Length of time required to complete a process performed by a specific microprocessor.

machine language A programming language directly executable by the microprocessor. In the strictest sense, it consists of 1's and 0's, however, it is common practice to refer to the hexadecimal representation of binary code as machine language. See hexadecimal loader.

macroassembler A computer program that converts a specified sequence of instructions into machine code. A macroassembler can define a large segment of frequently used code. However, unlike a subroutine, new code is generated each time it is used.

macrocode Instructions in a source code that is equivalent to a specified sequence of machine instructions.

macrocommand In programming, a source instruction that is equivalent to a specified sequence of machine instructions.

magic T A type of waveguide with two inputs and an output. Usually used in the front end of microwave receivers. Here the received signal and local oscillator frequencies are combined and sent to the system mixer. Sometimes referred to as a hybrid-T junction.

Magnetic lines of force around a bar magnet.

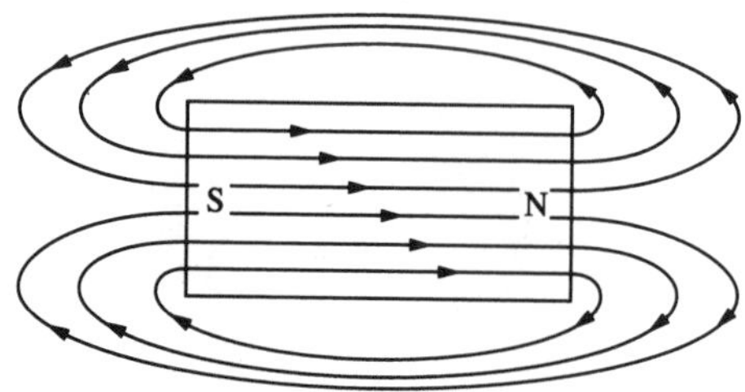

Magnetic attraction and repulsion.

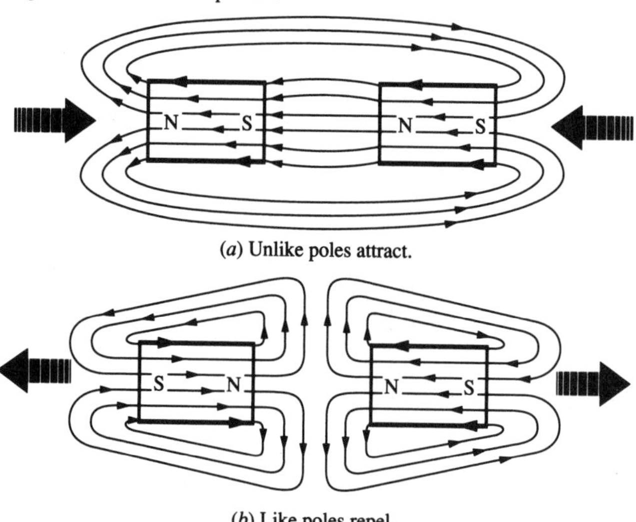

(*a*) Unlike poles attract.

(*b*) Like poles repel.

Magnet

magnet A body that has the property of attracting and repelling at a distance through the phenomenon known as magnetism.

magnetic amplifier A system where the action of a magnetic field is used to effect a power gain. A magnetic amplifier uses one or more saturable reactors to achieve this effect.

magnetic circuit An electrical configuration that uses magnetism and electricity for its operation.

magnetic deflection 1. In a cathode ray tube, a method of controlling the position of the electronic stream by using a magnetic field. 2. A method of controlling the path of an electron beam in a cathode ray tube. Magnetic deflection is produced by coils

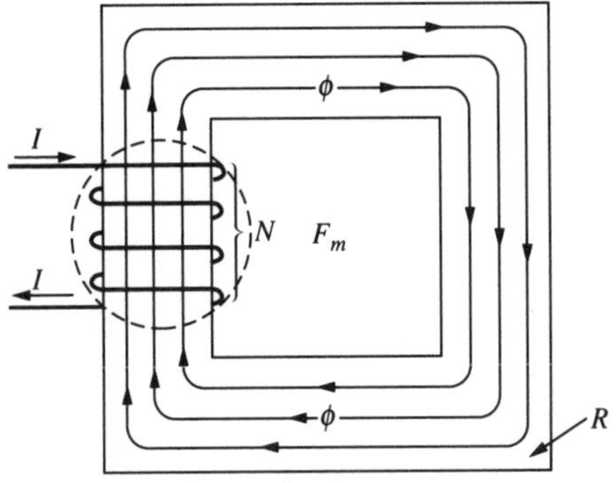

Magnetic circuit

placed around the neck of the tube. Electrical waveforms applied to these coils control where the electron beam will hit the face of the cathode ray tube giving the ability to form images that convey information.

magnetic domain The smallest magnetic area existing within ferromagnetic materials.

magnetic field An area defined by the effects of magnetism. A

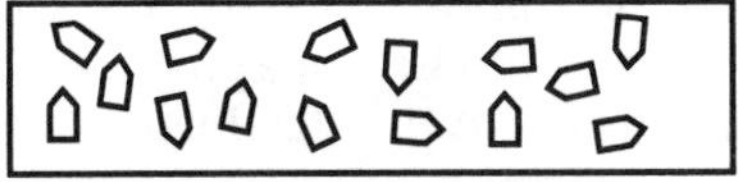

(*a*) The magnetic domains (N ◁ S) are randomly oriented in the unmagnetized material.

N (*b*) The magnetic domains become aligned when the material is magnetized. S

Magnetic domain

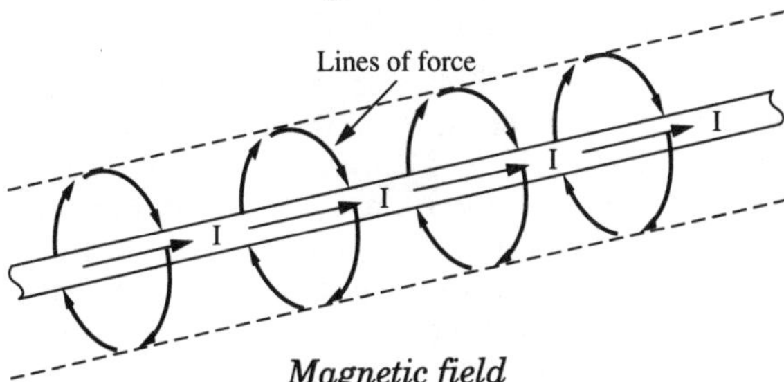

Magnetic field

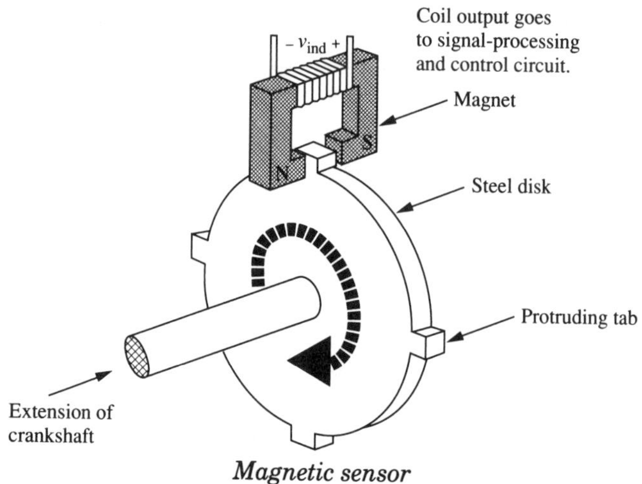

Magnetic sensor

helpful way to envision the phenomenon of magnetism.

magnetic flux lines Imaginary lines that indicate the strength and direction of a given magnetic field.

magnetic freezing The sticking of a relay because of residual magnetism in the coil.

magnetic memory 1. Any method of storing information that uses the property of magnetism. 2. The use of magnetic field patterns to store information. An example of magnetic memory is a floppy disk of a digital computer.

magnetic saturation The point at which any further increase in external energy does not produce any increase in a resulting magnetic field. In an electromagnet, magnetic saturation is achieved when any further increase in coil current does not produce any further increase in a resulting magnetic field.

magnetic sensor An electromagnetic device used for detecting some physical condition.

magnetic shield Any material used for the express purpose of protecting a given space from unwanted effects produced by magnetic fields. Usually constructed from a sheet of iron or other material that will affect a magnetic field.

magnetic switch A mechanical switch that is opened or closed by the action of a magnetic field.

magnetic tape Any tapelike material coated with a substance that responds to the effects of a magnetic field. For example, the magnetic tape used in sound recorders is plastic coated with a magnetic emulsion—usually ferrous oxide.

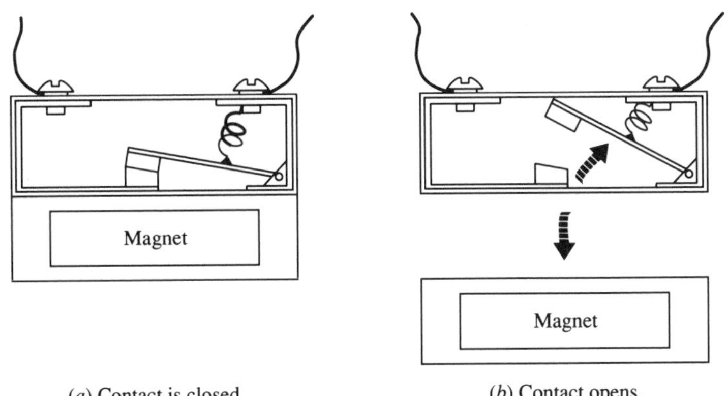

(*a*) Contact is closed.

(*b*) Contact opens.

Magnetic switch

Tracks

1 2 3 4 5 6 7 8 9

Tracks

Marker	ID	Data	Check sum

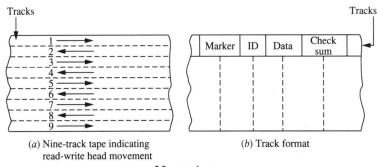

(*a*) Nine-track tape indicating read-write head movement

(*b*) Track format

Magnetic tape

magnetic thin film A layer of magnetic material usually less then 10,000 Å thick. Magnetic thin films may be used for storage of computer data.

magnetic wire Copper wire used for winding relay and transformer coils. Magnetic wire is usually insulated with an enamel coating.

magnetism An observable property where certain materials can exert a mechanical force at a distance on neighboring bodies of similar material and can cause voltages to be induced in conducting bodies moving relative to the magnetic fields.

magnetizing force Letter symbol H. In a material, the magnetomotive force per unit length of the material. Unit is the orested.

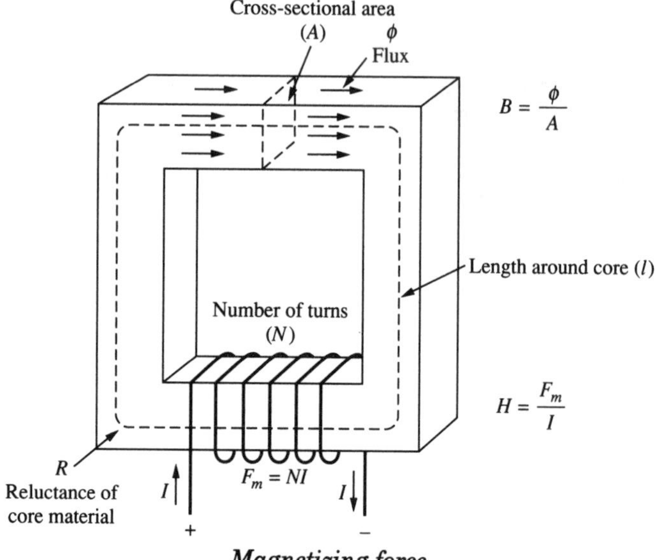

Magnetizing force

magnetometer An instrument for measuring the intensity and direction of a magnetic field.

magnetomotive force Letter symbol F. Measured in amp-turns, it is the pressure required to create a magnetic flux in a ferromagnetic material. Unit is the gilbert.

magnetoresistor A semiconductor device where its resistance is a function of an applied magnetic field.

magnetosphere A belt in the upper atmosphere composed primarily of helium gas.

magnetostriction The propriety of certain ferromagnetic metals that causes them to shrink or expand when influenced by a magnetic field. Conversely, these same materials will influence a magnetic field when subject to mechanical stress.

magnetron An electronic tube used to produce high-power-output frequencies in the ultra high frequency and super high frequency bands.

magnitude The value of a quantity.

magnitude comparator A circuit that compares the magnitude of two quantities in order to determine their relationship.

mainframe computer Referring to a large computer system that operates several terminals. The term mainframe referred to the large racks and panels of

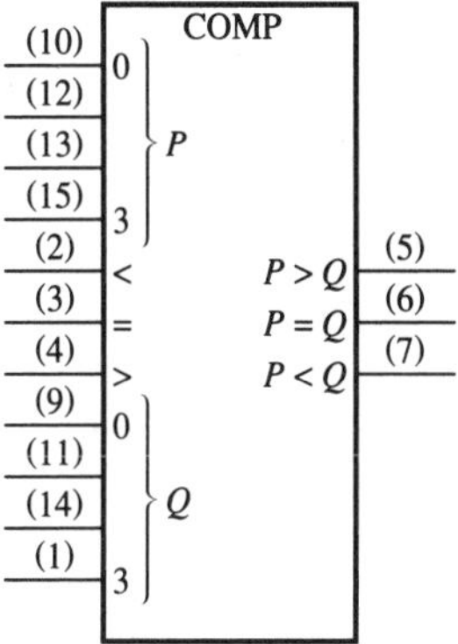

Magnitude comparator

vacuum tubes in the early computer models. These computers were stationary and not made to be moved about. Today, mainframe computers are solid-state machines intended to stay in one location and service many different terminals.

maintenance An activity designed to prevent system failures.

majority carrier The main carries in a semiconductor material. In n-type material, electrons are the majority carriers, while in p-type holes are the majority carriers.

male Shaped in such a manner as to fit into a hollow part.

male connector Pertaining to a mechanical device containing protrusion that will fit into another matching device.

manual Operated by the hand. A manually operated switch is a switch that must by controlled by the hand as opposed to an electrically operated switch that is controlled by some electrical action.

manual reset A term used to indicate required human action to restart a system or reset a device such as a relay.

MAR Abbreviation for memory address register. See memory address register.

marker generator An electronic instrument capable of producing a single stable frequency referred to as a marker frequency. A marker generator is commonly used with a sweep generator to produce one or more stable reference frequencies.

mask-programmable ROM A read-only memory that is programmed by the manufacturer. Refers to the method used to place a specified bit pattern into the device where a "mask" is used during the etching process that represents the desired bit pattern.

mask-ROM See mask-programmable ROM.

mass storage A term indicating the containment of large amounts of computer data. Mass storage is usually achieved on a secondary system such as magnetic tape.

master clock A very accurate timer, in computers, the primary source of timing signals.

master oscillator The main oscillator in an electronic system

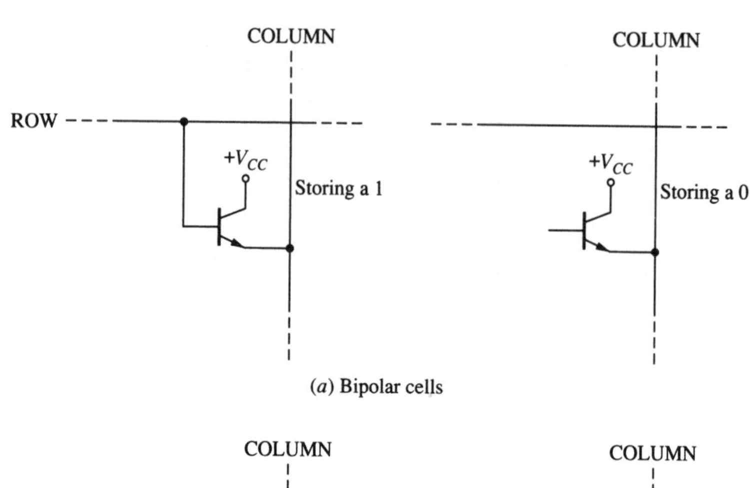

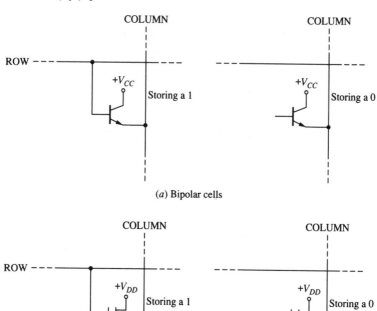

(*a*) Bipolar cells

(*b*) MOS cells

Mask ROM

that may have more than one oscillator or produce multiples of the master oscillator frequency.

master-slave flip-flop A bistable circuit consisting of two flip-flops treated as a single binary storage element. The advantage of such an arrangement is to prevent the output from changing while the input is changing.

master switch An ON/OFF device that is located electrically ahead of other similar devices.

matching transformer A transformer especially designed to match the impedance of the source to the load. Under ideal conditions, when the impedance of the source and load are matched, maximum power transfer will occur.

maximum output The largest average power output available without regard to the amount of distortion.

maximum power transfer In a complete circuit, maximum

Logic diagram for a basic master-slave S-R flip-flop.

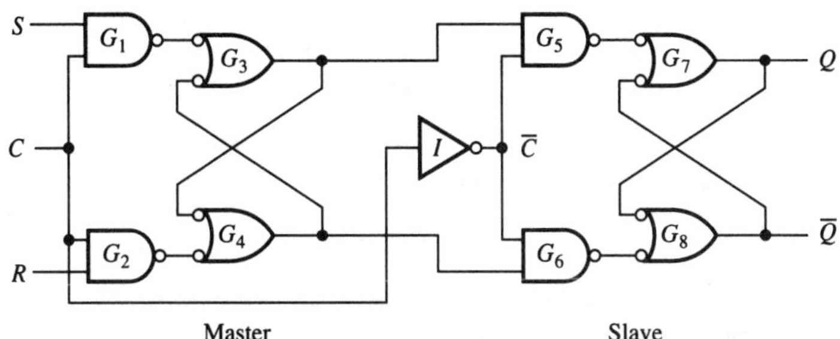

Truth table for the S-R master-slave flip-flop.

Inputs			Outputs		Comments
S	R	C	Q	$\overline{Q}$	
0	0	⎍	Q_0	$\overline{Q_0}$	No change
0	1	⎍	0	1	RESET
1	0	⎍	1	0	SET
1	1	⎍	?	?	Invalid

⎍ = clock pulse

Q_0 = output level prior to clock pulse

Master-slave

power is transferred to the load when the load resistance is equal to the output resistance of the circuit.

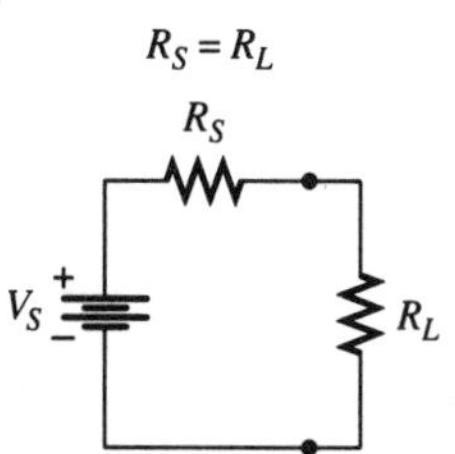

Maximum power transfer

maximum ratings Values specified for electrical devices that when exceeded may destroy the device.

maxwell In the CGS system, the unit of magnetic flux equal to 1 gauss per square centimeter.

MB Letter symbol for megabit. See megabit.

mechanical ohm A mechanical resistance, where a mechanical force of one dyne produces a linear velocity of one centimeter per second.

medium-scale integration Integrated circuits with packaging densities of up to 100 gates or similar circuits.

mega Prefix for one million. Mega = 1,000,000. As an example, a 4M-ohm resistor is 4,000,000 ohms.

megabit One million binary digits, letter symbol MB.

megahertz One million hertz. Means a frequency of 1,000,000 or 1×10^6 cycles each second.

megaohm One million ohms, abbreviated MΩ.

megavolt One million volts. One megavolt equals 1,000,000 volts or 1×10^6 volts.

megawatt One million watts. One megawatt equals 1,000,000 watts or 1×10^6 watts.

membrane switch A mechanical device used for making an electrical connection as long as pressure is applied to it. Usually a membrane switch is used to make up the keys on a hand held calculator. It consists of a plastic membrane raised slightly from the surface. When pressure is applied to the membrane (usually by the finger), it causes a small metal plate to be pressed down over two separate electrical conductors. Such action causes a completed circuit between the conductors. When the external pressure is removed, the membrane will raise back to its normal position causing electrical contact to be broken.

memory Any medium capable of storing information that may be used to cause some action through an electrical process. Common forms of memory are magnetic and optical disk storage, electrical RAM and ROM, as well as optical bar codes.

memory address register A storage place in the microprocessor that contains the location of data being accessed. The memory address register holds the address of the memory being accessed by the microprocessor.

memory dump In programming, the process of displaying the contents of a given sequence of memory locations.

memory expansion Increasing the size of memory by increasing the number of address locations, by increasing the data word size, or by doing both.

memory map A listing describing the contents of a specified range of memory locations. (See pg. 196 for illustration.)

memory-mapped I/O When computer input-output is addressed in the same manner as computer memory.

memory size A measurement of how bit patterns are stored in computer memory. The notation used as $A \times D$, where A gives the number of addresses (rows) and D gives the number of bits at each address (word size).

menu A list of selectable items. In computers, a menu offers the program user a selection of options from which to choose, where each option will evoke a particular program or other such action.

Meissner effect The levitation of a magnet above a superconductor.

mercury cell A small dry voltage source that provides 1.35 volts. A mercury cell is used in

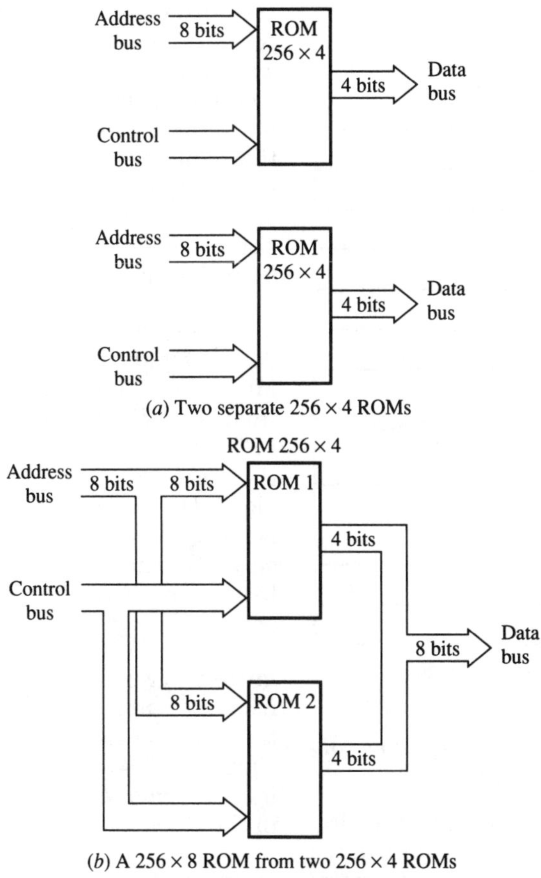

(*a*) Two separate 256 × 4 ROMs

(*b*) A 256 × 8 ROM from two 256 × 4 ROMs

Memory expansion

watches, calculators, and hearing aids.

mercury switch An electrical switch made up of a cylindrical container (usually glass or other insulating material) with a small pool of mercury. Tilting of the cylindrical container causes the mercury to make or break electrical contact.

mesh A circuit arrangement consisting of a set of branches forming a closed path in a network. A mesh circuit is constructed so that if any branch is omitted, the remaining branches will not form a closed path.

mesh analysis A method of solving for circuit currents of a circuit loop (mesh). The method allows a reduced set of equations compared to other methods of solution.

Address (hexadecimal)	Contents
0000	
0001	
0002	
0003	
0004	
0005	
0006	
0007	
FFFB	
FFFC	
FFFD	
FFFE	
FFFF	

Memory map

message A set of symbols containing information.

metal film resistor An electronic component constructed from a very thin layer of a metal alloy that is vacuum deposited on a substrate for the purpose of providing a specified amount of resistance.

metal-oxide semiconductor A solid-state circuit with a sandwichlike construction where a metal oxide acts as the dielectric insulator between the metal and the semiconductor. This process is used to construct a MOSFET or metal-oxide field-effect transistor whose gate is insulated from the channel.

metal-oxide semiconductor field-effect transistor The MOSFET differs in construction from the FET in that the MOSFET's gate is insulated from the channel by a layer of silicon dioxide (SiO_2). There are two basic types of MOSFETs, depletion only (D) and the depletion-enhancement (DE). See depletion area.

meter The unit of measurement for length in the MKS and the SI systems. One meter is equal to 1.094 yards in the English system and 100 centimeters in the CGS system.

metric notation The representation of certain common quantities that are used more often than others. For example, the metric prefix kilo (k) represents

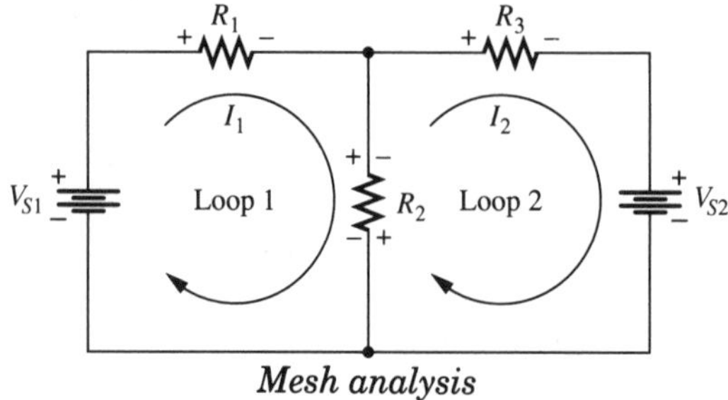

Mesh analysis

Power of Ten	Value	Metric Prefix	Metric Symbol
10^9	one billion	giga	G
10^6	one million	mega	M
10^3	one thousand	kilo	k
10^{-3}	one-thousandth	milli	m
10^{-6}	one-millionth	micro	μ
10^{-9}	one-billionth	nano	n
10^{-12}	one-trillionth	pico	p

Metric

1,000, and the metric prefix milli (m) represents 0.001.

MFM Abbreviation for modified frequency modulation. See modified frequency modulation.

MHz Abbreviation for megahertz. See megahertz.

mica A mineral with excellent resistance properties. Mica can be split into very thin sheets and has great resistance to both the conduction of heat as well as electricity.

mica capacitor A capacitor constructed by using mica as its dielectric.

microamp One microamp is 0.000001 or 1×10^{-6} amps.

microcode The instructions built into a microprocessor that determine a specific selectable process.

microcomputer A computer that uses a microprocessor as its central processing unit.

microelectronics That field of electronics that deals with the design, development, fabrication, and servicing of very small circuits. Microelectronics usually refers to integrated circuits.

microhenry One microhenry is 0.000001 or 1×10^{-6} henry.

microinstruction A single instruction in the microcode of a microprocessor.

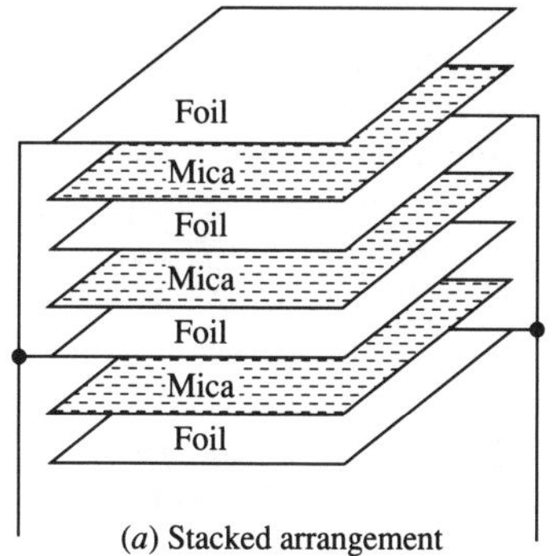

(*a*) Stacked arrangement

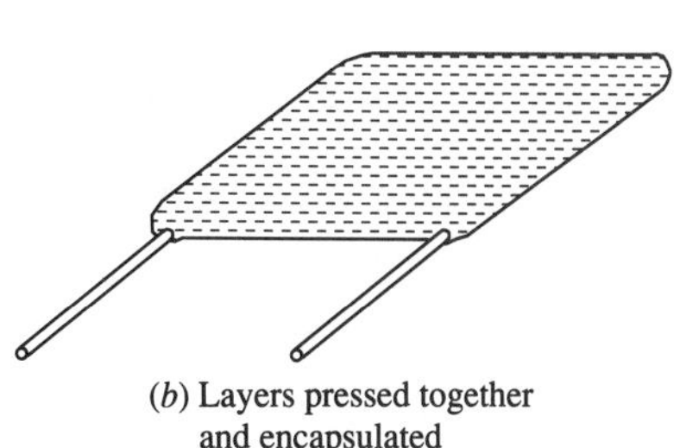
(*b*) Layers pressed together and encapsulated

Mica capacitator

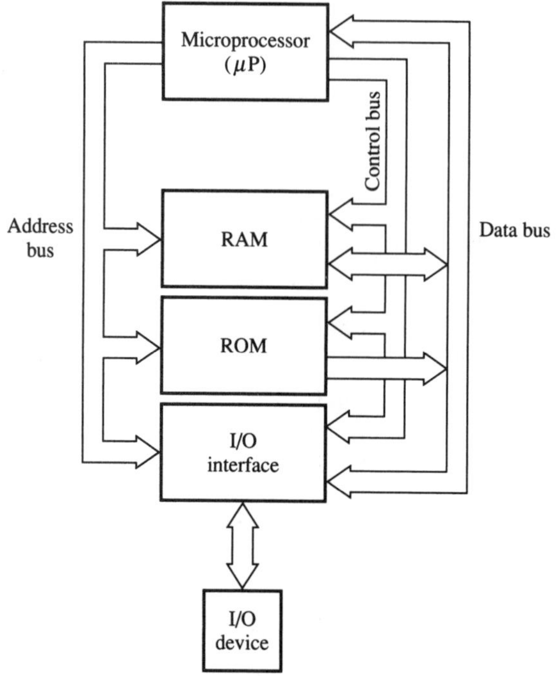

Microcomputer

micron A measurement of one-millionth of a meter. One micron equals 1×10^{-6} meters.

microphone A device that converts sound waves into equivalent electrical energies.

microphone boom A movable crane to which a microphone is attached.

microphonics The creation of electrical noise caused by the mechanical motion of internal parts within an electrical device.

microprocessor A single integrated circuit chip containing a fixed set of processes such as arithmetic, logic, and bit manipulation. Predictable processes are selected by specific logic patterns applied to the device. A program contains the logic patterns to produce a desired sequence of processes from the microprocessor.

microprocessor cycle The process of a microprocessor getting information from an external source (such as computer memory) and acting on that information. Referred to as fetch and execute.

microprogram The permanent instruction set inside the microprocessor used to perform a specific process.

microsecond One-millionth of a second. One microsecond equals 1×10^{-6} or 0.000001 seconds.

microvolt One-millionth of a volt. One microvolt equals 1×10^{-6} or 0.000001 volt.

microwatt One milionth of a watt. One microwatt equals 0.000001 or 1×10^{-6} watts.

microwave Literally means "small wave" from the fact that microwaves describe very high frequencies that have a small (micro) or short wavelength.

microwave oven A cooking device that uses microwave energies for the purpose of producing heat in food products. The main advantage of microwave ovens over that of conventional electrical ovens is that microwave energy is more efficient and will usually reduce the amount of cooking time. See microwave.

microwave spectrum Generally, the range of frequencies from 300 MHz to 300 GHz.

midrange 1. The frequency range between bass and treble. The audio frequency range of about 400 to 3000 Hz. 2. Specifying the middle of a range of frequencies. The frequency range of an amplifier lying between two points.

mike Slang for microphone.

military temperature range Electrical equipment and devices designed for military use to operate in the temperature range of −55 to 125°C.

milli The prefix designating one-thousandth. Milli = 1×10^{-3}.

milliamp One-thousandth of an ampere. One milliamp equals 0.001 or 1×10^{-3} amps.

milliammeter An electronic meter movement calibrated in milliamps.

millisecond One-thousandth of a second. One millisecond equals 0.001 second or 1×10^{-3} second.

millivolt One-thousandth of a volt. One millivolt equals 0.001 volt or 1×10^{-3} volt.

Millman conversion Converting a circuit to a simpler form using Millman's theorem. See Millman's theorem.

Millman's theorem A method of solving complex circuits that

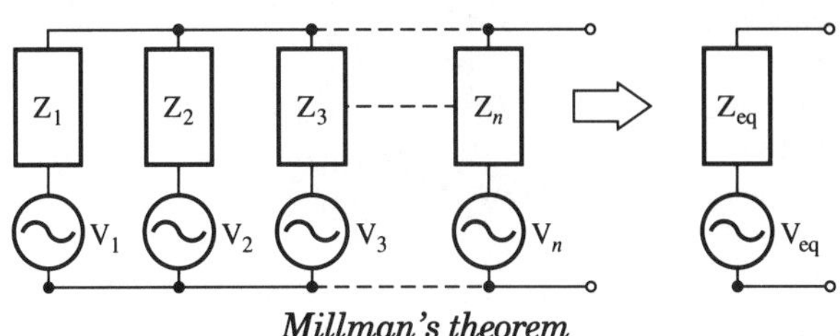

Millman's theorem

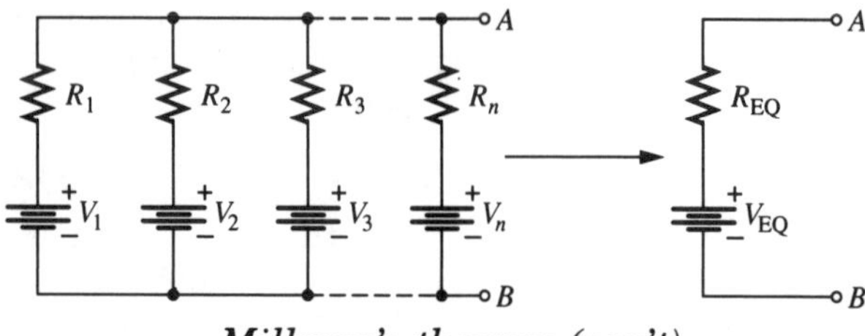

Millman's theorem (con't)

employs source conversions, thus permitting the solution of unknown variables in the circuit.

minority carrier The less predominant carriers in semiconductor material. In p-type material, the minority carriers are electrons; in n-type material, the minority carriers are holes.

minor lobe Any lobe in an antenna radiation pattern other than a major lobe. Minor lobes represent part of the total radiation pattern of an antenna.

mismatch The condition where the impedance of the load does not match the impedance of the source resulting in less than 100% transfer of power between the source and the load.

mixer An electrical device having two or more signal inputs with a resulting output. For example, the mixer in a superheterodyne receiver has the received station as one frequency and the local oscillator as the other. The resulting useful output is the difference between these two frequencies that is the frequency of the IF amplifiers.

MKS system The system of units that use the meter, kilogram, and the second as the fundamental units of measurement.

mmf Abbreviation for micromicrofarad (1×10^{-12} f).

mnemonic code A programming code where the commands are usually represented as three letters, such as CLA for "clear the accumulator."

mnemonics A device to help you remember something. Used in computer programming to assist in helping you remember what a process does. For example, the mnemonic CLA means "clear the accumulator." Primarily used in assembly language programming.

mobile receiver A communications receiver, such as a radio, in a movable vehicle such as a car.

modal dispersion In fiber optics, occurs when a light pulse transmitted through a fiber optic becomes washed out or flattened due to the properties of a multimode optical fiber. This undesirable effect can be reduced by using optical cladding with a refractive index smaller than that of the fiber.

modem Word derived from *mo*dulator/*dem*odulator. A modulator converts digital information into transmission signals. A demodulator converts the signals back to the original digital form.

modified frequency modulation A method of storing digital information on a magnetic disk. This method produces twice the density of disk information storage as with standard frequency modulation techniques.

mod number Referring to a counter, the number of states a counter will have under given conditions. For example, a 4-bit binary counter is mod 16 because it has 16 distinct states.

mod-12 counter A digital counter with 12 unique states.

modulate To modify. To modulate a signal means to change it in a meaningful way. As an example, amplitude modulation means to change the amplitude of a signal in such a manner that the changes represent some form of information. With AM transmission, the amplitude of the radio wave is modulated by the audio signal. These changes in amplitude are detected at the receiver and reconstructed to produce the orignal audio signal (which is usually voice or music).

modulated wave A carrier wave where some measurable characteristic, such as its amplitude, frequency, or phase is varied in accordance with another signal that represents information such as sound.

modulation envelope The continuous curve, drawn along the peaks of a carrier wave showing how the changes in amplitude of the carrier represent the waveform of the modulating wave.

modulation factor A measurement of the amount of modulation in an AM wave. The modulation factor is the ratio of half the difference between the maximum and minimum amplitudes to the average amplitude. Mathematically expressed as $m = B/A$, where m is the modulation factor (no units), B is the peak value of the modulating signal, and A the peak value of the unmodulated carrier. When multiplied by 100% the result is called the percentage of modulation.

modulation frequency The signal that causes modulation of a carrier wave.

modulation index A measurement of the degree of modulation in frequency modulation. The modulation index is the ratio of the frequency deviation to the frequency of the modulating wave. Expressed as $m = f_D/f_M$, where m is the modulation index (no units), f_D is the deviation of the FM wave (in hertz), and f_M is the frequency of the modulating signal (in hertz).

modulator Any device that creates the effects of modulation. A modulator causes an electrical

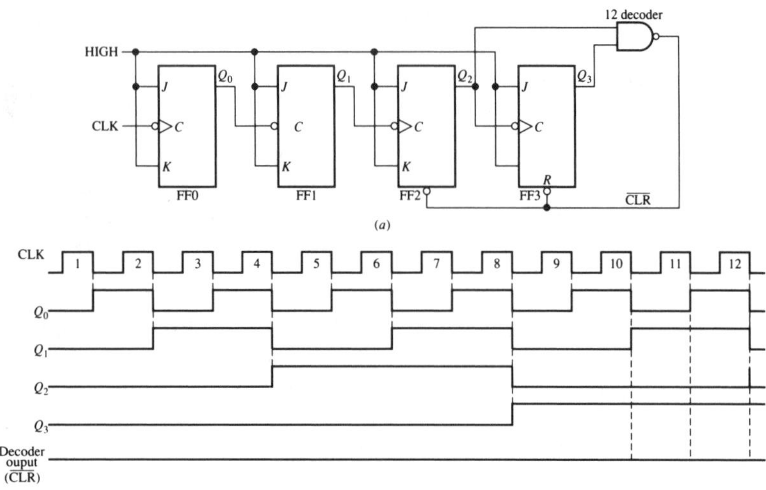

Modulus-12 counter

change caused by another signal called the modulating signal.

modulus When applied to counters refers to the maximum number of distinct states a counter can assume.

modulus-12 counter A binary counter with twelve distinct states. See binary counter.

molded inductor A type of fixed-value inductor. See fixed inductor.

molecule The smallest particle of a substance that still retains the property of that substance. For example, a molecule of water consists of two atoms of hydrogen and one atom of oxygen.

monaural Sound produced from one source. A monaural sound system has only one channel as opposed to a stereo system that has two channels (the left channel and the right channel).

monochromatic 1. Pertaining to a single color or radiation of a single wavelength. 2. Light consisting of a single color. Light consisting of a single frequency.

monochrome monitor A computer screen capable of producing shades of only one color.

monophonic See monaural.

monostable Pertaining to a two-state device (ON or OFF) that has one stable state in which it will normally stay and will stay only momentarily in the other state.

monostable multivibrator A two-state device that will stay in one condition for a predetermined amount of time and then auto-

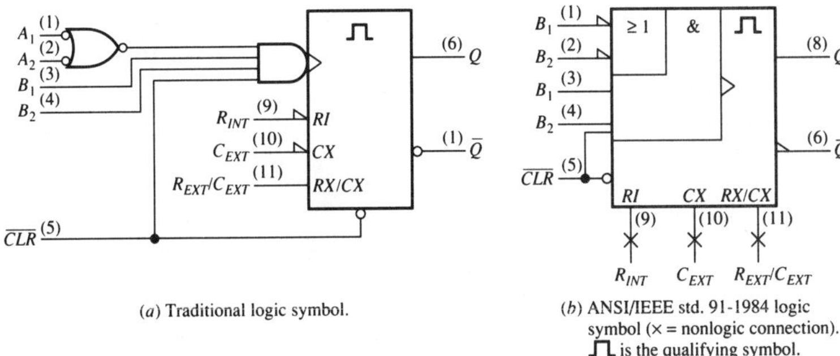

(*a*) Traditional logic symbol.

(*b*) ANSI/IEEE std. 91-1984 logic symbol (× = nonlogic connection). ⎍ is the qualifying symbol.

Monostable multivibrator

matically go back to its prior condition. See multivibrator.

monotone A single musical tone not varying in pitch.

Morse code A system of dots and dashes developed by Samuel Morse.

MOS Abbreviation for metal-oxide semiconductor. See metal-oxide semiconductor.

MOS memories Electrical computer memories manufactured from field-effect transistors.

MOSFET Abbreviation for metal-oxide semiconductor field-effect transistor.

most significant bit In a binary number, the significant bit contributing the largest quantity of the value of the number. For example, in the binary number 101_2 if the leftmost bit is considered the most significant bit, then it contributes a value of four to the number. The binary number $101_2 = 5$ to the base 10. See least significant bit.

most significant digit The single digit at the extreme left of a number. For example, in the number 439, the 4 is the most significant digit.

motherboard The largest single printed circuit board found in a personal computer. The motherboard has connectors that will accept other printed circuit boards such as disk drive and printer interface boards. The main circuit board of a personal computer.

motor A mechanical device capable of performing work by converting one form of energy into mechanical energy. An electrical motor converts electrical energy into mechanical energy.

motorboating Electrical interference on an audio system that sounds like the "putting" of a small boat motor. It is usually caused by problems with the power supply that converts ac to dc.

mouse A small device held in the hand and moved along a flat surface for the purpose of putting information into a computer. As the mouse is moved along the flat surface, patterns (such as arrows) may be moved on the computer screen. Switches in the form of buttons are provided on the device for the purpose of letting the computer know when a desired location on the screen has been reached.

moving coil meter A meter movement where a coil pivots between the poles of a permanent magnet.

moving coil pickup A transducer used to reproduce the sound contained on a phonographic recording. The moving coil pickup uses a coil to produce an electric output as it moves within the field of a permanent magnet.

moving coil speaker A transducer that converts electrical energy into sound. The moving coil speaker uses a coil of wire attached to the base of a paper cone. This coil of wire is in the magnetic field of a permanent magnet. The action of small electrical currents passing through the coil of wire causes mechanical motion of the coil and thus the paper cone, producing sound waves.

MPU Abbreviation for microprocessor unit.

ms Abbreviation for millisecond. See millisecond.

MSB Abbreviation for most significant bit. See most significant bit.

MS-DOS Acronym for Microsoft disk operating system. The disk operating system developed by Microsoft, Inc. Used by IBM PCs and compatibles.

MSI Abbreviation for medium-scale integration. See medium-scale integration.

MTS Abbreviation for multichannel TV sound. A system that includes the option of transmitting a TV signal with a stereo sound signal. First commercially televised program to use MTS was the 1984 Los Angeles Olympics.

mu English spelling for the Greek letter μ.

multibus system An interface bus system developed by Intel. A

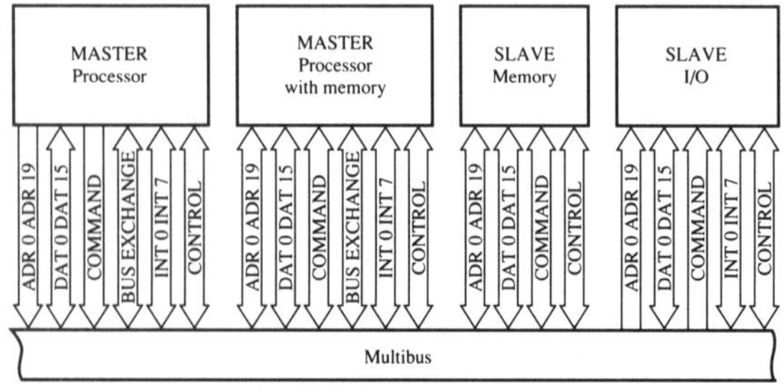

Multibus system

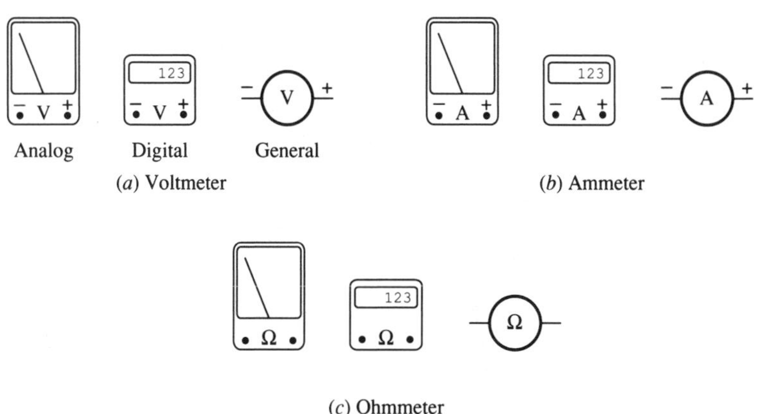

(*a*) Voltmeter (*b*) Ammeter

(*c*) Ohmmeter

Multimeter

major feature of this bus system is that several microprocessors (called masters) can be connected to the bus at the same time to cause a multiprocessing operation with other devices (slaves).

multimedia In computers, a computer-based presentation using two or more media, such as text, graphics, writing as well as video and audio. See hypermedia.

multimeter An electronic instrument for measuring resistance and ac or dc current and voltage.

multiple hop The action of a sky wave as it is reflected back and forth between the ionosphere and the earth. This can result in increasing the transmission distance of radio waves. See sky wave.

multiple primary transformer A transformer with more than one primary winding.

multiple range ammeter An ammeter with more than one current range.

multiple range voltmeter A voltmeter with more than one voltage range.

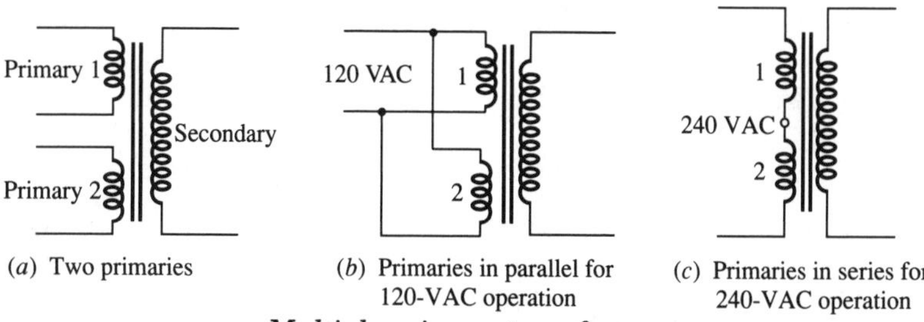

(*a*) Two primaries (*b*) Primaries in parallel for 120-VAC operation (*c*) Primaries in series for 240-VAC operation

Multiple-primary transformer

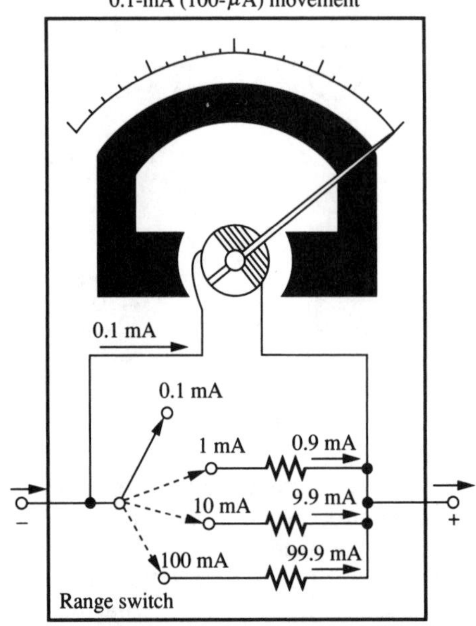

Multiple-range ammeter

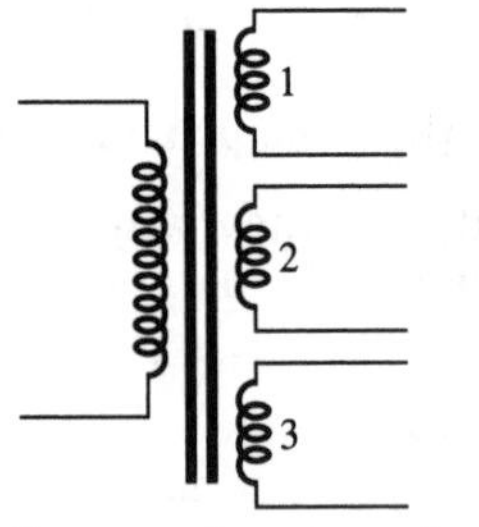

Multiple-secondary transformer

multiple secondary transformer A transformer with more than one secondary winding.

multiplex To carry out two or more functions at the same time in a related manner. A technique for transmitting two or more signals at the same time (such as in FM stereo).

multiplex stereo An electronic system for transmitting two channels on the same carrier frequency. Multiplex stereo is used in FM stereo transmission.

multiplexed I/O A digital system that uses the same bus for input and output to external devices. This is usually accomplished through the use of bidirectional tristate buffers. See bidirectional tristate buffer. (See pg. 208 for illustration.)

multiplexer An analog or digital device that can take several simultaneous inputs and convert them to a single output. See data

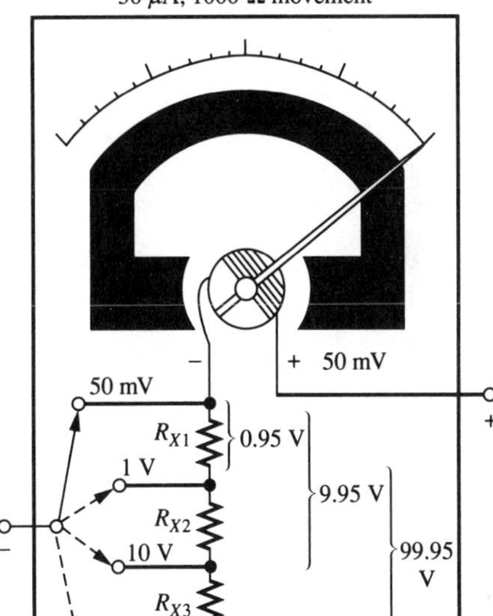

Multiple-range voltmeter

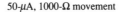

selector. (See pg. 208 for illustration.)

multiplexing Using two or more different signals through the same electrical element.

multiplier An electronic circuit that produces the product of two input signals. A frequency multiplier is a circuit that produces a harmonic frequency of its input signal.

multiprocessing In computers, the operation of more that one processing unit at the same time within a single system.

multiprocessor In computers, a computer capable of executing two or more programs at the same time through the use of two or more processing units under integrated control of programs or devices.

multistage Having more than one stage. A multistage amplifier has more than one stage of amplification. See stage.

multitasking The ability to perform two or more operations at the same time or what appears to be at the same time because of high processing speeds.

multivibrator A two-state output circuit capable of acting as an oscillator by generating its own signal or maintaining one stable state or maintaining one of two stable states.

music synthesizer An electrical device capable of producing

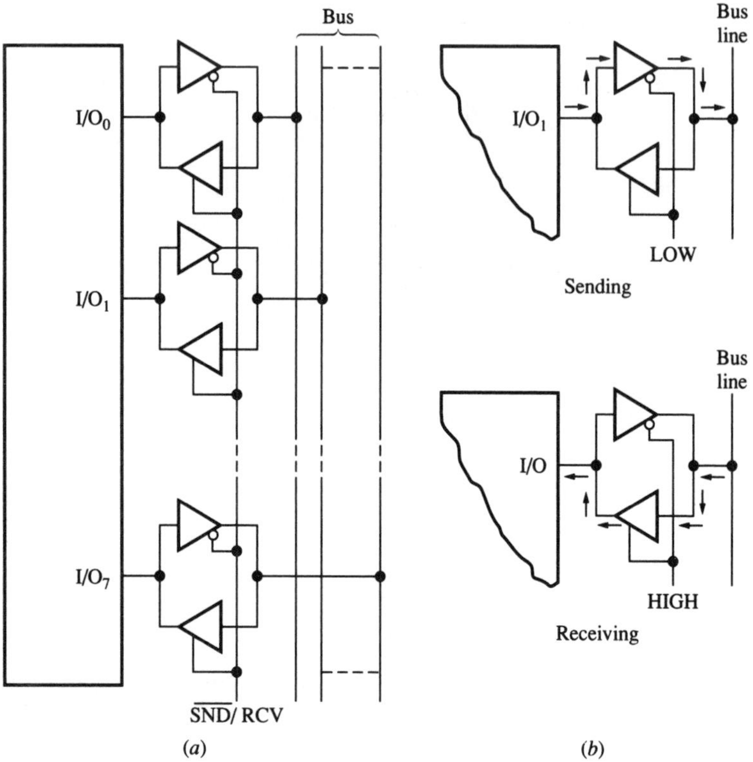

Bus

Bus line

I/O_0

I/O_1

I/O_7

$\overline{SND}$/ RCV

I/O_1

LOW

Sending

Bus line

I/O

HIGH

Receiving

(*a*) (*b*)

Multiplexed I/O

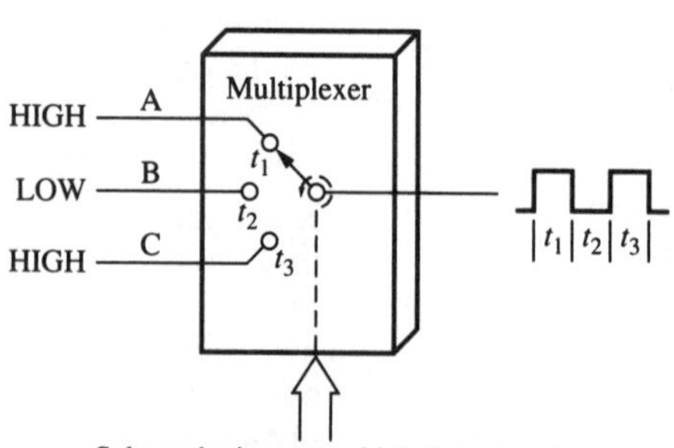

Multiplexer

the sound of music or replicating the sounds of musical instruments. Usually the combination of digital and analog circuits. See waveform synthesis.

mutual inductance The inductance effect caused by the magnetic coupling of two or more inductors.

mV Letter symbol for millivolt. See millivolt.

MV Letter symbol for megavolt. See megavolt.

MW Letter symbol for megawatt. See megawatt.

Mylar The E. I. Du Pont trade name for a highly durable, transparent plastic film of great strength. Used primarily as a base for magnetic tape.

Mylar capacitor A capacitor in which the dielectric is of Mylar material.

n Letter symbol for the prefix nano (1×10^{-9}).

nA Letter symbol for nanoamp. One nanoamp equals 0.000000001 or 1×10^{-9} amps.

NAND A Boolean function that is the inverse of the AND. The result of the function is FALSE only when all the variables within the function are TRUE. Otherwise, the function is always TRUE. Comes from NOT AND.

NAND equivalent Using NAND gates in proper logical combination to produce the other logic functions of NOT, AND, OR, and NOR.

(*a*) A NAND gate used as an inverter

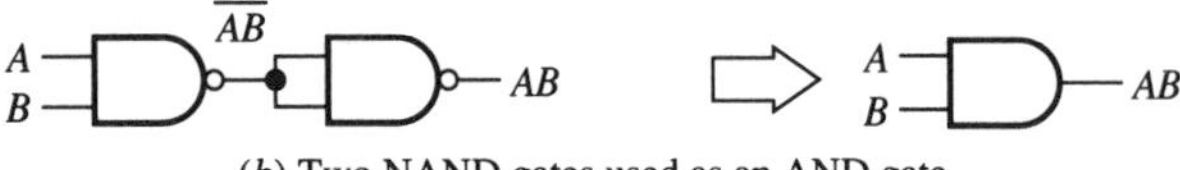

(*b*) Two NAND gates used as an AND gate

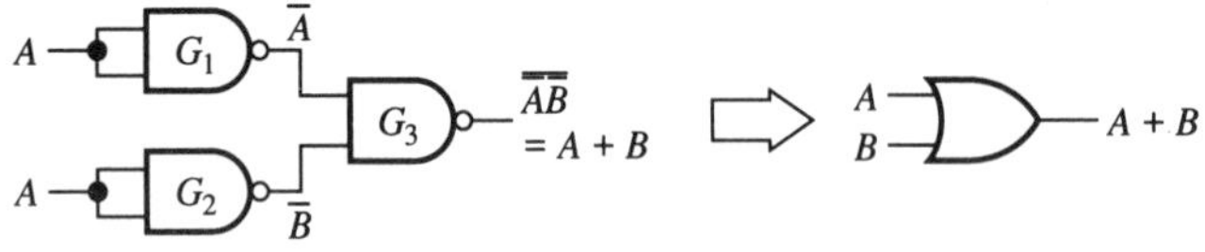

(*c*) Three NAND gates used as an OR gate

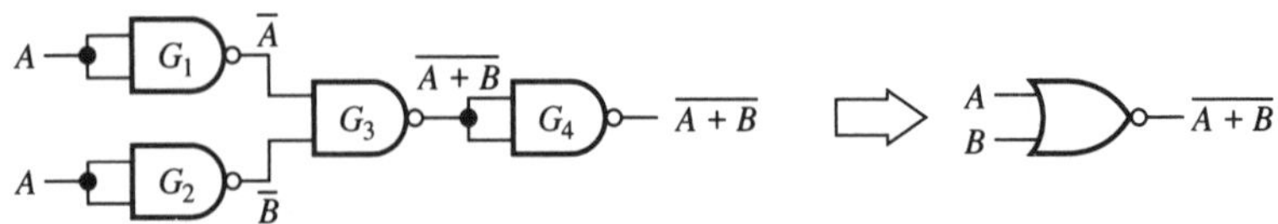

(*d*) Four NAND gates used as an NOR gate

NAND equivalent

A	B	$\overline{AB} = X$
0	0	$\overline{0 \cdot 0} = \overline{0} = 1$
0	1	$\overline{0 \cdot 1} = \overline{0} = 1$
1	0	$\overline{1 \cdot 0} = \overline{0} = 1$
1	1	$\overline{1 \cdot 1} = \overline{1} = 0$

NAND function

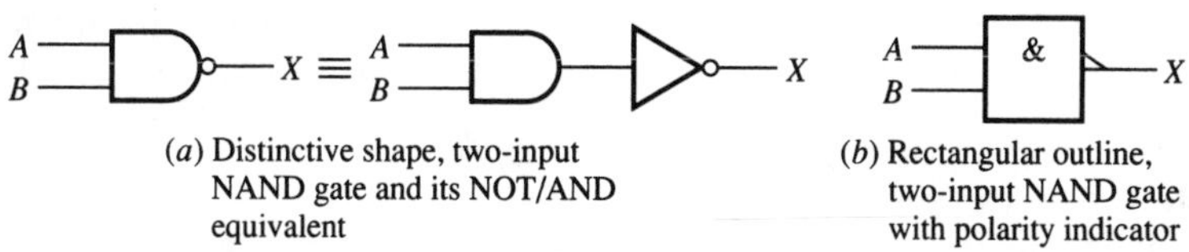

(*a*) Distinctive shape, two-input NAND gate and its NOT/AND equivalent

(*b*) Rectangular outline, two-input NAND gate with polarity indicator

NAND gate

NAND function A two-valued Boolean expression where the output is FALSE (or 0) only when all the inputs are TRUE (or 1). The NAND function is expressed as $X = \overline{A \blacksquare B}$. X is FALSE only if both A and B are TRUE; otherwise, X is TRUE.

NAND gate An electrical digital circuit consisting of one output and two or more inputs. Follows the logic of the Boolean NAND, where the output will be LOW only when all the inputs are HIGH.

nano Metric prefix meaning one-billionth or 1×10^{-9}. Letter symbol is n.

nanofarad One-billionth of a farad. May be expressed as 1×10^{-9} farads.

nanometer A measurement of 1×10^{-9} meters.

nanosecond One-billionth of a second. May be expressed as 1×10^{-9} seconds.

Naperian logarithm Also known as natural logarithms. A logarithm to the base of 2.7128.

narrowband FM Frequency modulation of the carrier where the modulation index is kept less than $\pi/2$. Narrowband FM produces a modulating signal where the resultant bandwidth depends mainly on the frequency of the modulating signal. In normal FM, the resultant bandwidth is also affected by the amplitude of the modulating frequency.

NASA Abbreviation for the National Aeronautics and Space Administration. NASA is an agency of the U.S. government for all scientific and some military space missions.

natural frequency The frequency a circuit will tend to oscillate at when supplied by a temporary force of energy. It is the lowest resonant frequency of a given circuit or device.

natural logarithm See Naperian logarithm.

negative-channel metal-oxide semiconductor Produces high packing densities. Since electrons are the current carriers, this results in greater speed and lower power consumption. Channel is N material.

negative charge One of the fundamental forms of electrical charge, the other fundamental form being a positive charge. In static electricity, two like charges repel, while two opposite charges will attract.

negative feedback 1. In an amplifier or similar circuit, the process of taking some of the output signal and feeding it back to the input out of phase. Causes the overall gain of the amplifier to be reduced, but results in greater system stability and better signal reproduction. 2. The process of taking part of an output signal and placing it back in the input 180 degrees out of phase. Negative feedback is used in amplifiers to help improve the amplifier's ability to produce a faithful reproduction of the input signal.

negative ground Having the circuit or system electrical reference point at the negative potential of the voltage source.

negative number Numbers that can be represented by points to the left of the origin (zero) on the horizontal axis or downward on the vertical axis of a graph.

negative peak clipper A circuit that clips the negative peaks of an input signal to a predetermined level.

negative resistance The condition where an increase in voltage causes a corresponding decrease in current, while a decrease in voltage causes a corresponding increase in current. These characteristics are used by the tunnel diode. See tunnel diode.

negative temperature coefficient A value that indicates a decrease in an electrical quantity with an increase in temperature. See positive temperature coefficient.

negative terminal The mechanical connection of a voltage source that has an excess of electrons. As opposed to the positive terminal, which has a deficiency of electrons.

nesting In computer programming a technique of placing smaller routines inside larger ones. Nesting may include nested loops where program loops are programmed inside other program loops.

net loss The totals of the gains and losses of a signal between the input and output terminals of a system, circuit, or device.

network A combination of electrical elements, systems, or other such interconnections. A computer network allows communications between otherwise separate computer terminals.

neutral As applied to electronics, a net charge of zero. A neutral surface has no excess negative or positive charges.

neutralize To design a circuit in such a manner that it will not self-oscillate.

neutralizing capacitor A capacitor used in a circuit to help prevent self-oscillations of the circuit. A neutralizing capacitor usually provides some form of negative feedback from the output of a circuit to its input.

neutron A particle in the nucleus of the atom that does not have any electrical charge. A neutron is neutral; it supplies mass to the atom but no charge.

newton A unit of measurement for force in both the SI and MKS systems. One newton is equal to 10^5 dynes in the CGS system of measurement.

nibble A binary group of 4 bits. See bit and byte.

nickel-cadmium cell A popular voltage cell that is easily recharged and can be used over and over again. It has a flat voltage discharge curve with an output of 1.25 volts. The cell is constructed with a nickel and oxide positive electrode and a cadmium negative electrode that are immersed in a solution of potassium hydroxide.

nm Letter abbreviation for nanometer. See nanometer.

NMOS Abbreviation for negative-channel metal-oxide semiconductor. See negative-channel metal-oxide semiconductor.

NO Abbreviation for normally open.

nodal analysis A method of solving for the voltages at the junctions (nodes) of a complex circuit. Nodal analysis uses determinants as a part of its solution.

node A connecting point in a circuit of two or more circuit elements; a junction where two or more current paths come together.

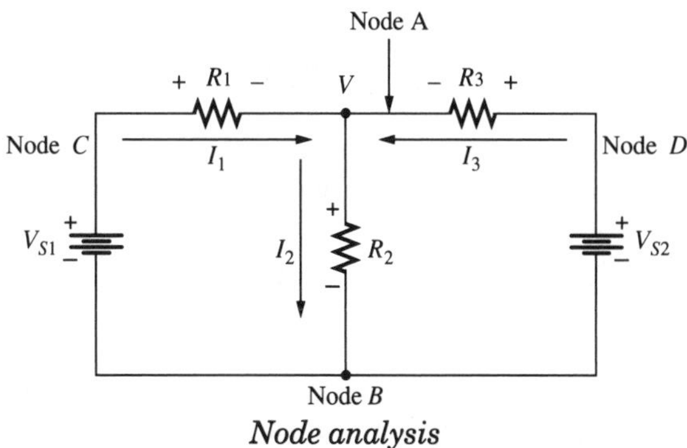

Node analysis

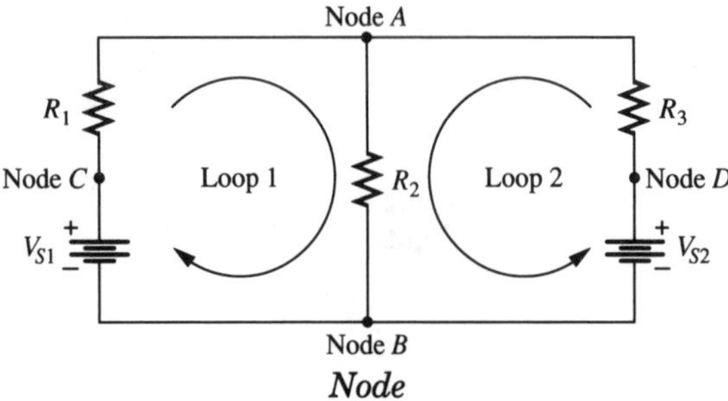

Node

noise Any undesirable electrical signal usually not intentionally produced. There are two sources of electrical noise. One is natural; the other is manufactured. Natural electrical noise is caused by weather conditions, energy from the sun, and radiation from space. Manufactured noise is caused by industry machinery, vehicles, and other communications equipment.

noise clipper A circuit that is used to help remove amplitude modulation noise from a given signal.

noise diode A special diode used to produce noise. The diode is operated at saturation and produces noise from the random emission of electrons.

noise figure A measure of the degradation of the signal caused by the receiving system. Expressed in decibels as $NF = (S_{in}/N_{in})/(S_{out}/N_{out})$, where NF is the noise figure, S_{in} is signal in, S_{out} is signal out, N_{in} is noise in and N_{out} is noise out. An ideal receiver has a noise figure of 0 dB.

noise filter An electrical circuit designed to reduce electrical noise. Noise filters are used in electrical power lines where they consist of a combination of capacitors and inductors.

noise free A space that does not contain electronic noise. A noise-free environment created for the purpose of testing ultrasensitive electrical equipment.

noise generator An electrical device used to generate noise for the purpose of testing circuits under noisy conditions.

noise margin The voltage differences between the input profile and the output profile of a digital integrated circuit. Expressed mathematically as $V_H = V_{OH,min} - V_{IH,min}$ and $V_L = V_{OL,max} - V_{IL,max}$, where V_H is the logic HIGH noise margin, V_L is logic LOW noise margin, $V_{OH,min}$ is the minimum HIGH input voltage, $V_{OL,max}$ is the maximum LOW output voltage, and $V_{IL,max}$ is the maximum LOW input voltage.

no load The condition when no power is demanded from the source. A no-load condition means that there is, in effect, an open circuit.

nominal value The published or specified value as opposed to the actual measured value of a device or system.

nonconductor An insulator. See insulator.

noninductive Having no properties of inductance.

noninverting amplifier An amplifier whose output signal is in phase with its input signal.

noninverting op amp See noninverting amplifier.

nonlinearity For a circuit means that a change on the input does not produce a corresponding change on the output. For a pulse, nonlinearity is any variation from a straight line drawn from the 10% to the 90% amplitude points.

nonperiodic Not repeating. A waveform whose pattern does not repeat itself. See periodic.

nonretriggerable A digital device (such as a one-shot) that will not respond to any additional trigger pulses while in its unstable state.

nonretriggerable one-shot A bistable digital circuit that will not respond to any additional trigger pulses from the time it is triggered into its unstable state until it returns to its stable state.

nonsinusoidal Any waveform that does not have the properties of a sine wave.

nonvolatile memory A storage medium for computer information that can retain information without the application of power external to the system.

NOR A two-state logic function whose results depend upon the conditions of two or more variables. The result is the inverse of the OR logic function; that is, its result will be FALSE when any

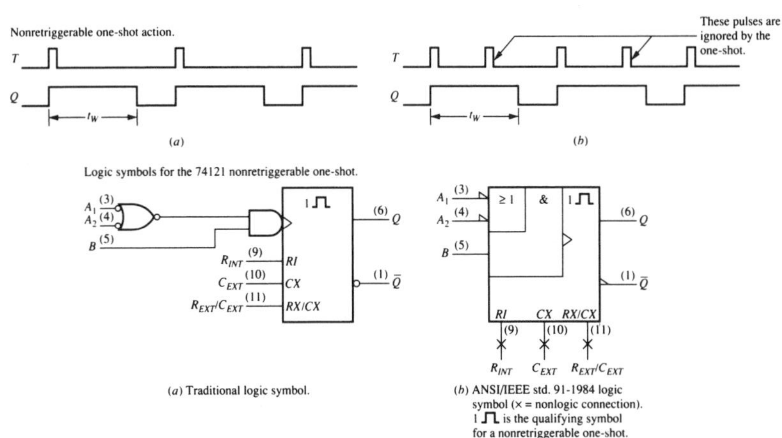
Nonretriggerable one-shot action.
T
Q
t_W
(a)
These pulses are ignored by the one-shot.
(b)
Logic symbols for the 74121 nonretriggerable one-shot.
A_1 (3)
A_2 (4)
B (5)
R_{INT} (9) RI
C_{EXT} (10) CX
R_{EXT}/C_{EXT} (11) RX/CX
(6) Q
(1) $\overline{Q}$
≥1
&
RI CX RX/CX
(a) Traditional logic symbol.
(b) ANSI/IEEE std. 91-1984 logic symbol (× = nonlogic connection). 1 ⎍ is the qualifying symbol for a nonretriggerable one-shot.

Nonretriggerable one-shot

(*a*) A NOR gate used as an inverter

(*b*) Two NOR gates used as an OR gate

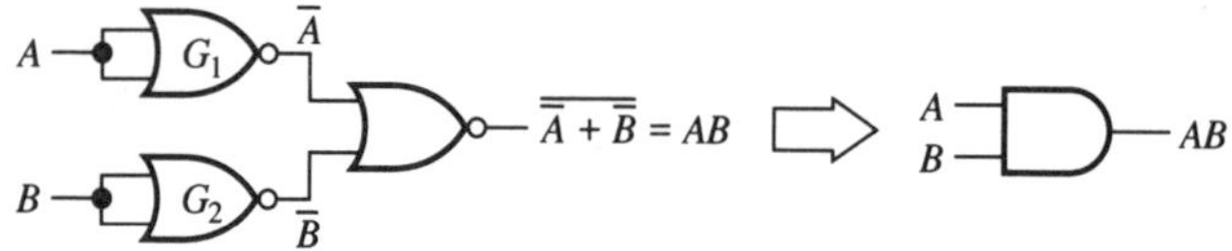

(*c*) Three NOR gates used as an AND gate

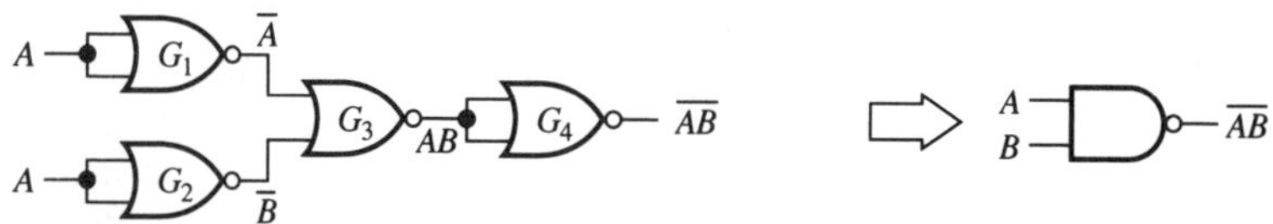

(*d*) Four gates NOR used as an NAND gate

NOR equivalent

one or more of its variables are TRUE.

NOR equivalent Using NOR gates in correct logical combinations to produce the other logic functions of NAND, NOT, AND, and OR.

NOR function The logical NOT OR. A NOR function is TRUE only if all of its inputs are FALSE, otherwise the function is FALSE. The NOR function is expressed as $X = \overline{A + B}$. X is TRUE only if A and B are both FALSE; otherwise, X is FALSE.

NOR gate A digital circuit that duplicates the NOR logic function. Contains one output with two or more inputs.

Norton equivalent An electrical circuit consisting of a current source in parallel with an impedance. The Norton equivalent circuit will model a more complex electrical circuit. Complex electrical circuits may be reduced to

A	B	$\overline{A+B} = X$
0	0	$\overline{0+0} = \overline{0} = 1$
0	1	$\overline{0+1} = \overline{1} = 0$
1	0	$\overline{1+0} = \overline{1} = 0$
1	1	$\overline{1+1} = \overline{1} = 0$

NOR function

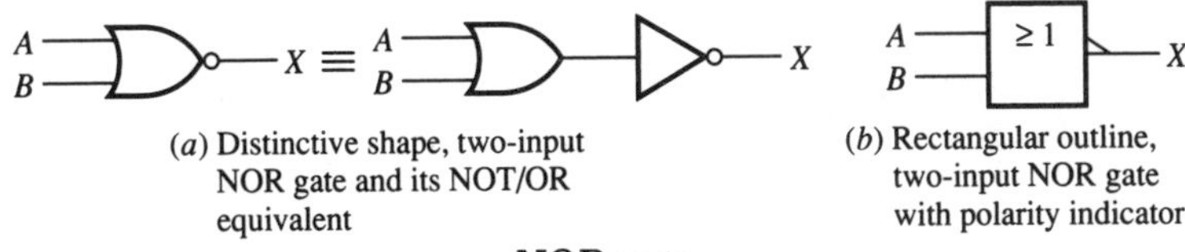

(*a*) Distinctive shape, two-input NOR gate and its NOT/OR equivalent

(*b*) Rectangular outline, two-input NOR gate with polarity indicator

NOR gate

their Norton equivalents, which are in turn easier to analyze under various loads than their original more complex circuits.

Norton's equivalent circuit A circuit representation of a Nortonized circuit. See Norton's theorem.

Norton's theorem States that the output terminals of any linear dc circuit can be represented as a current source in parallel with a fixed resistor.

NOT In digital circuits, the primary function is to change the logic level to the opposite logic level. Has an input and an output. If a HIGH is placed on the input, the output will be LOW. If a LOW is placed on the input, the output will be HIGH. In Boolean algebra, the NOT function is the complement (logical opposite). Thus, NOT 1 is 0, and NOT 0 is 1.

notch filter An electrical circuit designed to reject a specified range of frequencies. Called a notch filter because the graph of the bandpass represents a *notch* in the bandpass curve.

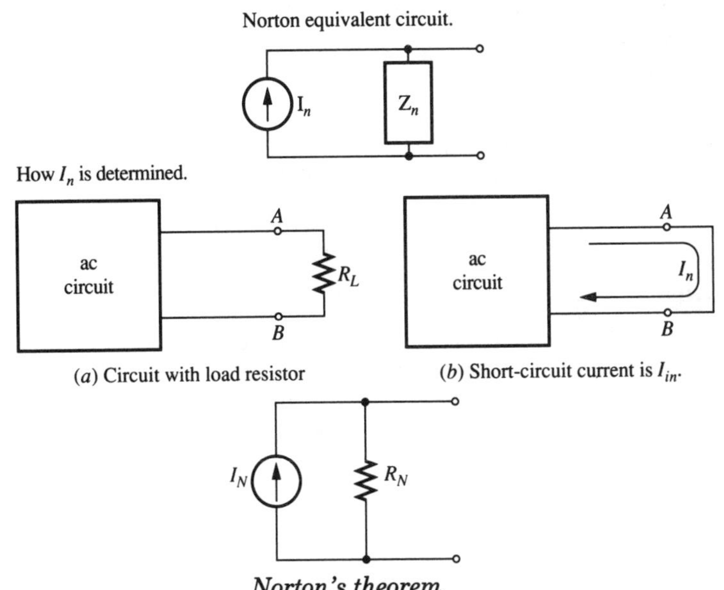

(*a*) Circuit with load resistor

(*b*) Short-circuit current is I_{in}.

Norton's theorem

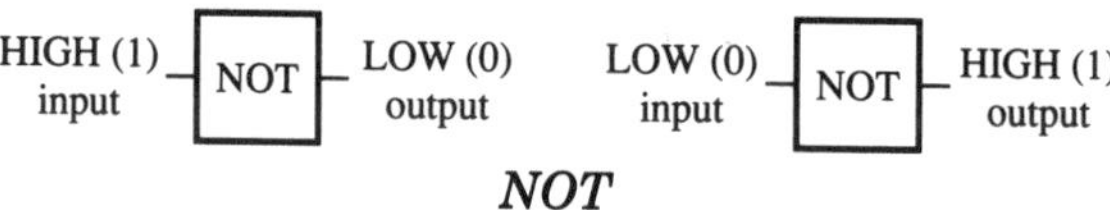

NOT

NOT function A two-valued Boolean expression that represents the inverse of the input. The expression $X = \overline{B}$ (where the overbar represents the inverse, opposite, or NOT) means that if B is TRUE (or 1), then X is FALSE (or 0), and if B is FALSE (or 0), then X is TRUE (or 1).

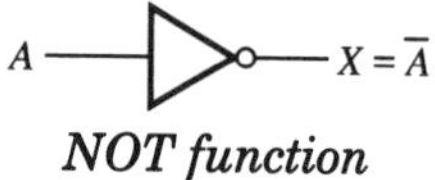

NOT function

npn transistor A bipolar transistor where the emitter and collector consist of material with an excess of electrons (n-material) and the base consists of material with a deficiency of electrons (p-material).

n-type semiconductor Semiconductor material, such as germanium or silicon, that has had impurities added to cause an excess of electrons. In n-type material, electrons are the current carriers.

nucleus The central part of the atom consisting of neutrons and positively charged protons. See Bohr model.

null A condition that represents a balanced, minimum, or zero output of a device or circuit.

number An agreed-to symbol that represents a specific quantity.

numerator The value of the top part of a fraction. For example, in the fraction 2/3, the 2 is the numerator. See denominator.

numerical aperture A measure of an optical fiber's light-gathering capabilities. $NA = \sin \phi$ where ϕ is the maximum acceptance angle for total internal reflection of light in the material and NA is the numerical aperture.

numerical control Systems that use computers to control physical processes, usually machinery used in some sort of manufacturing process.

Nyquist sampling theorem A theorem that states that the sampling frequency of a pulse-modulated system must be equal to or greater than twice the highest signal frequency to convey all the information of the original signal. See sampling frequency.

object code The code understood by the microprocessor, as opposed to source code, which is programming code understood by the programmer. Before a program is executed, it is converted, by another program, from source code to object code.

object-oriented programming A method of programming that is close to the very natural way items are normally viewed by people. It contains three main elements: encapsulation, which is a combining of a record with the procedures and functions that will manipulate it, thus creating a new data type called an object; inheritance, defining an object and then using this definition to build a hierarchy of descendant objects where each descendant inheriting will give access to all its ancestors' code and data; and polymorphism, which is the process of giving an action one name that is shared all along the hierarchy, with each object in the hierarchy implementing the action in a manner that is appropriate to that object.

octal The number system to the base 8. Has eight symbols: 0 through 7.

octave 1. In electronic frequency measurements, a change in frequency by a factor of 2. 2. Frequencies having a 2-to-1 relationship. Thus one octave above 100 Hz is 200 Hz.

octet simplification In Karnaugh mapping a method of simplifying a Boolean expression represented by the Karnaugh map by treating a group of eight horizontally and/or vertically adjacent cells as a single term.

odd harmonic A frequency that is an odd multiple of a fundamental frequency. Thus, 300 Hz is an odd harmonic of 100 Hz.

odd harmonics Odd multiples of a fundamental frequency.

odd parity See parity. (See pg. 221 for illustration.)

OFF An electrical condition in digital circuits signifying a FALSE, LOW, or condition of 0. Usually accomplished by the presence of 0 volts.

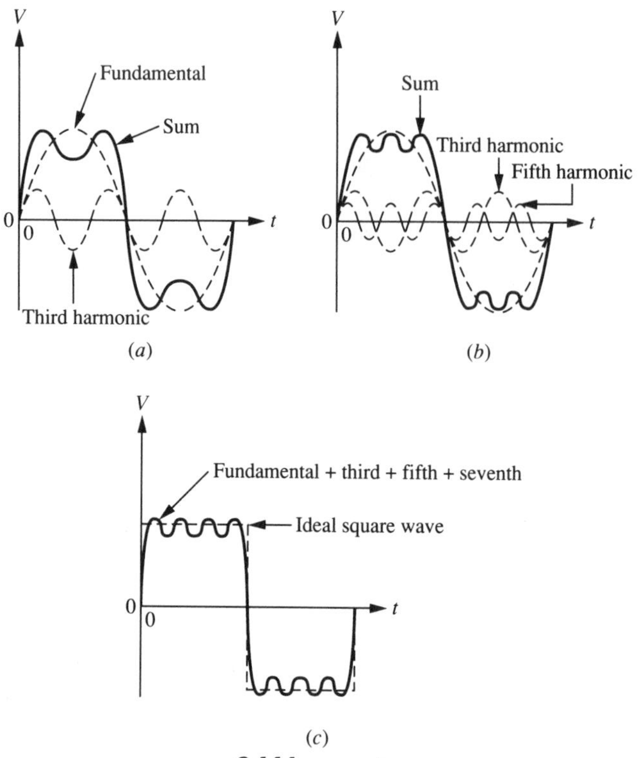

Odd harmonics

offset current In operational amplifiers, the difference in input currents required to bring the output voltage of the amplifier to zero.

offset voltage In operational amplifiers, the difference in input voltage required to bring the output voltage of the amplifier to zero.

ohm The unit of measurement of resistance. Symbol is the Greek letter omega (Ω). A resistance of one ohm will allow one ampere of current flow when a difference of potential of one volt is applied.

ohmic region The operation range of an electrical device where the device behaves as a resistor. The ohmic region of a device has a linear increase in current for a given increase in voltage.

ohmic resistance Direct current resistance.

ohmmeter An electrical instrument for measuring resistance. (See pg. 222 for illustration.)

Ohm's law The relationship among voltage, current, and re-

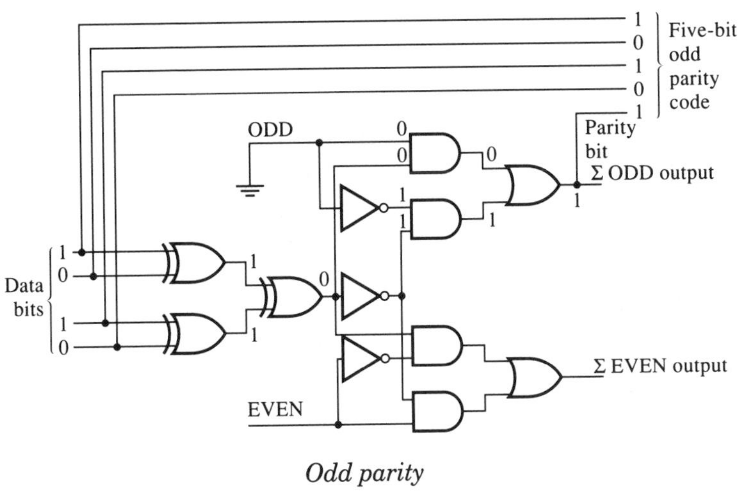

Odd parity

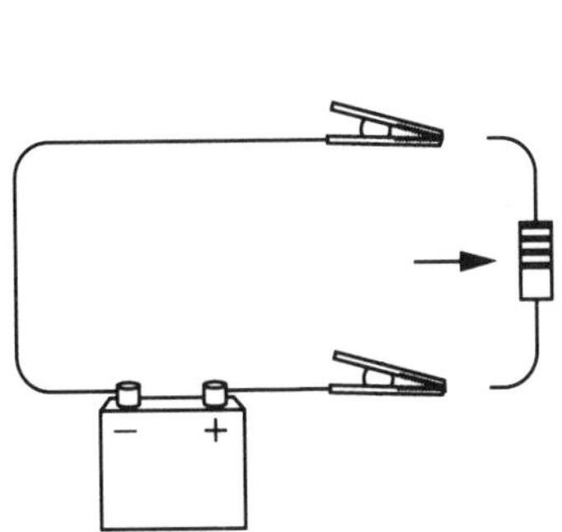

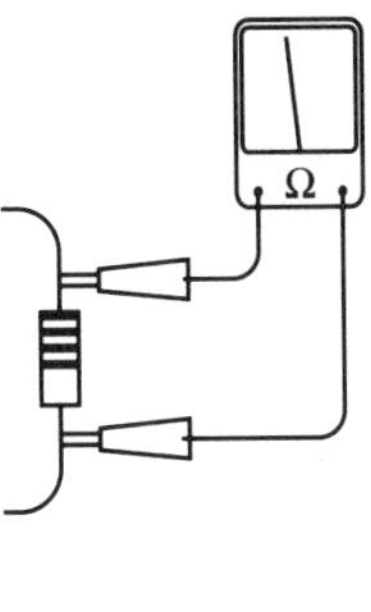

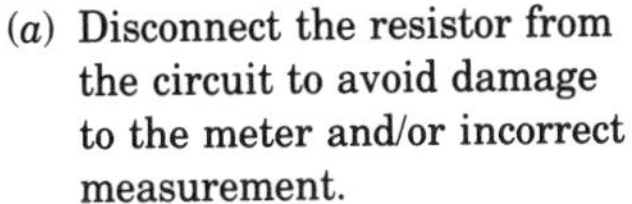

(*a*) Disconnect the resistor from the circuit to avoid damage to the meter and/or incorrect measurement.

(*b*) Measure the resistance. (Polarity is not important.)

Ohmmeter

Effect of changing the voltage with the same resistance in both circuits.

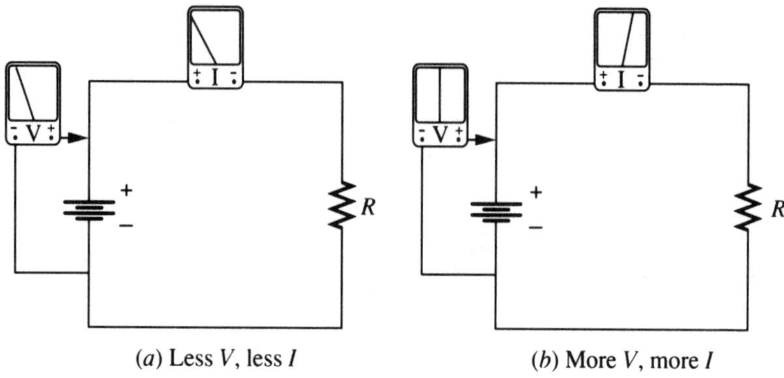

(*a*) Less *V*, less *I* (*b*) More *V*, more *I*

Effect of changing the resistance with the same voltage in both circuits.

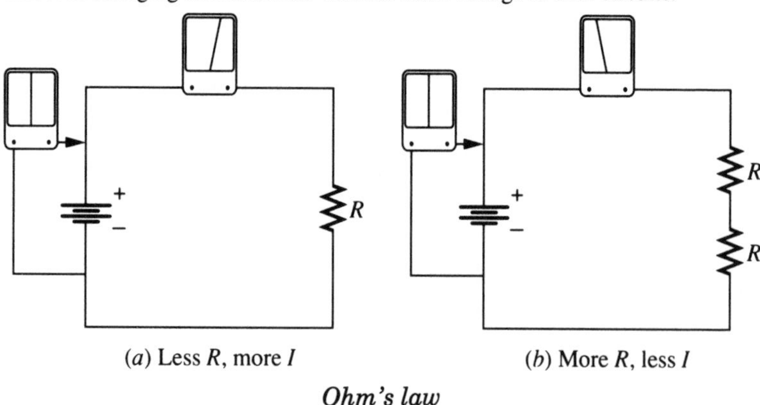

(*a*) Less *R*, more *I* (*b*) More *R*, less *I*

Ohm's law

sistance as stated by George S. Ohm. The basic form of Ohm's law states that voltage is equal to the product of current and resistance.

omnidirectional Having the same strength in all directions. Not favoring any one single direction.

ON An electrical condition in digital circuits signifying a TRUE, HIGH, or condition of 1. Usually accomplished by the presence of +5 volts.

one's complement notation Used in binary subtraction. The one's complement of a binary number is found by taking the complement of each bit of the number. Two's complement notation is more commonly used for subtraction.

one-shot 1. A flip-flop that is normally in an OFF condition and produces an ON time of a specified duration when its input is triggered by another signal. 2. A monostable multivibrator having only one stable state. When a one-shot is activated, it goes to its unstable state and stays there for a specified amount of time. It then will return back to its stable state.

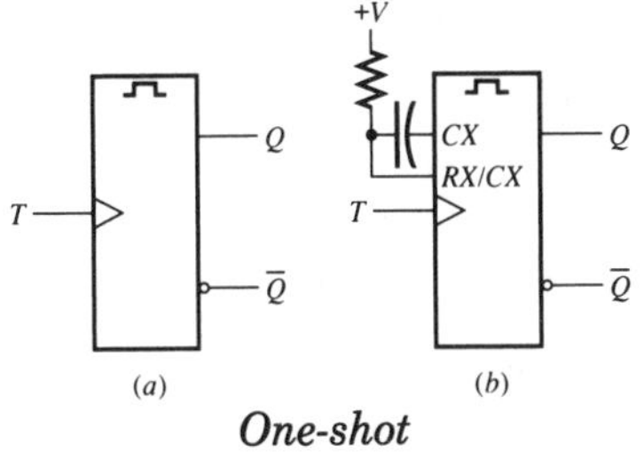

One-shot

on/off switch A mechanical device that is usually manually operated having two electrical positions, an open or a short. An on/off switch is usually used as a power switch to an electrical system.

opacity The ability of a substance to have light travel through it. The degree of nontransparency of a material.

op amp Abbreviation for operational amplifier. See operational amplifier.

open An electrical condition that does not allow any current flow. An open circuit is a break in the conduction path offering infinite resistance to the voltage or signal source.

open circuit A circuit that exhibits the properities of an open. See open.

open-circuit jack A female connector that is in the normally open position. When the matching male connector is connected, the normally open circuit will be closed.

open-circuit voltage The amount of voltage at a voltage source when no current is flowing from the source.

open collector logic A logic device whose output requires the addition of a pull-up resistor for proper operation.

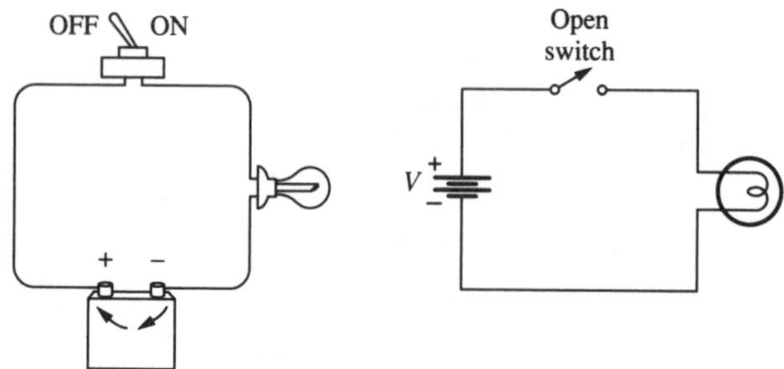

No current flows in an *open* circuit (switch is OFF or in the *open* position).

Open circuit

open loop An amplifier configuration in which there is no path for feedback.

open-loop gain Ratio of the output to the input signal of an amplifier when the output of the amplifier is not fed back to the input.

open transmission line A transmission line whose terminating end is an open circuit. If the length of the transmission line is an exact multiple of a quarter wavelength of the transmitted wave, then an open transmission line will appear as a short to the source.

operand In programming the data that are affected by the instruction.

operating life The amount of time that a device or system will function normally under normal operating conditions.

operating parameters The electrical characteristics of a device when it is actively used in a circuit.

operation In programming, a specific task the computer will do when given the proper instruction. In mathematics the action required of a stated mathematical process.

operational amplifier A high-gain wideband amplifier, usually found in integrated circuits, sometimes referred to as an op amp. An ideal operational amplifier has no frequency limitations and has infinite gain. Real operational amplifiers do not achieve this ideal, but in design applications these ideals are assumed. 2. Called operational because with proper external circuit components, these amplifiers can simulate mathematical operations such as addition, subtraction, multiplication, integration, and differentation.

optical communications Transferring information from one point to another through the use of visible electromagnetic energy. An example of optical communications is the use of laser beams to transmit information.

Parameter (times in ns)	TTL					CMOS
	7474	74LS76A	74L71	74107	74111	74HC112
t_{PHL} (CLK to Q)	40	20	150	40	30	31
t_{PLH} (CLK to Q)	25	20	75	25	17	31
t_{PHL} ($\overline{\text{CLR}}$ to Q)	40	20	200	40	30	41
t_{PLH} ($\overline{\text{PRE}}$ to Q)	25	20	75	25	18	41
t_s (set-up)	20	20	0	0	0	25
t_h(hold)	5	0	0	0	30	0
t_W (CLK, HI)	30	20	200	20	25	25
t_W (CLK, LO)	37	—	200	47	25	25
t_W ($\overline{\text{CLR}}/\overline{\text{PRE}}$)	30	25	100	25	25	25
f_{max} (MHz)	25	45	3	20	25	20
Power (mW/F-F)	43	10	3.8	50	70	0.12

Source: Compiled from Texas Instruments TTL data book, 1985.
Note: Values given are typical where available; otherwise, they may be maximum or minimum depending on availability.

Operating parameters

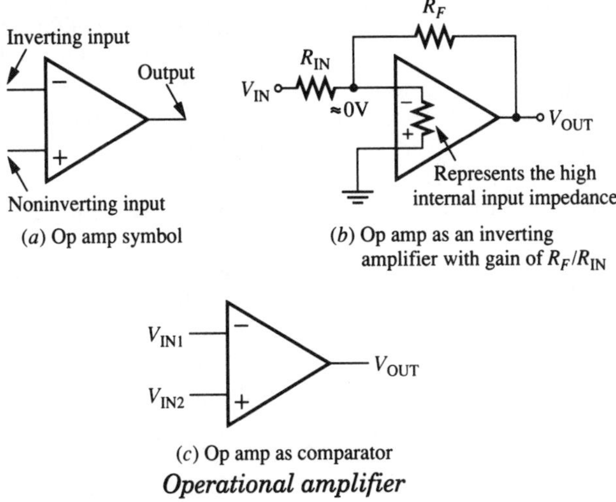

(*a*) Op amp symbol

(*b*) Op amp as an inverting amplifier with gain of R_F/R_{IN}

(*c*) Op amp as comparator

Operational amplifier

optical coupler An electrical device that transfers information from one circuit or system to another through the use of light. The main advantage of an optical coupler is that it offers electrical isolation. In this manner, a small 5-volt computer signal could control a large 120-volt power relay.

optical disk An optically reflective disk that contains digital information stored as surface changes that cause the reflection of a laser beam focused on the rotating disk to change.

optical fiber Transparent wirelike strands of material ca-

A	B	$A + B = X$
0	0	$0 + 0 = 0$
0	1	$0 + 1 = 1$
1	0	$1 + 0 = 1$
1	1	$1 + 1 = 1$

OR function

pable of having light transmitted within them. Optical fibers can be used to communicate over long distances with the use of light. Their advantage over that of solid wires is that by using light, much more information can be transmitted at the same time.

optical isolator See optical coupler.

optical memory The use of light to store and/or read computer information. Optical memory may use a laser beam to store and/or read data from an optical disk.

optical reader An electrical device that is capable of interpreting bar codes or other such information.

optical scanner In computers an electronic device capable of transferring printed images directly into the computer display. The scanner is swept over the image to be stored in the computer. Some optical scanners can also transfer printed text into the computer where it is treated as text accessible to other computer programs.

optical spectrum Electromagnetic radiation in the wavelength range of infrared, visible, and ultraviolet rays. This radiation has a velocity of 300,000,000 meters per second in a vacuum. The optical spectrum is wider than the visible spectrum. See visible spectrum.

optoelectronics The field of study that deals with the use of light and electricity as devices and systems.

OR function A two-valued Boolean expression where the single output is TRUE (or 1) if any of the inputs are TRUE (or 1). Otherwise, the output is FALSE (or 0). The OR expression is $X = A + B$. If either or both A and B are TRUE, then X is TRUE; otherwise, X is FALSE.

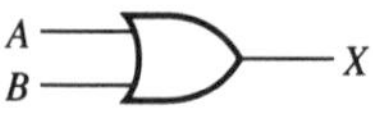

(*a*) Distinctive shape

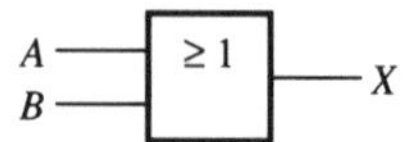

(*b*) Rectangular outline with OR (≥ 1) qualifying symbol

OR gate

OR gate An electrical digital circuit that performs the logical OR function.

ordinate The vertical line or lines on a graph.

oscillate To repeat a continuous waveform. A circuit is said to oscillate when it produces its own continuous waveform.

oscillator An electrical circuit capable of producing its own waveform, usually in the form of a sine wave. Oscillators can also produce a sawtooth waveform as in the case of the horizontal oscillator of an oscilloscope.

oscilloscope An instrument that has a cathode ray tube that is capable of displaying the actual shape of one or more electrical waveforms under test. An oscilloscope is usually used to reconstruct the amplitude and period of an observed waveform. Usually called a scope.

OS/2 The IBM Operating System 2. Used to define a line of computers using the IBM trademark.

outlet A connection that is used for delivering electrical power.

out of phase Two or more signals whose variations do not oc-

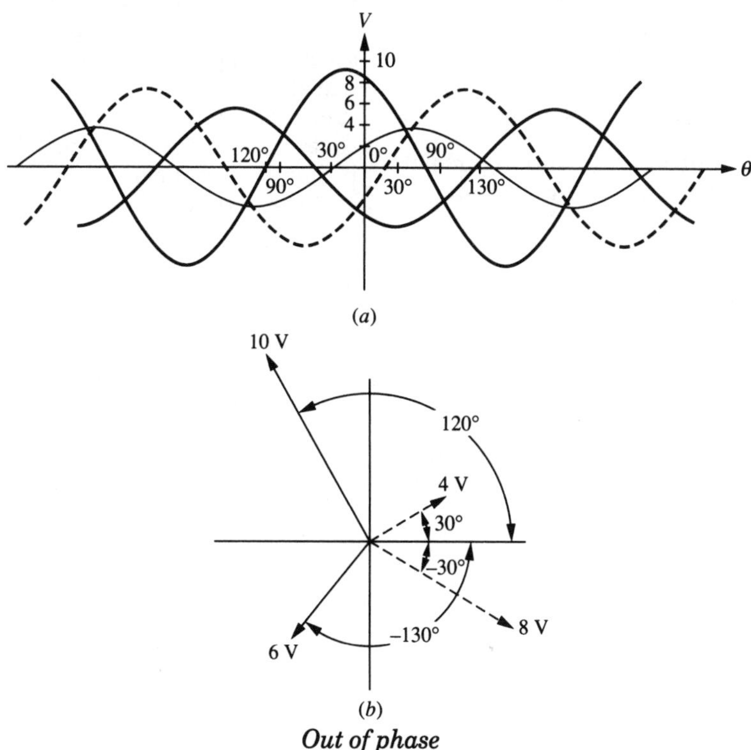

Out of phase

cur at the same point in time. An out-of-phase relationship may be represented by phasors.

output impedance The total opposition to current flow as measured at the output terminals of an electrical circuit, device, or system.

output profile For an electrical digital circuit, it is the specific voltage ranges of logic HIGH and logic LOW for that circuit's output. A typical output profile for TTL would be HIGH = +2.4 volts to +5.0 volts, LOW = 0.4 volts to 0 volts. This means that any voltage range on the output between +2.4 and +5.0 will be considered as a HIGH, while any output voltage between +0.4 volts and 0 volts would be considered as a LOW. See input profile.

output stage The last complete circuit of an electrical system.

overdriven amplifier An amplifier designed to be driven into cutoff and/or saturation.

overflow A condition, in a computer, indicating that the result of an arithmetic operation is not correct.

overload protection The process of keeping an electrical system safe by having it turn off when an internal voltage and/or current exceeds a given safe value.

overmodulate The process of modulating to the extreme results in the distortion of the information impressed on the carrier wave in the modulation process. An overmodulated signal will not be easy to understand or process at the receiver.

overmodulation See overmodulate.

overshoot 1. In a pulse waveform, the amount of changes that occur in the pulse after the initial transition of the pulse. Sometimes called rounding. 2. In a square wave, a sudden increase in amplitude just before the onset of the square wave.

overtone In a crystal oscillator, operating the crystal oscillator at a frequency that is a harmonic of the crystal frequency. This method allows crystal oscillators to operate at frequencies up to 100 MHz.

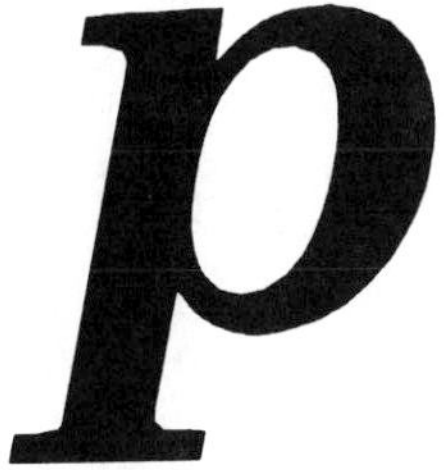

PA system Abbreviation for public address system. See public address system.

pair simplification In Karnaugh mapping, the process of simplifying a logic circuit by looking for cells that are adjacent either vertically or horizontally for the purpose of eliminating one variable from the Karnaugh map.

PAL Abbreviation for programmable array logic. See programmable array logic.

PAM Abbreviation for pulse amplitude modulation. See pulse amplitude modulation.

paper capacitor A fixed-value capacitor where the plates consist of a metal foil separated by oiled or waxed paper or similar dielectric.

parabola The shape formed when the location of all points from a given point and line are the same distance.

parabolic antenna An antenna consisting of a radiating and/or receiving element reflected into a reflector the shape of a rotated parabola. A parabolic antenna causes transmitted radiation to form into a beam. Used in line-of-sight microwave communications.

parallel binary adder A logic circuit capable of adding binary numbers where the numbers to be added are all placed into the adder at the same time.

parallel capacitors When capacitors are connected in parallel, their effective plate area increases, and the total capacitance

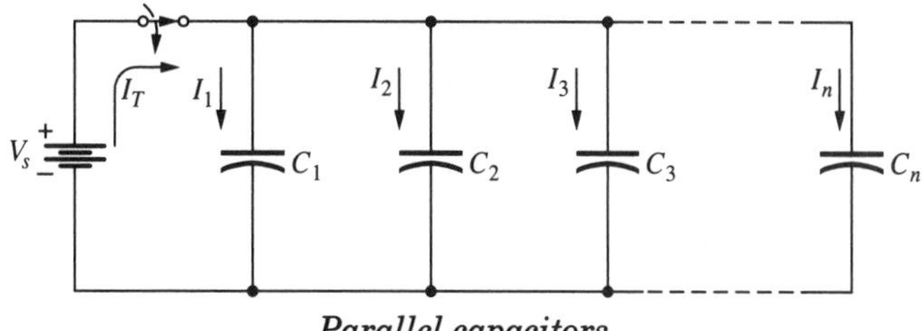

Parallel capacitors

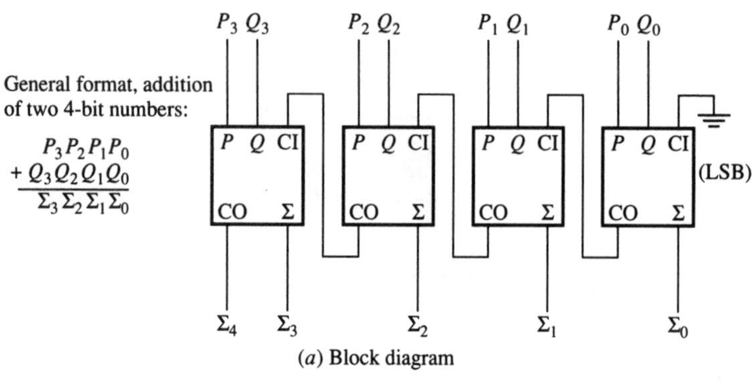

(*a*) Block diagram

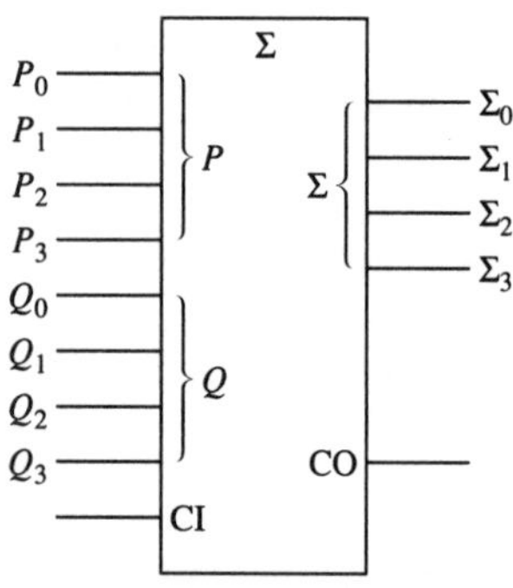

(*b*) Logic symbol

Parallel binary adder

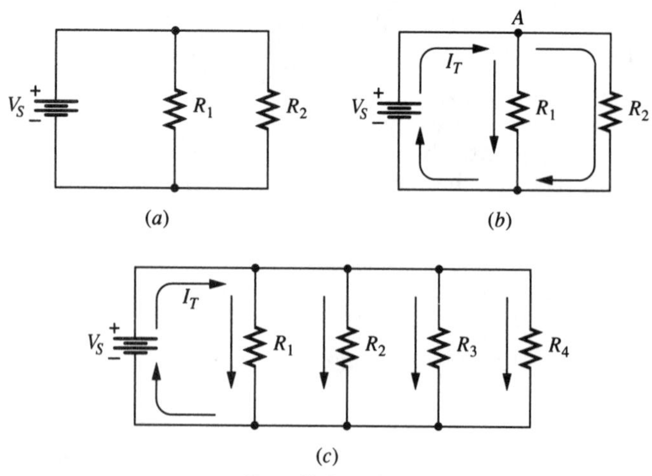

(*c*)

Parallel circuit

is equal to the sum of the individual capacitors.

parallel circuit A circuit arrangement where two or more components are connected across the same voltage source. Because of this arrangement, all components in a parallel circuit experience the same voltage, and there is now more than one path for current.

parallel data storage The method of copying to a digital storage device several bits of data at the same time.

parallel in A term used to describe the transfer of data into a digital device all at the same time. An example would be a parallel-in register where data are copied into the register in one clock pulse.

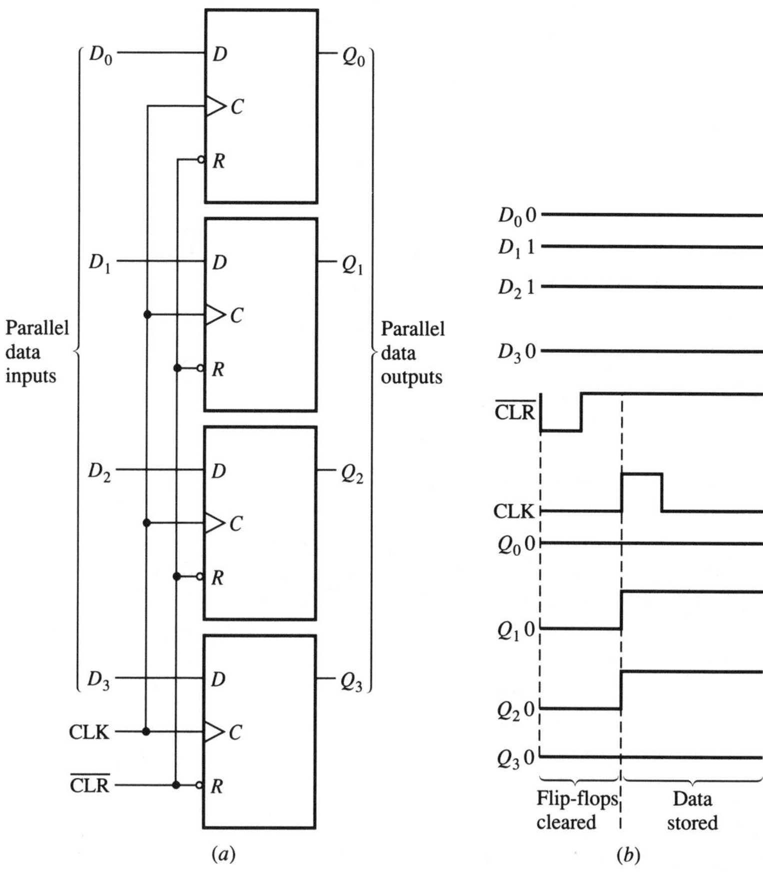

Parallel data storage

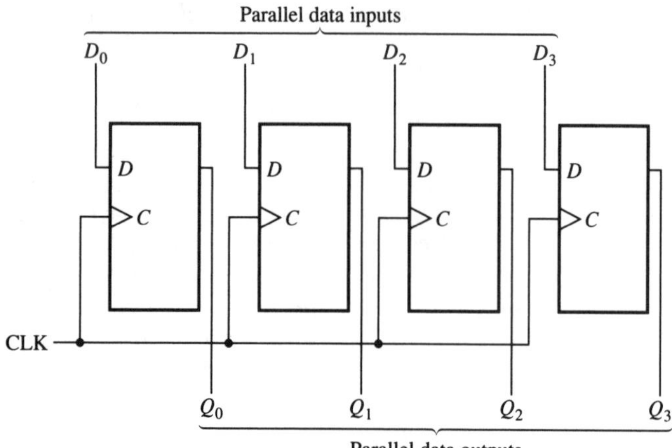

Parallel in parallel out

parallel in, parallel out A digital register that inputs data in parallel, all bits at the same time, and outputs its data in parallel, all bits at the same time.

parallel in, serial out A digital register that inputs data in parallel, all bits at the same time, and outputs its data serially, one bit at a time.

parallel out A term used to describe the transfer of data from a digital device all at the same time. An example would be a parallel-out register where data are copied from the register in one clock pulse.

parallel processing In computers, having more than one program being executed at the same time while the separate programs are capable of interacting with each other.

parallel *RC* circuit An electrical circuit with a capacitor and resistor connected in parallel to the power source.

parallel resonance In a parallel *LC* circuit, the condition when the inductive and capacitive reactances are equal. Under such conditions, the overall impedance of the circuit will be a maximum. This circuit will have a natural resonant frequency determined by the value of the inductor and capacitor.

parallel to serial A method of converting digital information from a parallel form to a serial form of data transfer.

parallel transfer In a digital system, the sending of a bit pat-

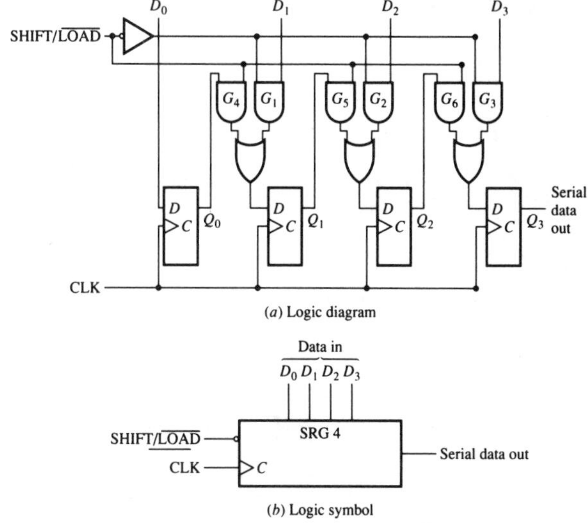

(a) Logic diagram

(b) Logic symbol

Parallel in serial out

tern over several lines at the same time. When transferring in parallel form, all the bits in a group are sent out on separate lines at the same time.

paramagnetic materials Materials that have a permeability that is slightly greater than that of free space.

parametric amplifier A special type of device used to increase the power output of microwave signals.

parasitic array Antenna elements not connected to the transmission or reception lines. Assists in developing a desired radiation pattern for the antenna. This is useful in making the antenna highly directional.

parasitic elements Antenna elements not connected to the

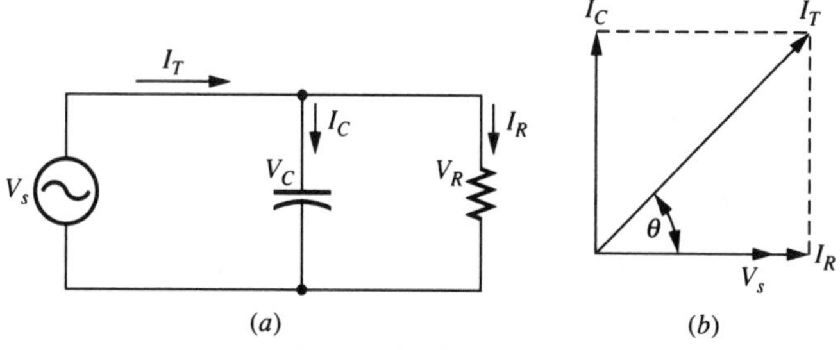

(a)

(b)

Parallel RC circuit

An ideal parallel LC circuit at resonance.

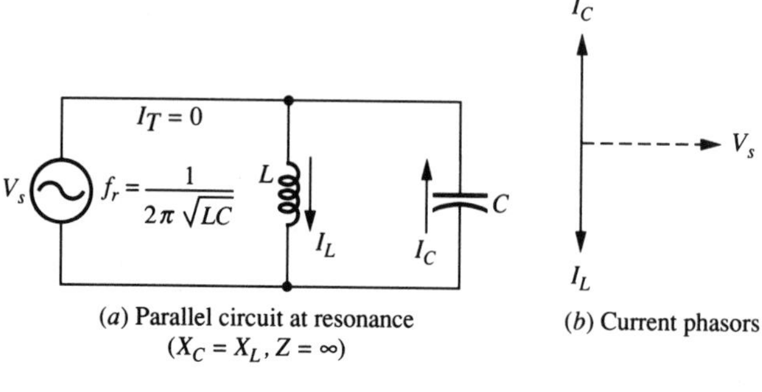

(*a*) Parallel circuit at resonance ($X_C = X_L$, $Z = \infty$)

(*b*) Current phasors

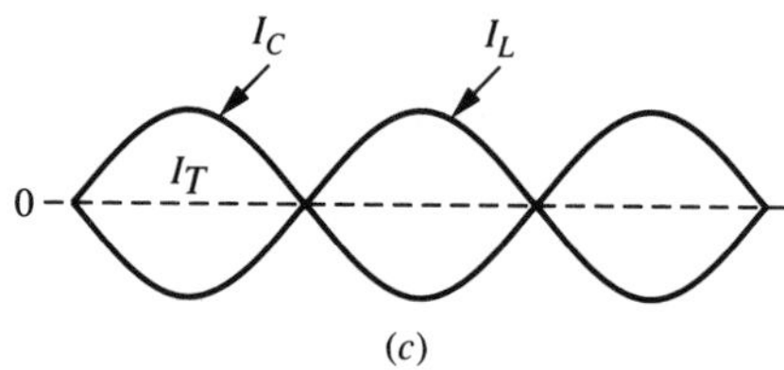

(*c*)

Generalized impedance curve for a parallel resonant circuit. The circuit is inductive below f_r, resistive at f_r, and capactive above f_r.

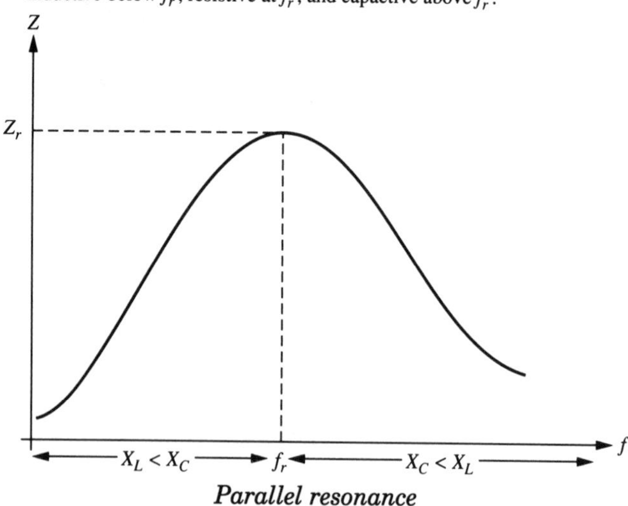

Parallel resonance

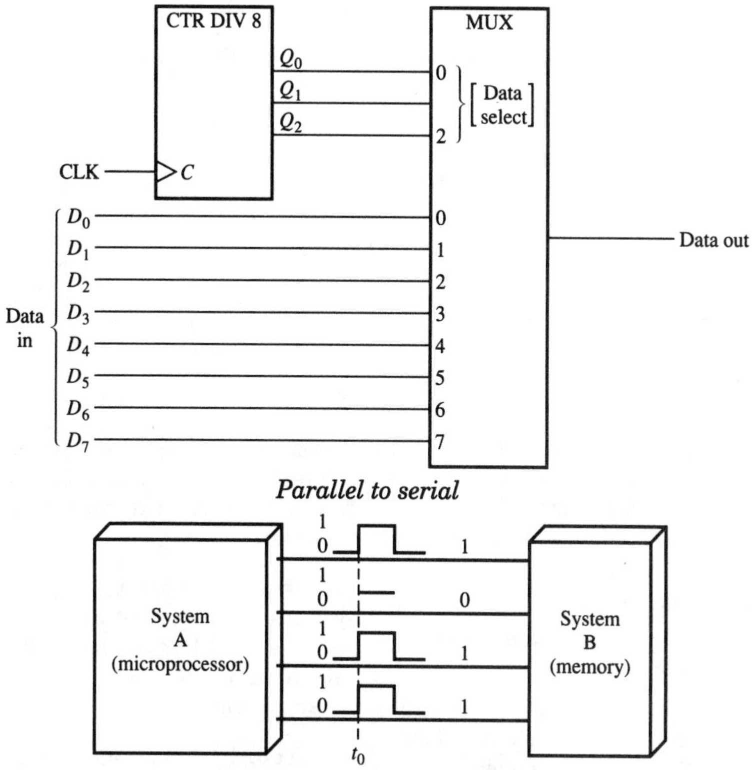

Parallel to serial

Parallel transfer of bits from A to B. All four bits are transferred simultaneously beginning at time t_0. Microprocessor to memory is an example.

Parallel transfer

transmission or reception lines. See parasitic array.

parasitic oscillations Undesirable signals caused by stray capacitance and inductance in a circuit. Parasitic oscillations are usually weak oscillations that may not be consistent; usually a high frequency that may change its characteristics when test probes are moved around the circuit.

parity Used for error checking when transferring information in a computer system. The number of bits for each group of data is always made odd (odd parity) or even (even parity); if the sum of the bits does not keep its parity, then an error has occurred.

parity bit The extra bit used in a word to cause odd or even parity. Used for error checking. See parity.

Even Parity		Odd Parity	
P	8421	*P*	8421
0	0000	1	0000
1	0001	0	0001
1	0010	0	0010
0	0011	1	0011
1	0100	0	0100
0	0101	1	0101
0	0110	1	0110
1	0111	0	0111
1	1000	0	1000
0	1001	1	1001

Parity

parity error An indication of an error when the sum of the parity check is not odd if it should have been odd, or even if it should have been even.

particle A tiny subdivision of matter such as an atom, electron, or photon.

Pascal A programming language invented by Niklaus Wirth, named in honor of the French mathematician Blaise Pascal. The language was intended to teach good programming habits such as program structure and the treatment of data. Became a popular applications language as well.

passband The continuous range of frequencies that are allowed to pass through a given circuit or system.

passive component An electrical component that does not require an external source of power and has no power gain characteristics such as resistors, capacitors, and inductors.

passive network Electrical connections of components that do not require an external source of power.

password In multicomputer systems, a unique set of characters assigned to each individual user to gain entry into the computer system.

patch cord A short mechanical connection made with a conductor between two points in a circuit that may be easily inserted or removed.

patch panel A smooth solid surface where connections are available to circuit outputs and inputs. These outputs and inputs may be interconnected by the use of patch cords. See patch cord.

PBNC Abbreviation for push button normally closed. See push button normally closed.

PBNO Abbreviation for push button normally open. See push button normally open.

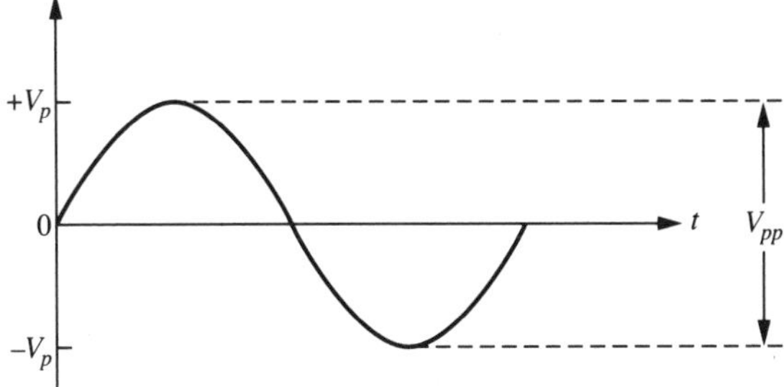

Peak to peak

PC Abbreviation for personal computer. Sometimes lowercased, used to mean printed circuit, as in pc board.

PCM Abbreviation for pulse code modulation. See pulse code modulation.

PDM Abbreviation for pulse duration modulation. See pulse duration modulation.

peak In waveform analysis, the amount of excursion from the resting point to a maximum value, usually that of a sine wave. In a sine wave, the peak value is exactly one-half the peak-to-peak value.

peak clipper A circuit that clips either the positive or negative peaks of an input signal. See positive peak clipper, negative peak clipper.

peak detector An electronic circuit that determines the maximum amplitude of an input signal.

peak inverse voltage The amount of reverse-biased voltage across a diode. For example, a rectifier diode used in a power supply has a peak inverse voltage equal to the peak value of the voltage it is rectifying.

peak to peak In waveform analysis, meaning the amount of excursion from the minimum value to the maximum value of a waveform, usually that of a sine wave.

peak-to-peak voltage The total voltage variation of a waveform from its minimum or negative value to its maximum or positive value.

peak value The maximum value of a wave form above its resting place. In a sine wave, the maximum value is the peak value of the sine wave as measured from its zero value.

peak voltage The maximum positive or negative value of a voltage waveform.

pentavalent In solid-state electronics, meaning five. A pentavalent impurity (an element with five valence electrons) such as arsenic, phosphorus, and antimony is added to pure silicon to create an n-type material.

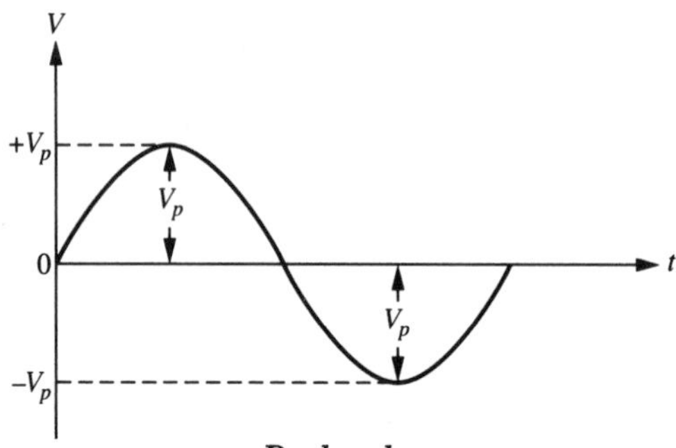

Peak value

pentode A five-element vacuum tube consisting of a cathode, control grid, screen grid, supressor grid, and plate.

percentage input regulation Specifies how much change occurs in the output voltage for a given change in input voltage. Expressed as a percentage change in the output voltage for a one-volt change in input voltage. Usually used with power supplies.

percentage load regulation For a power supply, specifies the amount of change in the output voltage over a range of load current values, usually from no load (no current) to full load (full current).

percentage modulation Used to measure the amount of modulation of a carrier. In AM, the percentage modulation is defined as $m_p = B/A \times 100\%$, where m_p is the percentage modulation, B is the peak value of the modulating signal, and A is the peak value of the unmodulated carrier. In FM, percentage modulation is defined as $M_{\mathrm{MFM}} = (f_{D(\mathrm{actual})}/f_{D(\mathrm{max})}) \times 100\%$, where M_{MFM} is the percentage of FM modulation, $F_{D(\mathrm{actual})}$ is the actual frequency deviation of FM carrier (in hertz), and $f_{D(\mathrm{max})}$ is the maximum allowable frequency deviation of FM carrier (in hertz).

percentage ripple The ripple factor × 100%. See ripple factor.

period 1. The amount of time it takes to complete one complete cycle. 2. The amount of time required to complete one cycle of a waveform, measured in seconds.

periodic To repeat. A periodic waveform is a waveform that continually repeats itself such as a clock pulse or a sine wave.

peripheral In computers, any device that is external to the actual computing parts of a computer. For example, a printer and disk drives are peripheral devices.

peripheral interface adapter A digital device used to parallel input and output data.

permanent magnet A material that remains magnetized for

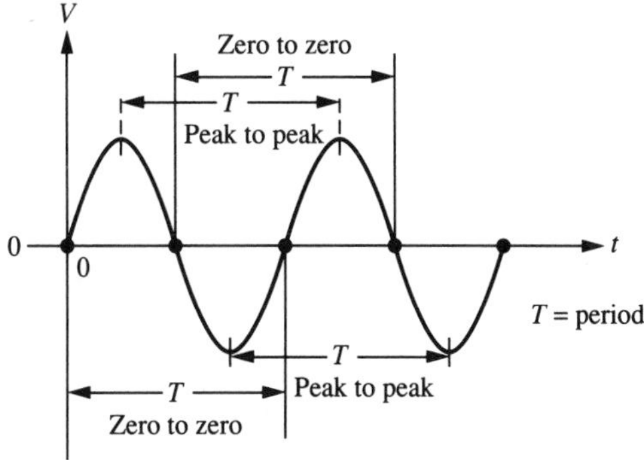

Period

long periods of time with no external means.

permanent magnet speaker A paper cone speaker that uses a permanent magnet for its magnetic field. A small coil of wire, attached to the center of the paper cone, has the electrical currents, that represent sound, moving through it. These currents in turn set up a changing magnetic field around the coil that interacts with the permanent magnetic field and causes the paper cone to move air thus producing sound.

permanent memory Storage for information where the information cannot be changed without destroying the medium containing the information.

permeability Symbolized by the Greek letter μ. A measure of the ease in which magnetic flux can be established in a material. Permeability is measured in webers per ampere-turn-meter.

persistence In a cathode ray tube, the amount of time the phosphor coating will glow after the electron beam strikes it.

personal computer A computer made for use by nonprofessional programmers. To be used by the general public or businesses without needing the employment of a professional programmer. Personal computers have a variety of applications, from entertainment to handling business tasks for small companies.

pF Letter symbol for picofarad. See picofarad.

PF Abbreviation for power factor. See power factor.

phase The measurement of angular displacement of a waveform, usually measured in degrees.

phase angle The time difference measured in degrees or radians between two or more sine waves. Phase angles may be represented by a vector diagram. See vector diagram.

Phase reference.

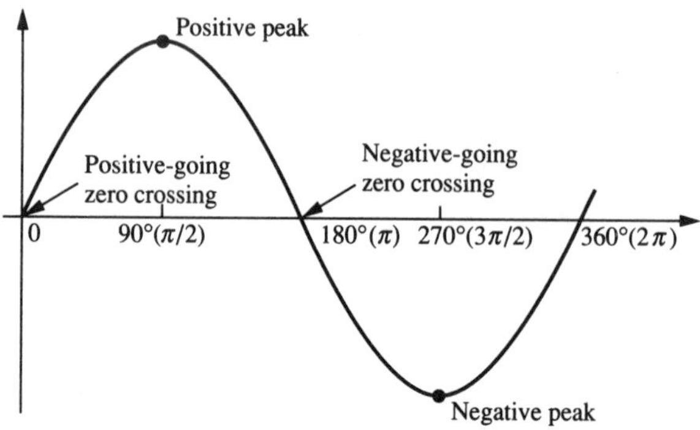

Illustration of a phase shift.

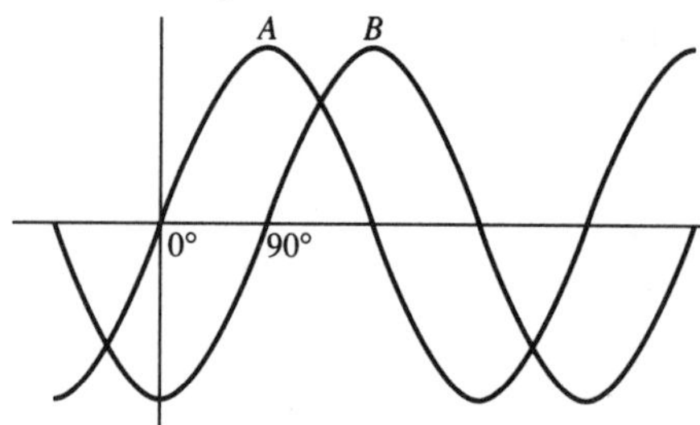

(*a*) *A* leads *B* by 90°, or *B* lags *A* by 90°.

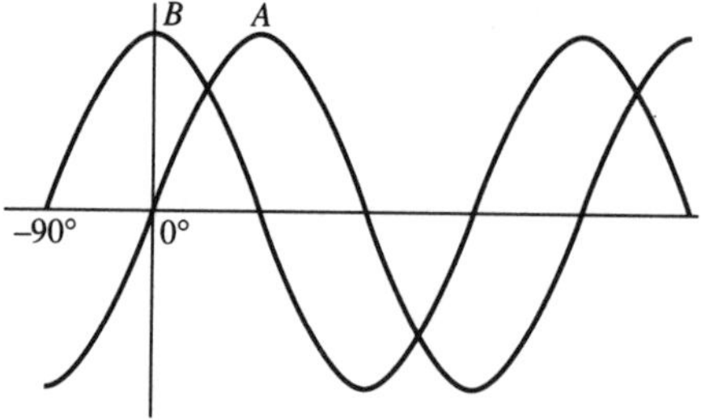

(*b*) *B* leads *A* by 90°, or *A* lags *B* by 90°.

Phase

phase detector A circuit where the voltage output is proportional to the phase difference between two input signals.

phase difference The time difference, measured in degrees or radians, between two waveforms.

phase inverter An electrical circuit that changes the phase of an input signal by 180 degrees.

phase lock A technique for making the phase of an oscillator signal follow the phase of a reference signal. Doing this allows the frequency of the oscillator to follow the frequency of the reference signal.

phase-locked loop An electrical circuit consisting of a phase detector, a low-pass filter, and a voltage-controlled oscillator (VCO). When a phase-locked loop is locked, only the phase between the VCO and incoming frequency is different, and the frequency will be exactly the same. Phase-locked loops are used in radio telemetry from satellites that have signals buried in noise.

phase-locked receiver A communications receiver that utilizes a phase-locked loop. Results in a low signal-to-noise ratio. Also known as a Doppler detector because it is useful in detecting the change in frequency due to the Doppler effect caused by an object, such as a satellite, moving at a very high velocity relative to the receiver.

phase margin The additional amount of phase shift that can be allowed before instability occurs in an amplifier.

phase measurement The measuring of the phase relationship between two or more signals. Usually done with the use of a dual-trace scope.

phase modulation The process of changing the phase of the

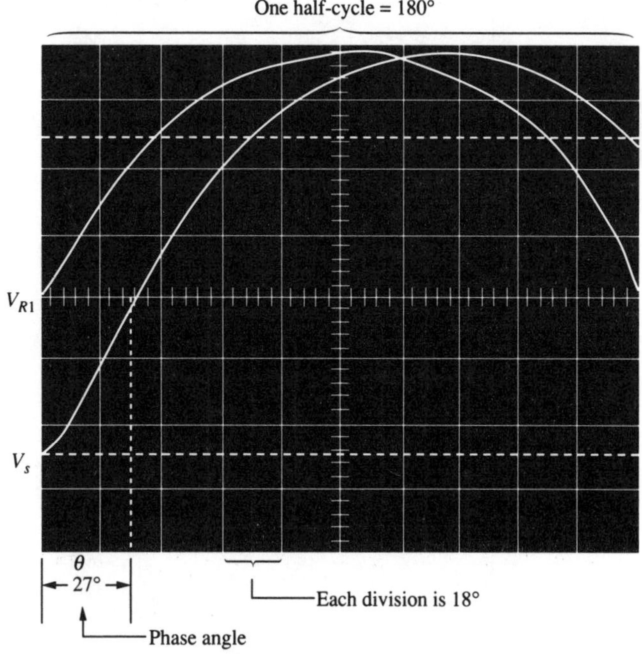

Phase measurement

carrier signal. In phase modulation, the changes in the phase of the carrier represent the information being transmitted. In color television, phase modulation is used to transmit the color information.

phase shift The amount of change in the phase of a signal. For example, if the output signal of an amplifier is 180 degrees out of phase with its input signal, the amplifier is said to have a phase shift of 180 degrees.

phase-shift oscillator A circuit that generates its own signal. Uses three *RC* networks, each of which produces a phase shift of 60 degrees, for a total phase shift of 180 degrees. Frequency of resultant sine wave is determined by the values of the resistors and capacitors used to shift the phase.

phase splitter An electrical circuit that produces two or more output waves from an input wave where the output waves are different in phase from the input wave.

phasor A method of representing the angular displacement of

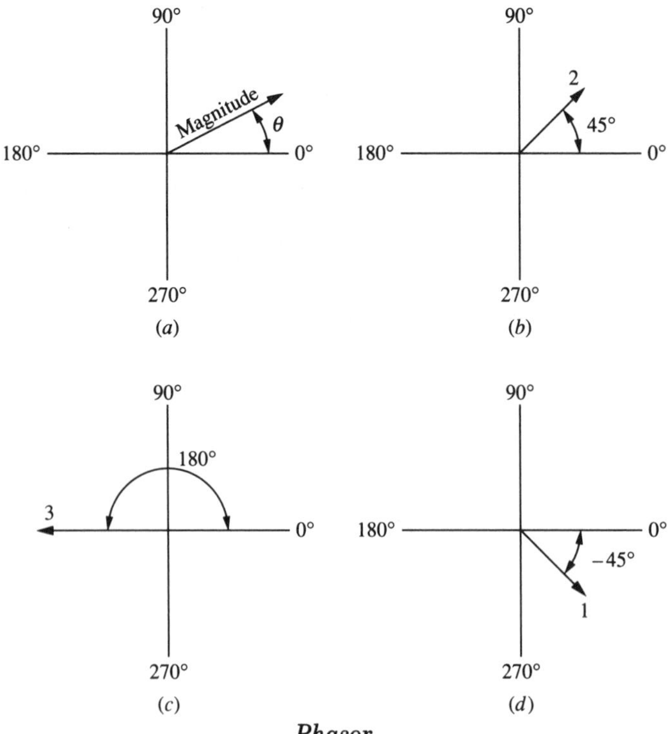

Phasor

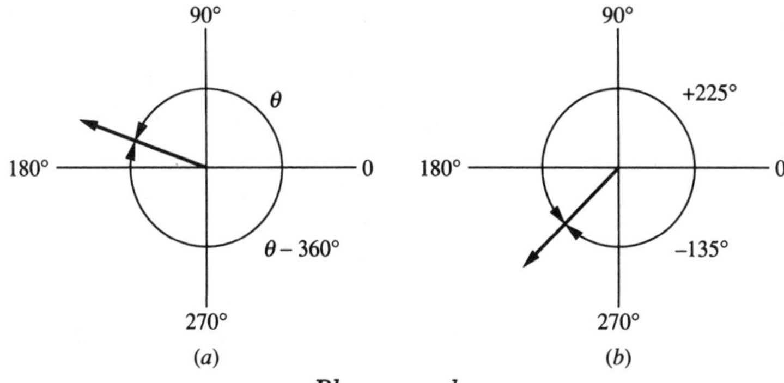

Phasor angles

a waveform. A graphic means of measuring quantities that have both magnitude and direction.

phasor angles A measure of the position of a phasor represented in polar coordinates. Measured counterclockwise from the zero degree line.

phasor diagram The graphical representation of a phasor. The length of the phasor represents the magnitude while its displacement angle represents its direction.

Phillips screw A screw that has a cross slot on its head. A Phillips screw is different from the conventional single-slot headscrew.

phone jack A receptacle that is used for the connection of a telephone or telephone equipment to an outside line.

phonemes The basic sounds that make up a language. English has about 40 phonemes 16 vowel phonemes and 24 consonant phonemes.

phonetic alphabet The use of words for each letter of the alphabet. Using a phonetic alphabet

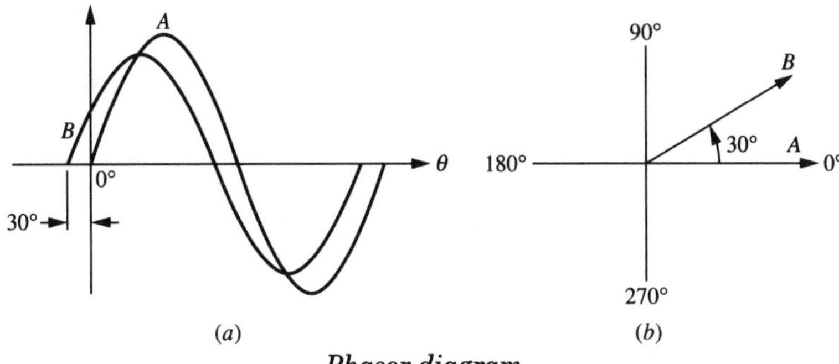

Phasor diagram

helps clarify the letter being used such as DELTA for D and ECHO for E.

phosphor A material that emits light when receiving the proper amount of energy. The inside face of a CRT is coated with phosphor and emits light when struck by an electron beam.

photoconductive The property of a material where its ability to conduct depends upon the amount of radiation (such as light) that it receives from an independent source.

photoconductivity The ability of certain material to show less resistance when exposed to light.

photoconductor An electrical device whose resistance is determined by the amount of light striking its surface. The greater the light, the less the resistance.

photodetector An electrical device that is capable of changing its electrical characteristics from exposure to light.

photodiode A diode whose electrical characteristics are affected by the amount of light on its surface. A pn-junction semiconductor device normally operated in the reverse-biased direction where the reverse leakage current is directly proportional to the amount of light striking its junction. Used as a device for measuring light levels.

photoelectric effect The phenomenon of light converting directly to the emission of electrons. An example of the photoelectric effect is the emission of electrons from the surface of a metal.

photoelectrons Electrons released as a result of photoelectric activity.

photons Particles of light. Light can be thought of as existing as tiny particles called photons. This is part of the duality of light where it can be thought of as particles (photons) or waves.

photoresist A chemical substance rendered insoluble by light exposure.

photoresistor A device whose resistance value is sensitive to the amount of light and is intended to be used for this effect.

photosensitive Material that will react in some manner when exposed to light.

phototransistor A solid-state pnp or npn device that reacts to light. A phototransistor has its collector current increase when light strikes its base. The base of this type of transistor is exposed to light through a small lens.

photovoltaic The property of a device where light striking its surface will cause a corresponding output voltage. A photovoltaic cell is sometimes referred to as a photocell.

pi The symbol π from the Greek alphabet that represents the constant 3.14159. . . .

PIA Abbreviation for peripheral interface adapter. See peripheral interface adapter.

pico The metric prefix meaning 10^{-12}.

picofarad A measure of capacitance equal to 10^{-12} farad.

picosecond A measure of time equal to 10^{-12} seconds.

Pierce oscillator A circuit capable of generating its own signal. A Pierce oscillator uses a crystal as its frequency-determining component. The whole circuit consists of very few components when compared to other oscillator circuits.

piezoelectric effect See crystal.

pi filter An electrical circuit consisting of two parallel connected capacitors separated by a resistor or an inductor. A pi filter is used to reduce the amount of ac ripple on the output of a power rectifer circuit. It gets its name from the fact that the network arrangement looks like the Greek letter π

pinch off In a field-effect transistor, the amount and polarity of voltage applied to the gate that will cause the drain current to become zero. The pinch-off voltage causes the depletion region to extend all the way across the channel, thus cutting off the channel current.

pin diode A semiconductor device consisting of a heavily doped p and n regions separated by an undoped semiconductor region. When reversed biased, the pin diode exhibits an almost constant capacitance. Usually used as a dc-controlled microwave switch where its resistance can be controlled by the amount of applied voltage when forward biased, making it a good modulating device.

pi network An electrical circuit made up of three separate components where each component is connected to the other so as to take the form of the Greek letter π.

pins 1. The connection leads of an integrated circuit package. 2. Electrical contacts of a mechanical connector.

pirating The act of illegally copying software for personal use or for sale or use by others to avoid or reduce the purchasing expense of the original software. Pirating is punishable by severe fines and/or imprisonment.

PIV Abbreviation for peak inverse voltage.

pixel Picture element. The smallest display element that can be controlled by a computer on the screen of its monitor.

PLA Abbreviation for programmable logic array. See programmable logic array.

place value The representation of values by a positional value system. For example, the symbol 5 represents different numerical values depending upon its placement with respect to a decimal point, such as 5., 50., 500., or 5000.

planetary electron An electron bound to an atom. A planetary electron is confined to an orbit or shell of the atom.

Plank's constant Symbolized by the letter *h*. A constant that represents a ratio of the energy of radiation to its frequency. Equal to 6.547×10^{-27} ergs-second. Named in honor of the physicist Max Plank.

plasma Gas that is ionized and may therefore be affected by electrical and magnetic fields.

plate The anode of a vacuum tube. That part of the vacuum tube that attracts electrons from the cathode and is connected to the most positive voltage associated with the vacuum tube.

PLD Abbreviation for programmable logic device. See programmable logic device.

PLG Abbreviation for programmable logic gate. See programmable logic gate.

PLL Abbreviation for phase-locked loop. See phase-locked loop.

PLS Abbreviation for programmable logic sequencer. See programmable logic sequencer.

plug That part of two mating connectors that is free moving when not attached to the mating half.

plumbing See waveguide.

PM Abbreviation for phase modulation. See phase modulation.

PMOS Abbreviation for positive-channel metal-oxide semiconductor. See positive-channel metal-oxide semiconductor.

pn junction In semiconductors, the area between the p-type material and the n-type material. The pn junction is void of current carriers because it forms a depletion area.

pnp transistor 1. A solid state device consisting of two pn junctions and three elements called the emitter, base, and collector, where the emitter and collector are of p-type material and the base separating the two is of n-type material. 2. A bipolar transistor where the emitter and collector consist of material with a deficiency of electrons (p-material) and the base consists of material with an excess of electrons (n-material).

polar components The representation of a phasor in polar form where the magnitude and then the direction are given. If the angle is θ, then the vertical

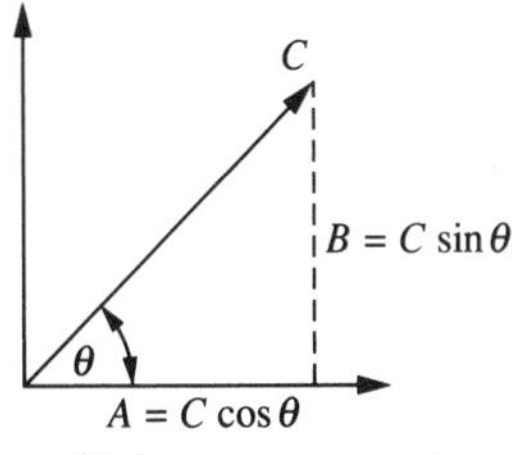

Polar components

component is represented by $C(\sin\theta)$ and the horizontal component by $C(\cos\theta)$, where C is the magnitude of the phasor.

polar form The representation of magnitude and direction by expressing the value of the magnitude and the angle of direction from the positive horizontal axis. For example, $12\angle 30°$ is the polar form representation of a magnitude of 12 units at an angle of 30 degrees rotated counterclockwise from the horizontal.

polarity Having a voltage direction signified by positive (+) and negative (−) that determine the direction of current flow.

polarized capacitor An electrolytic capacitor which has a defined voltage direction due to its internal construction. Inserting a polarized capacitor into the circuit with incorrect polarity can result in damage to the capacitor as well as the circuit.

polarized plug A connector made in such a manner that it can be inserted into its mating connector in only one way, thus preventing an incorrect connection.

Polish notation A mathematical notation, usually used with calculators, that does not use parentheses. Originated by the Polish logician J. Lukasiewicz. Because his name is difficult for

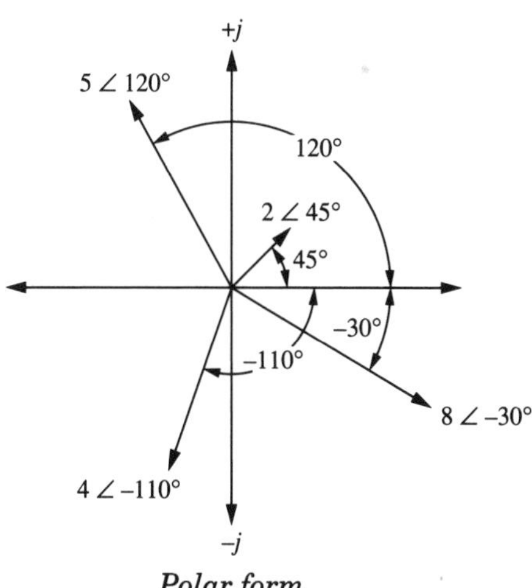

Polar form

English speaking people to pronounce, it is referred to simply as Polish notation.

polygon A closed figure made with straight lines.

polyphase Describing electrical circuits that have two or more phases. Two-, three- and six-phase circuits are common.

polyphase motor An electrical motor wound in such a manner as to be operated from a polyphase system.

port 1. The connecting hardware that allows communications between a digital system and an external device, for example, a printer port on a microcomputer. 2. A mechanical access to an electrical system.

POS Abbreviation for product of sums. See product of sums.

positional notation A method of representing the value of a number by the relative position of the characters used to represent the number. For example, positional notation allows the representation of a number through the use of only 10 symbols in the decimal number system. Thus, 2 is a value of two, while 20 is a value of twenty and 200 is a value of two hundred.

positional number system A systematic method for representing values, where any value may be represented as a sequence of multiples of successive powers of a given base. As an example in the base 10 number system, $250 = (2 \times 10^2) + (5 \times 10^1) + (0 \times 10^0)$.

positive-channel metal-oxide semiconductor Produces high packing densities. Channel is p-type material.

positive charge One of the two fundamental types of electrical charge. Opposite to that of a negative charge. In static electricity, like charges repel while unlike charges attract.

positive feedback In an amplifier or similar circuit, the process of taking some of the output signal and feeding it back to the input in phase. Used to create an oscillator where the circuit will generate its own signal whose frequency will largely be determined by the values of circuit components.

positive ion An atom having a net positive charge due to the deficiency of one or more electrons.

positive number Numbers that can be represented by points from the origin of a graph (0) to the right of this origin on the horizontal or upward on the vertical axis of a graph.

positive peak clipper A circuit that clips the positive peaks of an incoming waveform to a predetermined level.

positive temperature coefficient A value that indicates an increase in an electrical quantity

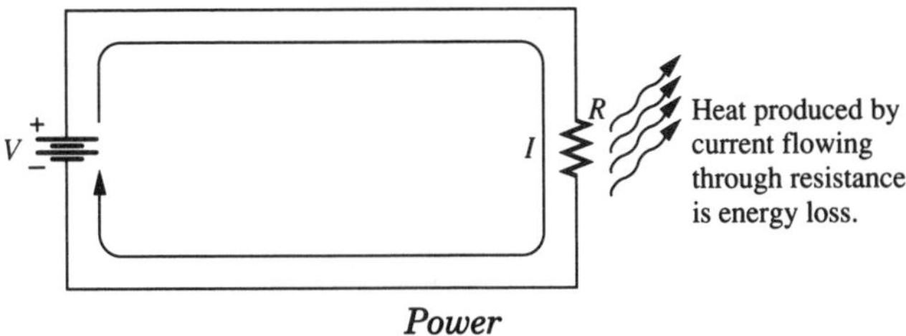

Power

with temperature. See negative temperature coefficient.

potential The difference in voltage between two points in a circuit.

potential barrier In semiconductor material, the pn junction forms a region of positive and negative ions that tend to prevent conduction across the region. For silicon, it takes 0.7 volts to overcome the potential barrier, 0.3 volts for germanium (at 25 °C).

potential difference The voltage created between two points in an electrical system; a voltage that exists across two points of an impedance.

potential energy Energy produced by mass because of its ability to do work as a result of its position. A ball sitting on the edge of a table has potential energy because of its ability to fall to the floor.

potentiometer A variable resistor consisting of a fixed resistance value along which a wiper arm may be moved. The ends of the fixed resistor value are connected to a voltage source and the output voltage is taken from the wiper arm, thus creating a variable voltage divider. As the position of the wiper arm is changed, so is the voltage at the wiper arm terminal.

pound The unit of measurement for force in the English system of measurement. One pound is equal to 4.45 newtons in the MKS or SI systems.

power A measurement of how quickly energy is delivered. Measured in watts, it is defined as $P = E/t$ where P is power (in watts), E is energy (in joules), and t is time (in seconds). Electrical power is also defined as $P = IE$, where I is the current (in amps) and E the voltage (in volts).

power amplifier A circuit designed to increase the power of a signal. For example, an audio power amplifier is designed to deliver maximum electrical power to the speakers. Power amplifiers do not usually have a signal voltage gain greater than one but have large signal current gains.

power bus In microprocessor terminology a group of conductors that supply electrical power to the microprocessor. The power

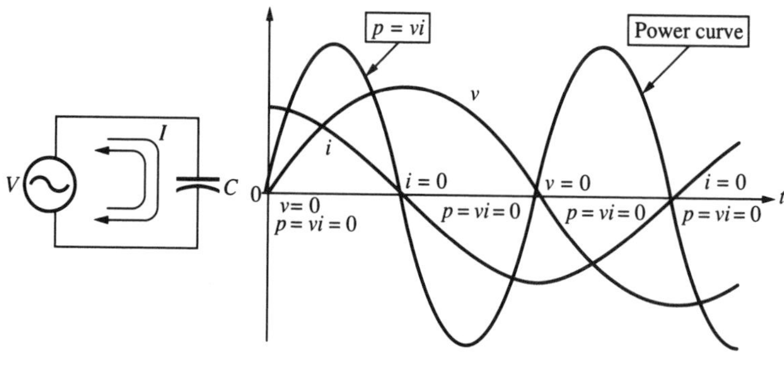

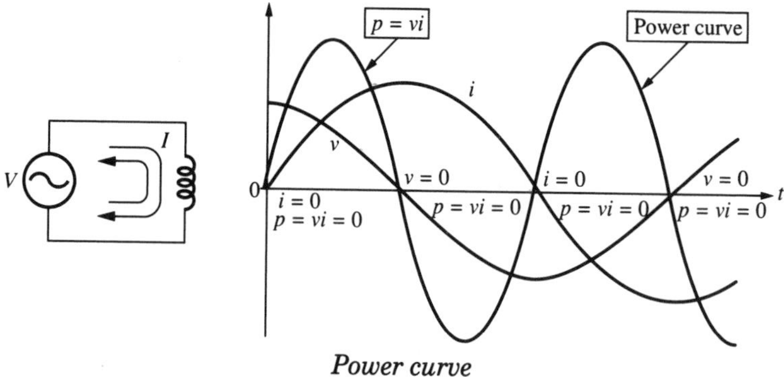

Power curve

bus does not contain any data. See bus.

power cord An insulated connector used to supply electrical energy to electrical systems. A power cord usually has a standard male connector at its terminating end that is used to plug into a power source receptacle.

power curve The resulting curve when the product of the sine curve for current and the sine curve for voltage is taken. The power curve for a purely resistive circuit will always have a positive area indicating that power is absorbed by the load (in the form of heat). The power curve for a purely reactive circuit (such as a pure capacitive circuit) will have equal positive and negative areas showing that no net power is absorbed by the load.

power distribution A method of allowing the transmission of electrical power to one or more

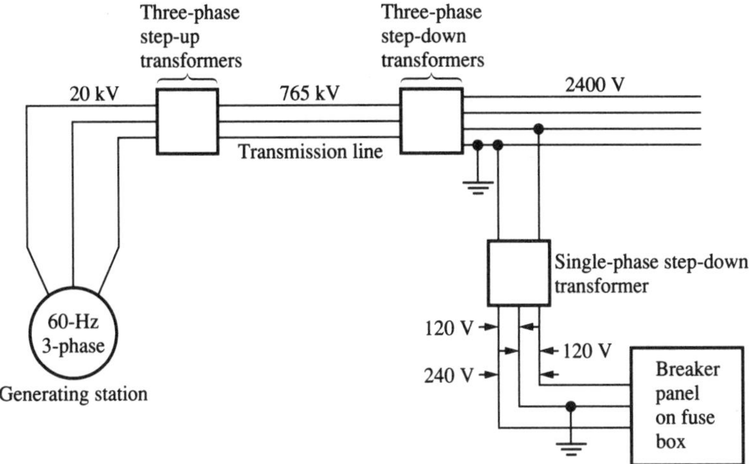

Power distribution

systems that require the power for their operation.

power factor Indicates how much power is actually being delivered to the load in an ac circuit. Mathematically, the power factor is PF = cos θ, where θ is the phase angle of the circuit current and voltage. When these are in phase, θ = 0 and the power factor becomes 1, meaning that the product of IE is the actual power.

power gain A ratio of the output power of a device to its input power. Expressed as $P_{gain} = P_{out}/P_{in}$, where P_{gain} (no units) is the power gain, P_{out} is the output power (in watts), and P_{in} is the input power (in watts). Note that if the output power is less than the input power (as in the case with resistive networks), the power gain will be less than one.

power line isolation A method of isolating the power source from electrical equipment through the use of a transformer. Power line isolation helps to reduce the hazards of severe electrical shock.

power output In an amplifier, the amount of power, measured in watts, delivered to a load by a power amplifier or similar circuit.

power rating The value of the maximum power that an electrical device can handle without damage.

powers of 10 Expressing numbers using an exponent with the base 10. The exponent indicates the number of decimal places to the left or right of the decimal place. As an example, the number 42.8 is expressed as $(4 \times 10^1) + (2 \times 10^0) + (8 \times 10^{-1})$.

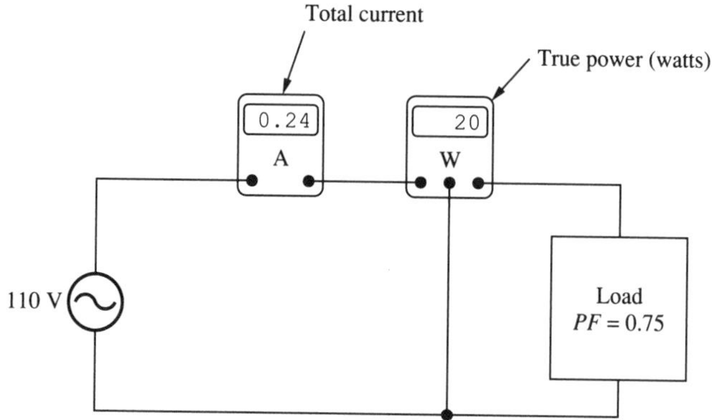

(*a*) A lower power factor means more total current for a given power dissipation (watts). A larger source is required to deliver the watts.

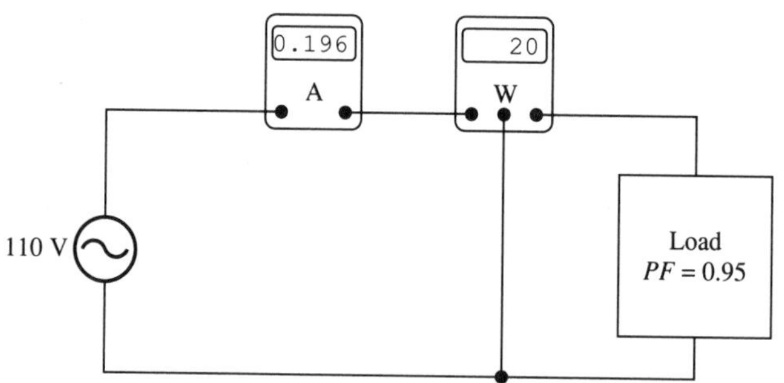

(*b*) A higher power factor means less total current for a given power dissipation. A smaller source can deliver the same true power (watts).

Power factor

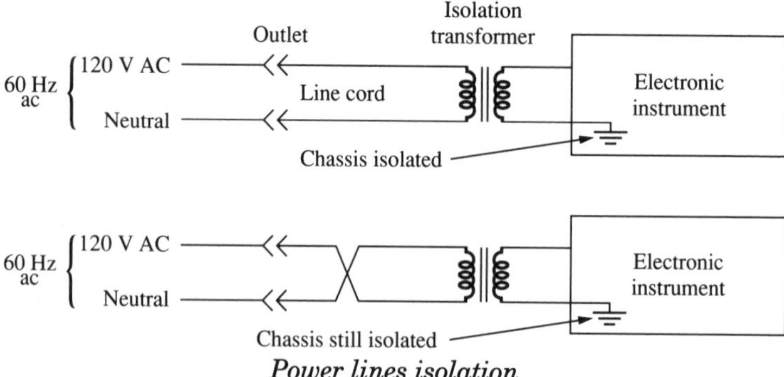

Power lines isolation

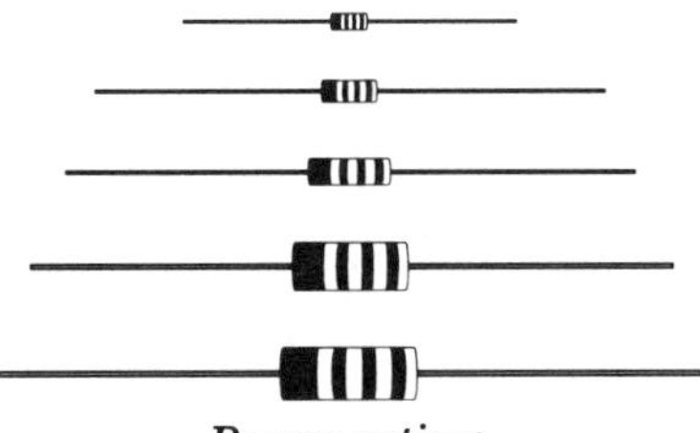
Power rating

powers of 2 The raising of the base 2 to any integer power starting at 0. Powers of 2 are useful in converting between decimal and binary numbers.

power source See power supply.

power supply A circuit that supplies the dc voltages required for the operation of an electronic system. Usually a power supply will convert the 120-VAC line

Count	Decimal Number	Binary Number
zero	0	00000
one	1	00001
two	2	00010
three	3	00011
four	4	00100
five	5	00101
six	6	00110
seven	7	00111
eight	8	01000
nine	9	01001
ten	10	01010
eleven	11	01011
twelve	12	01100
thirteen	13	01101
fourteen	14	01110
fifteen	15	01111
sixteen	16	10000
seventeen	17	10001
eighteen	18	10010
nineteen	19	10011
twenty	20	10100
twenty-one	21	10101
twenty-two	22	10110
twenty-three	23	10111
twenty-four	24	11000
twenty-five	25	11001
twenty-six	26	11010
twenty-seven	27	11011
twenty-eight	28	11100
twenty-nine	29	11101
thirty	30	11110
thirty-one	31	11111

Powers of 2

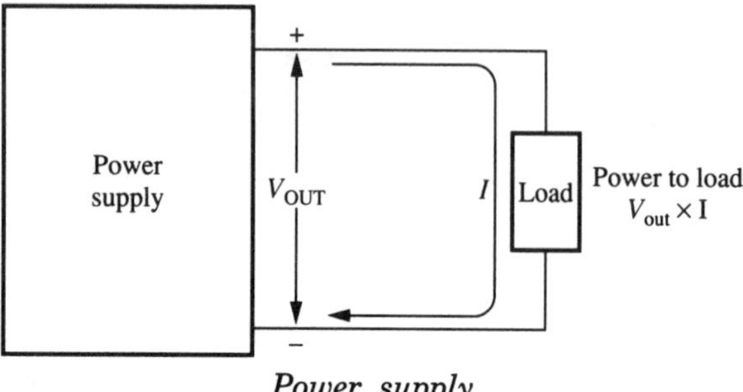

Power supply

voltage to usable dc voltages for a given system. As an example, the power supply of a computer may convert the 120-VAC line voltage to +5 VDC, −5 VDC, +12 VDC, and −12 VDC.

power transformer A transformer used to change the line voltage to a larger or smaller value, where the new voltage value will now be rectified and filtered into dc for use by an electrical circuit or system.

power transistor A transistor designed to handle large amounts of electrical power without damage. A power transistor usually requires mounting to a heat sink to help it dissipate safely the heat generated by the electrical power within the transistor.

p-p Abbreviation for peak to peak. See peak to peak.

PPI Abbreviation for programmable peripheral interface. See programmable peripheral interface.

PPM Abbreviation for pulse position modulation. See pulse position modulation.

precedence of operations The order in which arithmetic or Boolean operations are to be performed. As an example, in arithmetic operations, multiplication is done before addition and subtraction. In Boolean operations, Boolean multiplication (AND) takes precedence over Boolean addition (OR).

preemphasis A method of using an electrical circuit to emphasize some frequency components. Usually used in a transmitter to help reduce the noise characteristics of the signal.

preemphasis network An electrical circuit in FM transmitters to increase the modulating index for higher frequencies resulting in less distortion due to FM noise. See preemphasis.

preshoot A change of amplitude of the opposite polarity that precedes a pulse.

primary voltage Voltage applied to the primary windings of a transformer.

primary winding The input to a transformer. In a power trans-

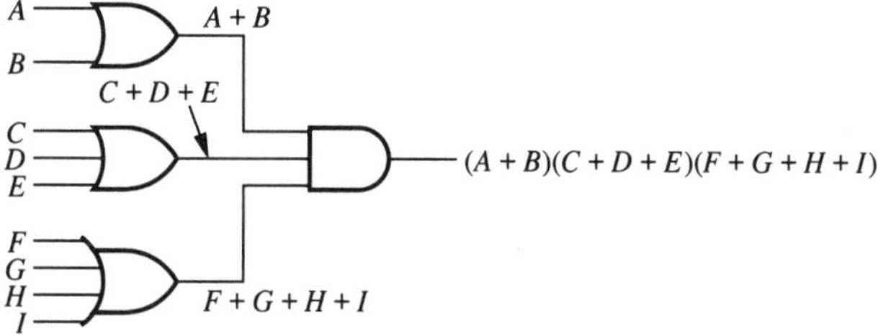

Product of sums

former, the primary winding is the one that receives the electrical energy from the ac power source.

printed circuit Electrical circuit connections made by conductive strips bonded to the surface of an insulated flat surface. A printed circuit board can be made by a photoetching process on a copper-clad board. The board is coated with a light-sensitive material, exposed to light focused through a negative containing the desired pattern. An acid bath is used to remove the unwanted copper leaving behind the desired interconnection pattern.

priority encoder A special type of encoder that senses if two or more inputs are active and then gives an output corresponding to the highest value of the input. See encoder.

probe A device used for sampling some property of an electrical system. The leads on a voltmeter are called test probes because they are used for the purpose of testing (reading) the value of a certain voltage.

procedure A precise, step-by-step process used to solve a problem. In programming it is called an algorithm. See algorithm.

processor That part of a computer that is capable of receiving data and instructions, manipulating the data according to the instructions, and transferring the results for storage and/or observation or further action.

product line In PLD notation, a line that represents multiple inputs to an AND gate.

product of sums 1. A logic connection of two or more OR gates whose outputs are fed into a single AND gate. POS logic for 2 two-input OR gates feeding a two-input AND gate is $X = (A + B)(C + D)$, where X is the output of the single AND gate and A, B, C, and D represent inputs to the OR gates. 2. A Boolean expression consisting of ORed terms that have been ANDed together.

program In computers, a sequence of instructions that invokes standard processes built into the instruction set of the computer. No matter how the program is written by the programmer, it must somehow always be

Simplified example of an FPLA.

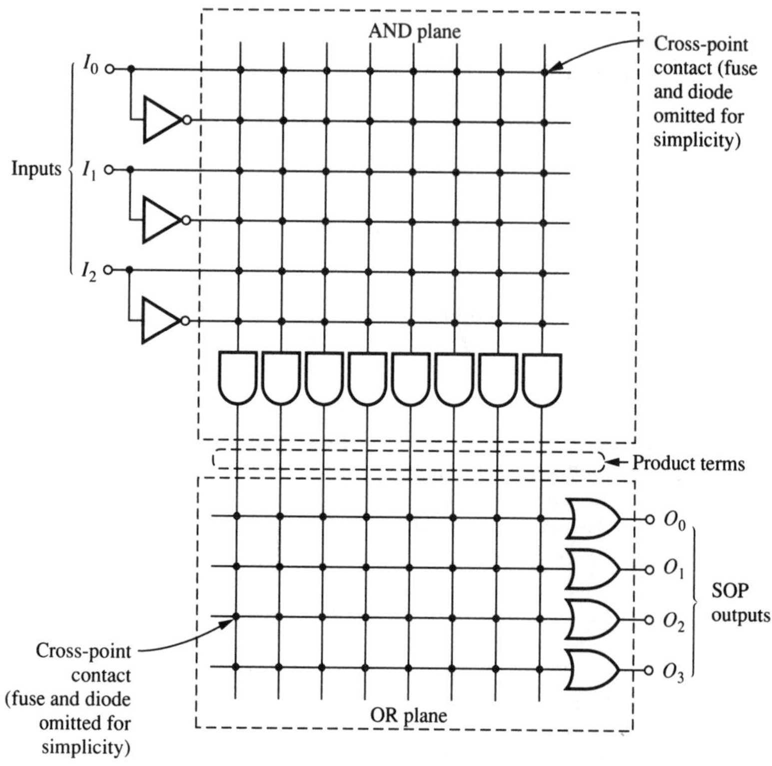

Programmable logic devices

reduced to a bit pattern of 1's and 0's that evoke a predetermined process built into the processor by the manufacturer.

programmable array logic A form of programmable logic devices. It has a programmable AND plane with a fixed OR plane.

programmable controller An electronic device similar to a computer that can be programmed to control predetermined operations that interact with the environment. Programmable controllers are used in automated manufacturing processes.

programmable counter A counter that has its MOD number controlled by a changeable bit pattern.

programmable logic array A two-level AND/OR product of sums logic network, where both

The FPLA programmed for the specified functions.

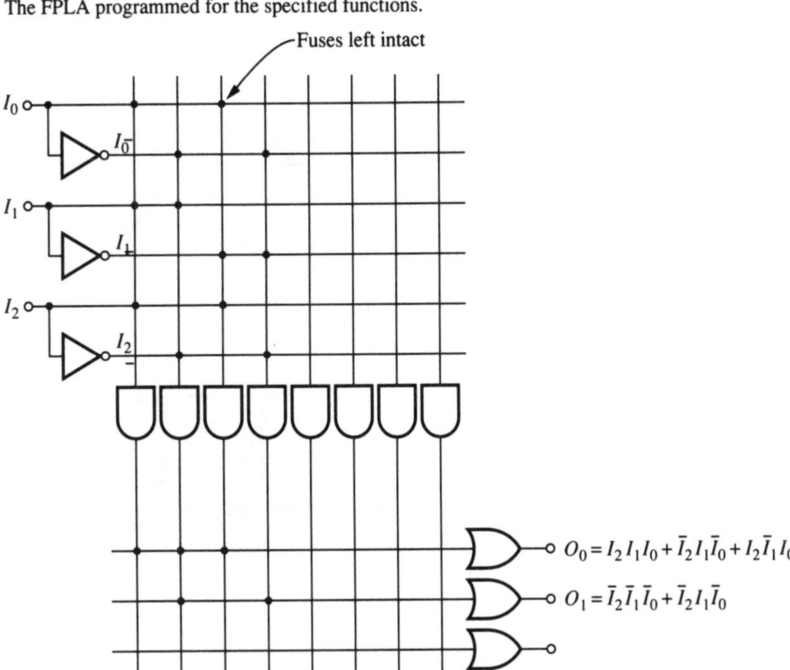

Programmable logic devices (con't.)

the AND connections and the OR connections can be programmed. Used in developing various logic network combinations from a single standard logic chip.

programmable logic device (PLD) Similar to a programmable read-only memory because it is fuse programmable. A PLD may be programmed to simulate Boolean functions of several gates and flip flops.

programmable logic gate A logic network that can replicate any of the fundamental gates (AND, OR, NAND, NOR, XOR) with the setting of a bit pattern that determines the type of gate it will replicate.

programmable logic sequencer Similar to programmable array logic (PAL) except that flip flops are incorporated in the device.

programmable peripheral interface A digital interface device with two or more I/O ports that may be programmed for var-

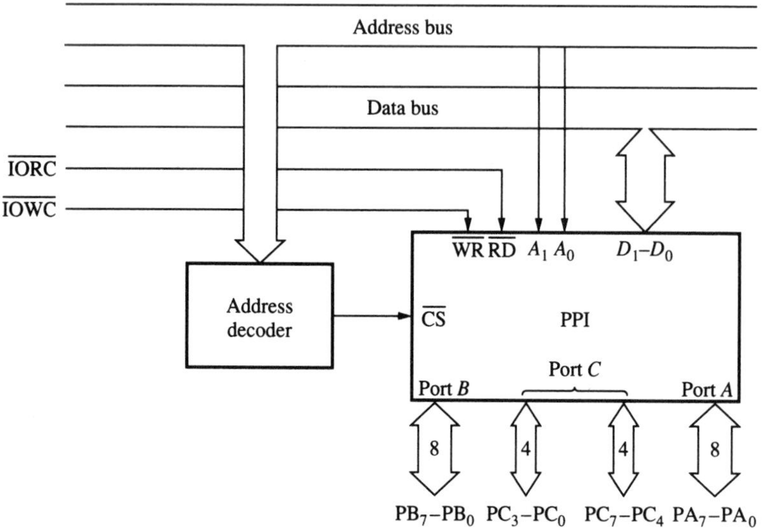

Programmable peripheral interface

ious combinations of input and output ports or bidirectional ports. These devices are commonly used in personal computers.

programmable read-only memory A programmable ROM that can be programmed in the field by the user. Can only be programmed once. This differs from a factory ROM where the programming cannot be changed by the user. See EPROM.

programmable ROM See programmable read-only memory (PROM).

programmable unijunction transistor Abbreviated PUT. A four-layer device like an SCR where the anode gate is brought out to control the device. The triggering voltage of the PUT is programmable and can be controlled by an external circuit.

programming The process of preparing a list of predetermined instructions that will cause specific processes, built-in processes to take place within the computer. If the computer is connected to other devices, such as a monitor, a printer, and so on, then these instructions may also affect these devices if the computer is capable of such processes.

project engineer An engineer who is responsible for the design,

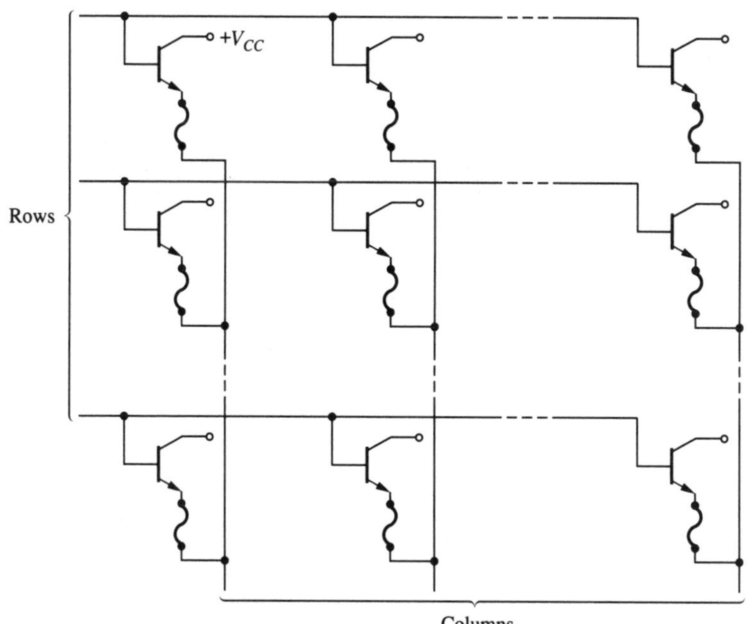

Programmable ROM

operation, and maintenance of a specified set of activities leading to a desired outcome. For example, a project engineer is responsible for the development of a new disk drive system for a computer.

PROM Abbreviation for programmable read-only memory. See programmable read-only memory.

propagation The outward flow of electromagnetic energy into the surrounding space.

propagation delay The amount of time it takes after an input signal has been applied for the resulting output change to occur.

propagation loss The reduction of an electromagnetic signal as it is transmitted from one point in space to another.

propagation time The amount of time it takes for an electromagnetic signal to travel from one point in space to another.

proton The positively charged atomic particle found in the central nucleus. See Bohr model.

prototype Used in the design and development of systems as an intermediate construction to be tested and modified as needed. From what is learned in working with the prototype, the final sys-

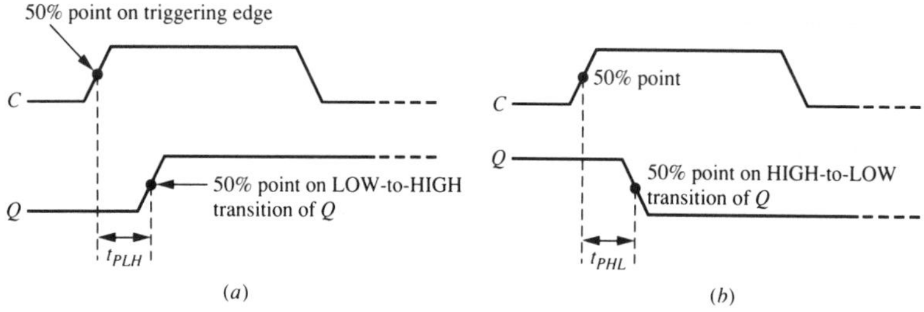

Propagation delay

tem can be constructed to meet required design specifications.

prototype board A plastic board consisting of small holes aligned in columns and rows for the purpose of constructing a temporary circuit. The small holes will accept the wire ends of electrical devices such as resistors, transistors, and dual in-line packages. Small wire clips on the bottom of the board make electrical connections between rows or columns of the mechanical holes.

proximity switch An electronic switch that reacts when an object is close to, not necessarily in contact with, it. Proximity switches are used in alarm systems.

PRT Abbreviation for pulse repetition time. See pulse repetition time.

ps Letter symbol for picosecond. See picosecond.

pseudorandom An apparent output of random numbers created by an algorithm; thus, the resulting numbers are not truly random.

PSpice A commercially developed computer program for the purpose of analyzing electric and electronic circuits. Essentially, the software enables you to create a computer breadboard of a circuit for testing and refinement without having to build with the hardware.

PTM Abbreviation for pulse time modulation. See pulse duration modulation.

p-type material Semiconductor material, such as germanium or silicon, that has had impurities added to cause a deficiency of electrons. Such a deficiency of electrons is said to contain holes. Holes are thought of as positive current carriers in order to help explain the operation of semiconductor devices.

public address system A system consisting of methods for producing sound for a large audience. A PA system usually consists of one or more micro-

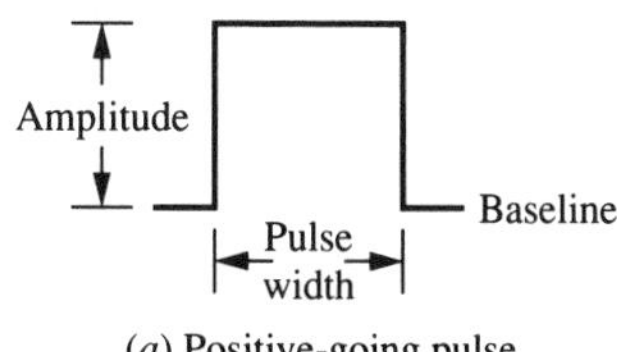

(*a*) Positive-going pulse

(*b*) Negative-going pulse

Pulse

phones, audio-frequency amplifiers, and loudspeakers.

pull-down resistor A resistor connected from the output of an electrical circuit to ground or a negative voltage. Pull-down resistors are used to reference the output of a logic circuit to 0 volts. Can also be used to lower the output impedance of a circuit.

pull-up resistor A resistor used to limit the current in a digital circuit. An example is a resistor used with a switch so that the switch may supply 0 volts or +5 volts to an output. The pull-up resistor is usually connected between the +5 volt line and the switch output. Commonly found in digital circuits. A pull-up resistor is used in logic circuits to hold the output voltage equal to or greater than the input transition level of the device. See open collector logic for another application.

pulsating dc Electrical currents that flow in one direction in an ON and OFF pattern. The resulting wave when ac is rectified.

pulse A momentary change in an electrical characteristic of a circuit. Usually a momentary change in voltage. In most digital circuits, a pulse consists of a momentary transition between 0 volts and +5 volts.

pulse amplifier A wideband amplifier capable of amplifying a square wave with a minimum of distortion.

pulse amplitude modulation A method of transmitting information where the amplitude changes of pulses (usually zero to some positive or negative value) represent the amplitude of the information.

pulse code modulation A method of transmitting information where a code, exemplified by a sequence of pulses (usually zero to some positive or negative value), represents the amplitude of the information.

pulse duration The amount of time as measured between the two 50% amplitude points of a pulse.

pulse duration modulation A method of transmitting information where the changes in width of pulses represent the amplitude of the information.

pulse duty factor The ratio of the amount of pulse on time to the pulse off time.

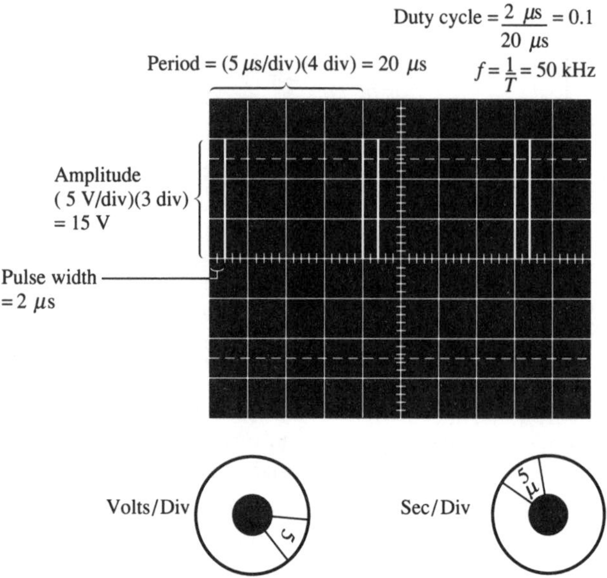

Measurement of the pulse output for its maximum amplitude and maximum frequency

Pulse measurement

pulse fall time The amount of time it takes for a pulse to go from 90% of its amplitude to 10% of its amplitude.

pulse generator An electrical device for developing a series of controlled pulses.

pulse interval See pulse spacing.

pulse length See pulse duration.

pulse length modulation See pulse duration modulation.

pulse measurement The process of measuring any of the electrical parameters of a pulse waveform. Usually performed through the use of the oscilloscope.

pulse modulation Using a series of pulses where a characteristic of the pulses is changed in order to convey information. Forms of pulse modulation are PPM, PDM, PAM, and PCM.

pulse position modulation A method of transmitting information where the relative positions of pulses (usually zero to some positive or negative value) represent the amplitude of the information.

pulse repetition frequency The rate in hertz at which pulses are transmitted. The term is usually used in radar applications.

pulse repetition time The amount of time between pulses. A term used in radar transmitters

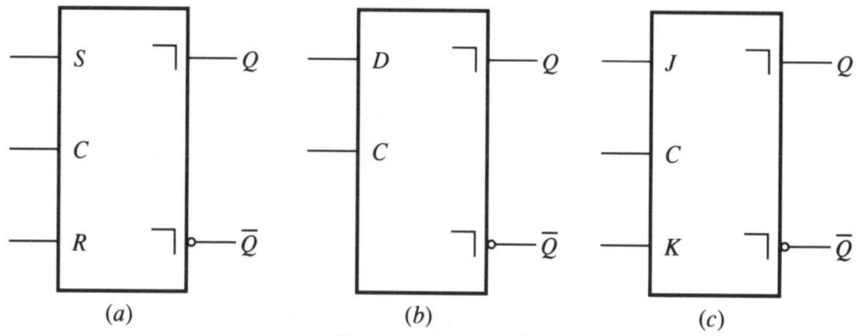

Pulse triggered

where the pulse repetition time is the amount of time between transmitted pulses.

pulse rise time The amount of time it takes a pulse to go from 10% of its amplitude to 90% of its amplitude.

pulse spacing The amount of time between pulses, measured in seconds.

pulse stretcher An electrical circuit designed to extend the duration of a pulse.

pulse train A succession of pulses usually at equal time intervals.

pulse triggered A pulse-triggered flip-flop means that data are entered into the device on the leading edge of the clock pulse. However, the output of the flip-flop does not reflect this input state until the trailing edge of the same clock pulse.

pulse-triggered master-slave flip-flop A flip-flop whose inputs are active at one level of the clock and whose outputs will reflect the input condition only at a different level of the clock. Can be thought of as containing two separate flip-flops, where the input flip-flop controls the output flip-flop. The input flip-flop is called the master and the output flip-flop the slave.

pulse width The amount of time between the 50% rise and fall maximums of a pulse.

push button normally closed A simple spring-loaded SPST switch that is kept in the closed position by its spring. Usually constructed in the shape of a small movable push button that must have mechanical pressure applied to it to open the switch.

push button normally open A simple spring-loaded SPST switch that is kept by its spring normally opened. Usually constructed in the shape of a small movable push button that must have mechanical pressure applied to it to close the switch.

push-pull Refers to an electrical circuit consisting of two separate but similar circuits that each work on a different half cycle of an input signal.

push-pull amplifier Two amplifier stages that operate on different half cycles of an input signal. Each amplifier operates on a 180-degree phase difference of the input signal.

PUT Abbreviation for programmable unijunction transistor. See programmable unijunction transistor.

Pythagorean theorem In a right triangle, the relationship of the longest side of the triangle (the hypotenuse) to the other two sides as given by $h^2 = a^2 + b^2$, where h is the hypotenuse and a and b are the other two sides.

Setting the pulse width of a 74121.

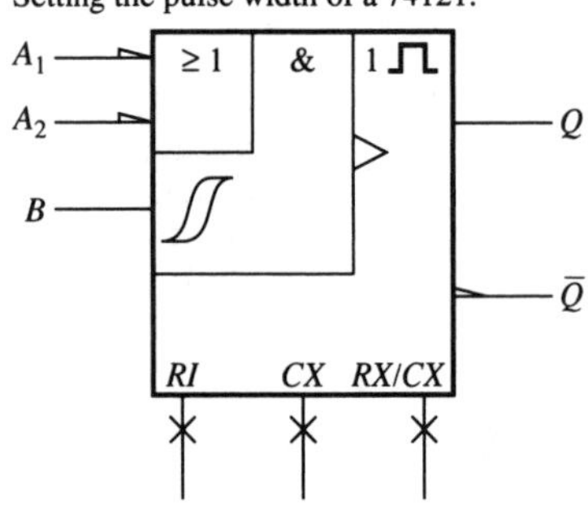

(*a*) No external components ($t_w \cong 30$ ns)

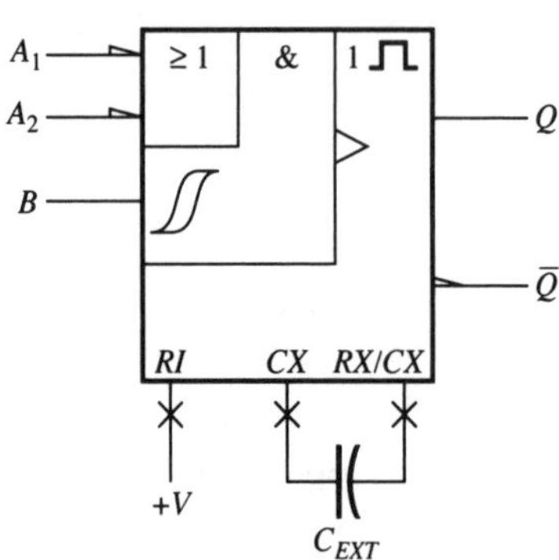

(*b*) Internal *R* and C_{EXT}

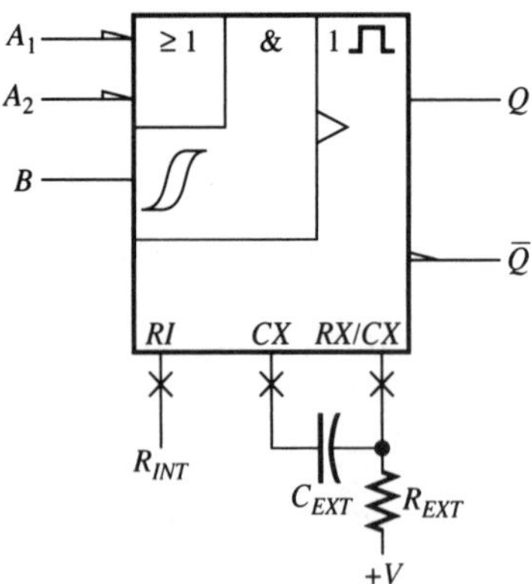

(*c*) R_{EXT} and C_{EXT}

Pulse width

q

Q 1. Letter symbol for electrical charge. 2. Quality of an electrical circuit as measured by the ratio of the reactance to the resistance. For example, the Q of an inductor is X_L/R_L, where X_L is the inductive reactance of the inductor and R_L is the dc resistance of the inductor.

***Q* factor** A measure of the quality of a tuned circuit. The higher the Q, the narrower the bandwidth. $Q = BW/f_r$, where Q is the Q of the tuned circuit (no units), BW is the bandwidth as measured at the 3-dB points (in hertz) and f_r is the resonant frequency of the circuit (in hertz). Q may also be expressed as X_L/R and serve as a measure of the quality of an inductor at a given frequency. $X_L = 2fL$, where X_L is the inductive reactance (in ohms), f is the applied frequency (in hertz), and L the value of the inductor (in henrys).

***Q*-meter** An instrument for measuring the Q of a tuned circuit.

***Q*-multiplier** A filter that has a steep response curve, either bandpass or band-reject.

***Q* output** The reference used to name the output of a flip-flop. In flip-flops with complementary outputs, one output is called the Q output and the other the $\overline{Q}$ (NOT Q) output.

***Q* point** The dc operating point of an amplifier. The Q point is established by external circuit components and characterizes the class of operation of the amplifier.

quad Four. In electronics some meaningful combination of four elements such as an integrated circuit consisting of four NAND gates would be a quad NAND gate.

quad latch A group of four flip-flops usually in a single IC package.

quadrant 1. In the Cartesian coordinate system, the X- and Y-axis divide the plane into four separate areas called quadrants. In a like manner, the complex plane is also divided into four areas called quadrants. The quadrants are numbered in a counterclockwise rotation where the first quadrant lies between

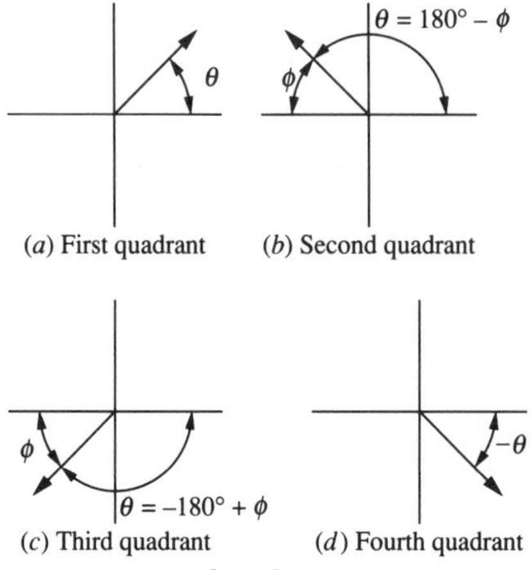

Quadrant

the right of the vertical axis and above the horizontal axis. The first quadrant has both X and Y positive, the second quadrant has $-X$ and $+Y$, the third $-X$ and $-Y$, and the fourth $+X$ and $-Y$.

quadraphonic Applying to the generation of sound from four distinct channels or the simulation of such sound generation requiring four distinct speaker systems.

quadrasonic The arrangement of four loudspeakers to give the listener four distinct points of sound.

quadrature The condition of two periodic functions separated by a quarter of a cycle. Phase relationship state of 90 degrees.

quadriphonic See quadrasonic. Also spelled quadraphonic.

quad simplification In Karnaugh mapping, a method of simplifying a Boolean expression represented by the Karnaugh map by treating a group of four vertically or horizontally adjacent cells as a single Boolean term.

quality control The activities required to produce a uniform quality product. Includes the control in workmanship and materials as well as testing standards.

quality factor See Q factor.

quantity A term of some numerical value.

quantization The process of digitizing an analog waveform by breaking up the amplitudes of the wave into discrete digital changes. Similar to the process produced by an analog to digital converter. See A-to-D converter.

quantization error Produced by quantization of an analog wave. It is the difference between the actual value of the analog wave and the resulting digitized values.

quantize 1. To convert a continuously changing variable, such as a waveform, into discrete steps of change, where there are no "in-between" values. The process of performing such a conversion. See quantization. 2. To break into small units.

quantizing noise Noise that represents an error or distortion introduced by PCM when the modulating signal is not an exact value of the resulting code. Quantizing noise occurs when the modulating wave lies somewhere between two quantizing points in PCM.

quantum A discrete amount of energy. The concept that energy is produced in discrete amounts rather than a continuous change.

quarter-wave antenna Any transmitting or receiving antenna where its electrical length is one quarter of the wavelength being transmitted or received.

quarter-wave stub A transmission line equal in length to one quarter of the wavelength of the frequency it is designed to transmit. It is shorted at the terminating end with the result that it appears as a parallel resonant circuit to the source at the designed frequency. Because of this, the source sees a maximum voltage and a minimum current and thus appears as an open to the source.

quarter wavelength The distance between two periodic wave intervals that corresponds to 90 degrees.

quartz A mineral (silicon dioxide) occurring in nature that has piezoelectric properties. When properly cut, quartz will vibrate at a prescribed frequency when electrically excited. This action results in extremely stable frequencies.

queue 1. A file or line of elements. A microprocessor will use digital circuits that are called a queue to increase its processing speed. In this case, the queue is used to store or line up instructions from memory for immediate processing. 2. A line of items waiting for servicing in a system where items added at one end are removed at the other end. Used in microprocessor storage of instructions and data taken from memory.

quick-brake fuse A specially made fuse that will open rapidly when its current rating is exceeded. Constructed with a spring, so as to hasten the fuse brake.

quiescent At rest. Circuit condition when no input signal is being applied; power is, however, still applied to the circuit.

quiescent point The point on the characteristic curves of an amplifier when no input signal is being applied.

R Letter symbol for resistance, resistor, or reluctance.

race condition An undesirable condition in a logic network where the differences in propagation times through two or more signal paths can produce an incorrect result.

raceway A physical channel used for holding cables and wires in an electrical system.

radar Comes from the word "radio." Descriptive of a system for determining the location of objects in space where a radio wave is transmitted and reflected

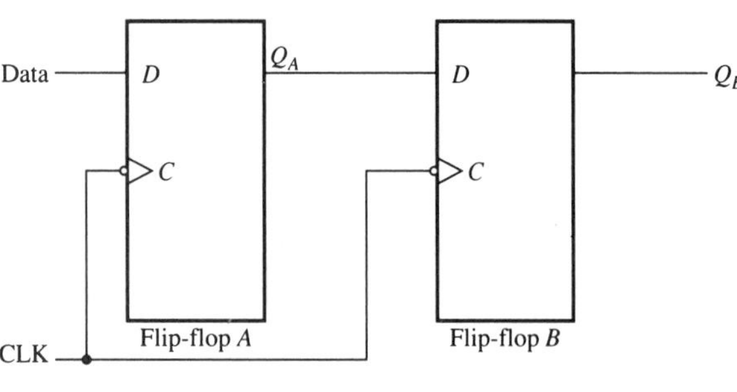

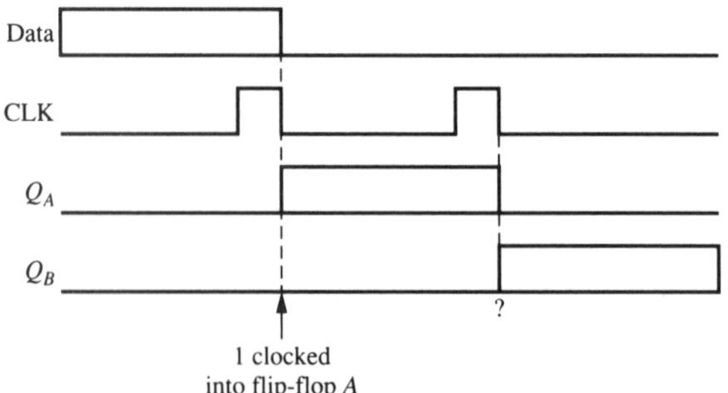

Race condition

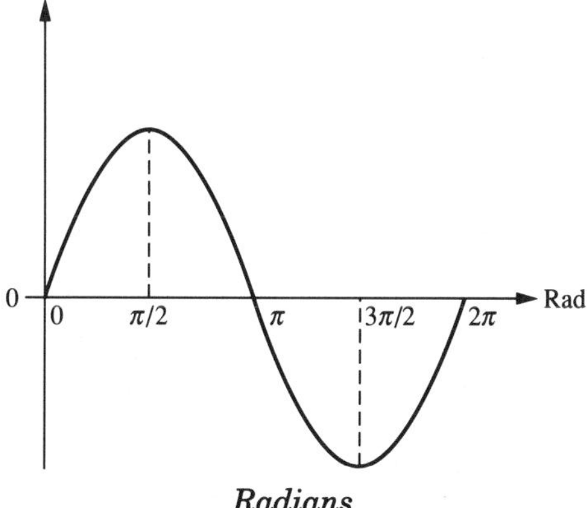

Radians

from the object back to the transmitter. By knowing the amount of time it takes for the complete round trip of the signal, the object's distance can be determined.

radial lead A component lead that extends from its sides rather than from its ends (where it is then called an axial lead).

radian The angle of a circle arc where the distance of the arc is equal to the radius of the circle. One radian = 77.3 degrees. To convert degrees to radians, use radian = $\pi/180 \times$ degrees.

radiance A measurement of the strength of a light source. Used in laser technology. Mathematically expressed as $RA = P/wA_{ap}$, where RA is the radiance (in watts per steradian-square centimeters), w is the solid angle (in steradians) and A_{ap} is the aperture area of laser output (in square centimeters).

radian frequency As used in electronics is the product of the frequency in hertz and 2π. Thus $w = 2\pi f$, where w is the radian frequency.

radiant energy Energy transmission in the form or electromagnetic radiation such as radio, heat, light waves, and X rays.

radiation The emission of electromagnetic waves or subatomic particles from a source or substance.

radiation pattern The shape of the electromagnetic field produced by an antenna supplied with electrical energy. The shape of the pattern is greatly influenced by the shape and wavelength of the antenna as well as the phase relationships of the pattern.

radiation resistance Of an antenna, defined as the ratio of the electrical power converted into electromagnetic radiation and the rms value of the antenna current.

radio A device capable of receiving radio waves, removing the intelligence contained in the waves, and reproducing it in a usable form.

radioastronomy A branch of astronomy that analyzes the non-visible radiation from celestial objects.

radiobiology The study of the effect of high-energy radiation on living matter.

radiocommunications A general term used to describe communications via electromagnetic waves. Includes radio and television.

radio direction finder A device used to detect the presence and origin of a radio transmission.

radio frequencies Frequencies of electrical energy that will produce electromagnetic radiation. Usually from 10 kHz and higher.

radio frequency choke An inductor used in radio frequency circuits that consists of a core of air or powered iron.

radio frequency generator An instrument capable of generating radio frequencies. It is used in the maintenance, troubleshooting and alignment of communications equipment.

radio frequency oscillator An electrical circuit that generates changing electromagnetic energy at frequencies normally used in some form of electronic communications.

radio frequency transformer A transformer used to handle radio frequencies. Usually used to help impedance match one RF stage to another.

radio horizon The point where direct transmission of a radio signal is no longer possible along the surface of the earth. Usually affected by atmospheric conditions.

radio jamming The act of interfering with a radio transmission by overpowering it with another transmission.

radiology The branch of science that deals with the treatment and diagnosis of diseases through the use of radiant energy.

radiometric measurements Measurements made within the full spectrum of the electromagnetic energy with the optical spectrum. See optical spectrum.

radio signal See radio wave.

radio wave Descriptive of electromagnetic radiation used to communicate between a transmitter and a receiver.

radix Also referred to as the base of a number system. The radix represents the total number of unique symbols used in the number system to represent value. For example, the radix of the decimal number system is 10 because 10 unique symbols are used: 0, 1, 2, 3, 4, 5, 6, 7, 8, and 9.

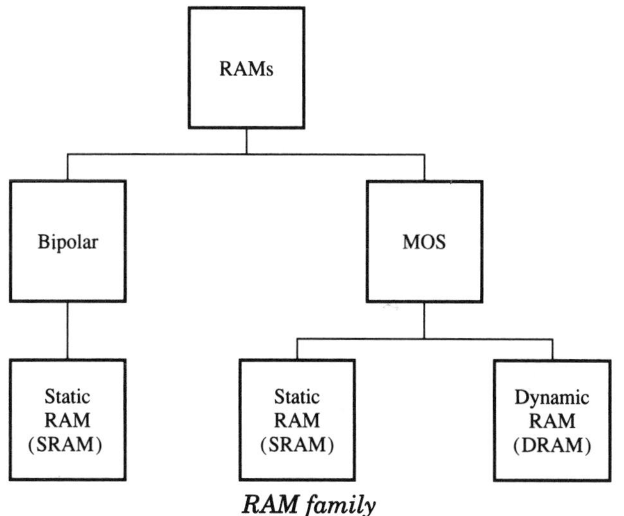

RAM family

RAM Literally means randomly accessible memory but is used to mean volatile, randomly accessible, electrical memory. See read-write memory.

RAM family The grouping of integrated circuit technologies for random access memory chips (RAM chips).

ramp A linear increase or decrease in voltage or current.

random Not having any set or predictable pattern.

random access In digital systems, retrieving and storing data in a parallel fashion. In this system, the amount of time it takes to access data does not depend upon its location in memory. See sequential access.

random failure Any system or component failure that is unpre-

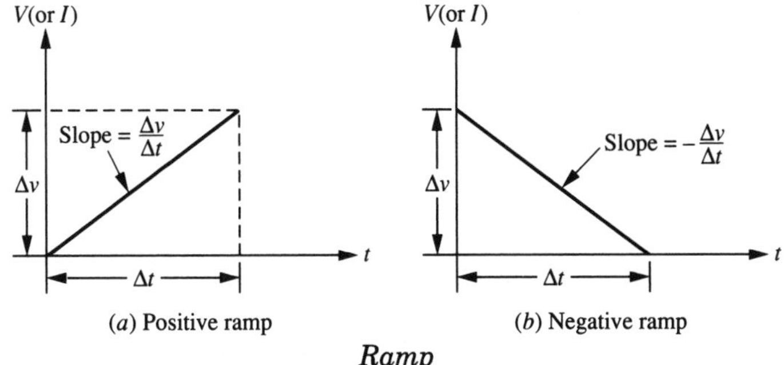

(*a*) Positive ramp (*b*) Negative ramp

Ramp

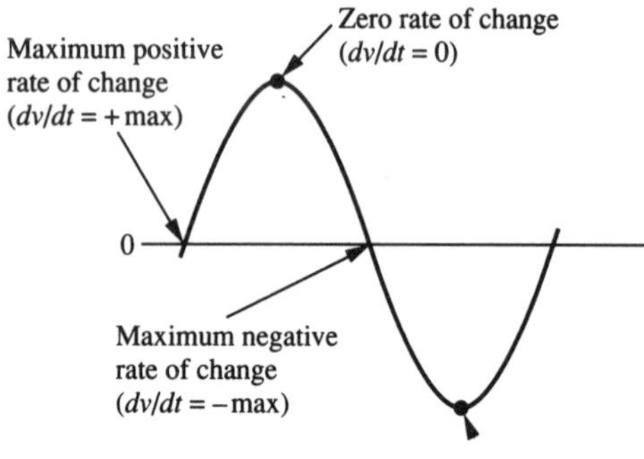

Rate of change

dictable in terms of its time of occurrence.

random noise generator An electrical instrument capable of producing a succession of random signals over a wide range of frequencies.

random sample A method of selecting items where the item selected is completely unpredictable.

rated voltage The voltage at which a device is designed to operate under normal conditions.

rate of change The amount of change that takes place at a particular point on a curve. The rate of change of a sine wave at its maximum value is zero.

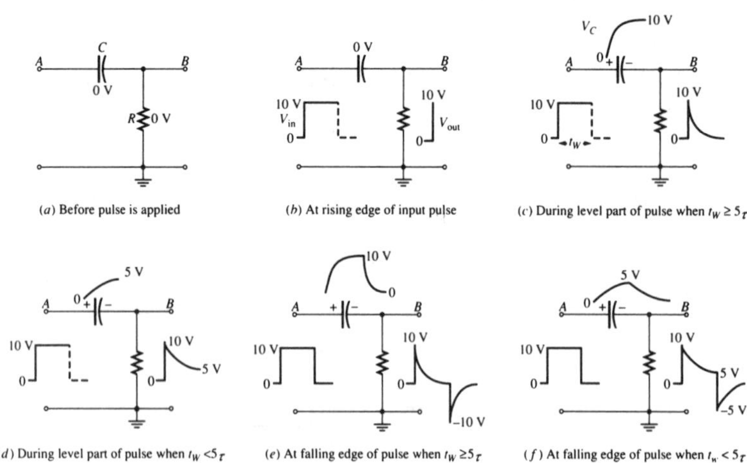

(a) Before pulse is applied
(b) At rising edge of input pulse
(c) During level part of pulse when $t_W \geq 5\tau$
(d) During level part of pulse when $t_W < 5\tau$
(e) At falling edge of pulse when $t_W \geq 5\tau$
(f) At falling edge of pulse when $t_W < 5\tau$

RC differentiator pulse response

ratio The value obtained as a result of dividing one number by another.

ratio detector An electrical circuit used in FM receivers to detect the audio signal from the carrier frequency change. A ratio detector is insensitive to amplitude variations of the signal and is thus good at noise reduction.

ray A straight line representing the direction of travel of an electromagnetic wave such as light.

RC Abbreviation for resistance-capacitance network. See *RC* network.

***RC* constant** The time constant of a resistor-capacitor combination that is equal to the product of the two values the result of which is measured in seconds.

***RC* coupling** Abbreviation for resistance-capacitance coupling. See resistance-capacitance coupling.

***RC* differentiator** An *RC* circuit that approximates the mathematical function of differentiation. A simple *RC* differentiator consists of a series connection of a resistor and a capacitor with the output taken across the resistor.

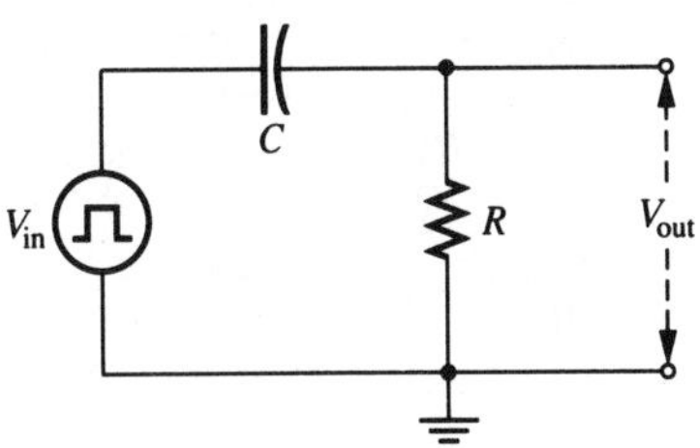

RC differentiator

***RC* differentiator pulse response** The effects on a pulse waveform due to the electrical actions of an *RC* differentiator.

***RC* differentiator time constant** The time constants involved in an *RC* differentiator circuit. Equal to the value of the resistor and the capacitor.

***RC* filter** A high-pass filter network consisting of a resistor and a capacitor.

***RC* integrator** An *RC* circuit is a circuit that approximates the mathematical function of integration. A simple *RC* integrator consists of a series connection of a resistor and a capacitor with the output taken across the capacitor.

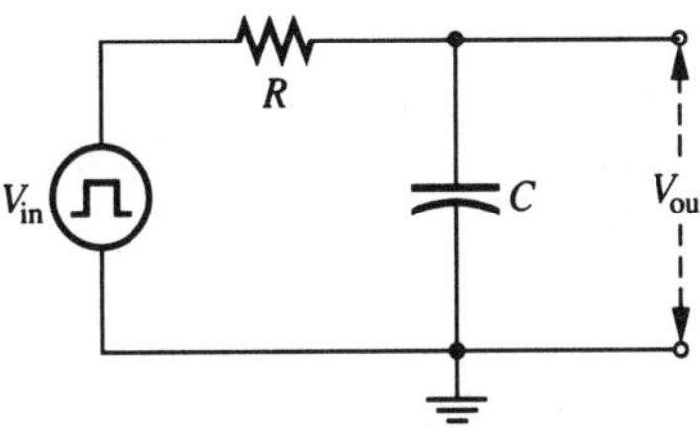

RC integrator

***RC* integrator pulse response** The effects on a pulse waveform due to the electrical actions of an *RC* integrator. (See pg. 275 for illustration.)

***RC* integrator time constant** The time constants involved in an *RC* integrator circuit. Equal to the value of the resistor and the capacitor.

***RC* lag network** A series *RC* circuit where the output voltage

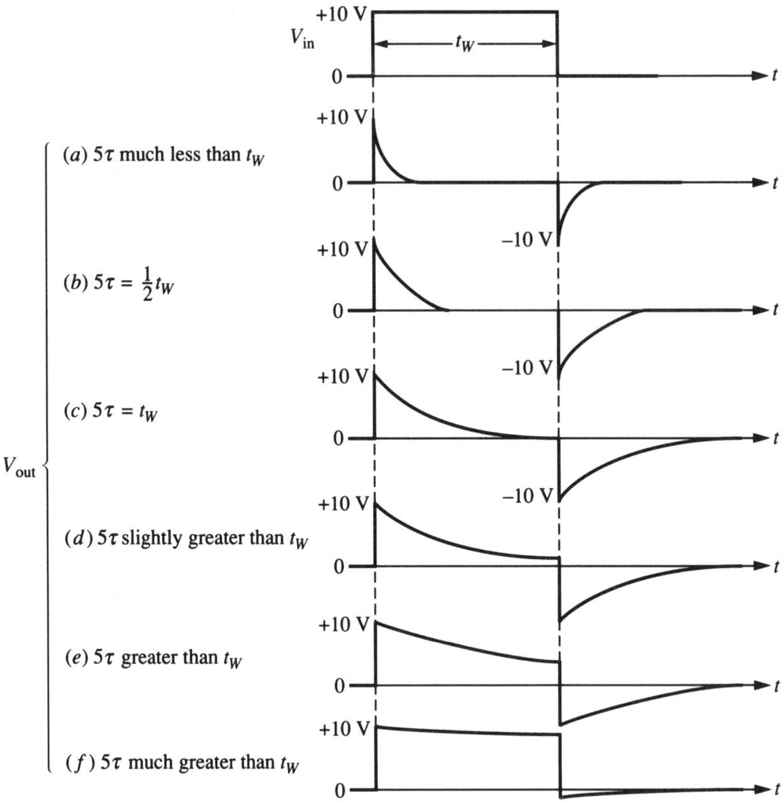

RC differentiator time constant

is taken across the capacitor. (See pg. 275 for illustration.) Such a network will cause the output voltage to lag the input voltage by an amount determined by the source frequency and values of the resistor and capacitor.

***RC* lead network** A series *RC* circuit where the output voltage is taken across the resistor. Such a network will cause the output voltage to lead the input voltage by an amount determined by the source frequency and values of the resistor and capacitor.

***RC* network** An electrical circuit consisting of resistors and capacitors. *RC* networks respond to frequency changes and are usually used as high-pass or low-pass filters. *RC* networks are also found in timing circuits where the values and value ratios of the capacitors and resistors deter-

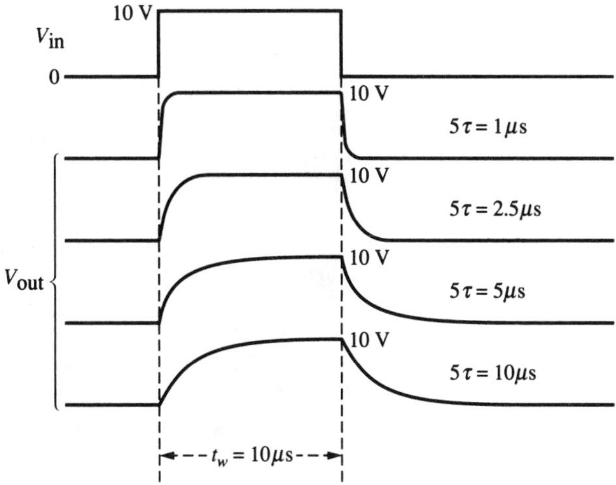

RC integrator pulse response

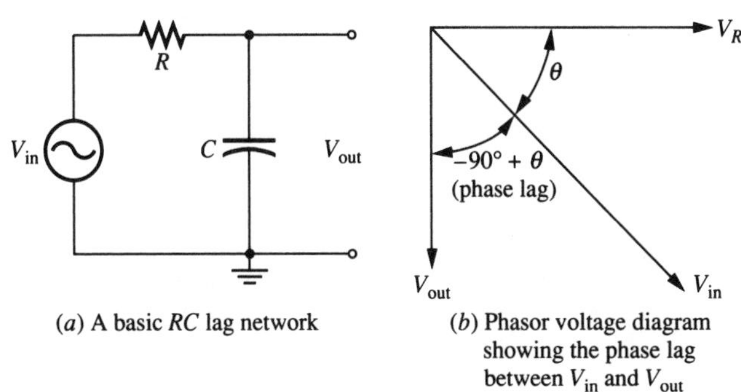

(*a*) A basic *RC* lag network

(*b*) Phasor voltage diagram showing the phase lag between V_{in} and V_{out}

RC lag network

mine the frequency and duty cycle of the resultant timing waveform.

***RC* oscillator** An electrical circuit capable of generating its own signal where the resulting frequency is determined by the values of the resistors and capacitors used in the circuit. If the resulting signal is a square wave, the ratios of the resistors to the capacitors will determine the duty cycle.

RDF Abbreviation for radio direction finder. See radio direction finder.

reactance Measurement of the opposition to alternating current for an inductor or a capacitor.

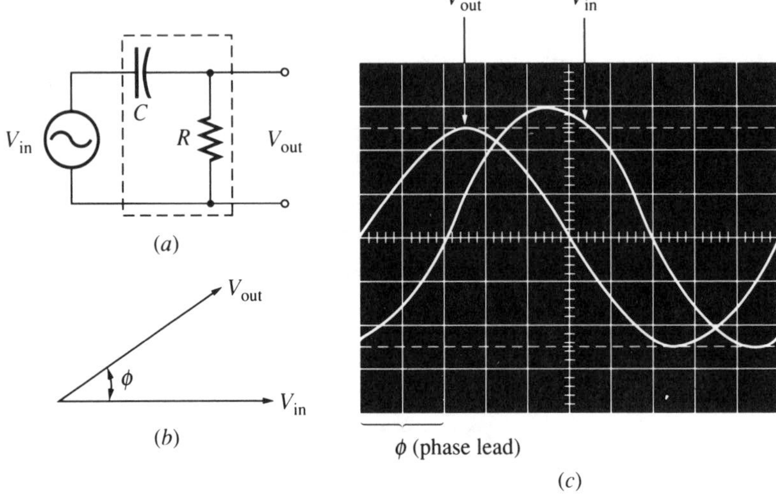

RC lead network

Symbol (X) and measured in ohms. Capacitive reactance (X_c) is equal to $1/(2\pi fC)$ and inductive reactance (X_l) is equal to $2\pi fL$, where f is the frequency (in hertz), C the capacitance (in farads), and L the inductance (in henrys).

reactive The opposition to sinusoidal current by a capacitor or an inductor.

reactive power The product of the reactive voltage times the reactive current. Sometimes referred to as wattless power. Unit of measurement is the var (standing for volts-amps reactive).

read In microprocessor terminology the process of copying data from a memory location into the microprocessor. See write.

read-only memory Memory that contains a preset bit pattern that cannot easily be changed without destroying the medium containing the memory.

read operation 1. The process of a microprocessor getting information from memory. 2. The activation of a read function. See read.

read-write cycle In a computer, the operation of reading and writing data to and from the microprocessor. The timing diagram of a read-write cycle shows the relationships among various control, data, and address lines.

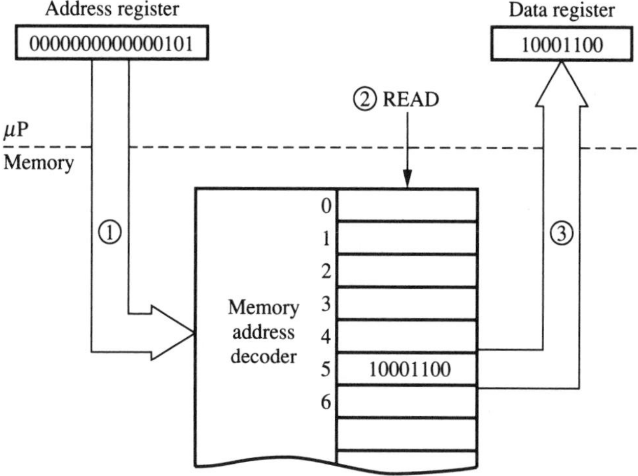

① Address 5_{10} placed on address bus.

② READ signal applied.

③ Contents of address 5_{10} in memory placed on data bus and stored by data register.

Read operation

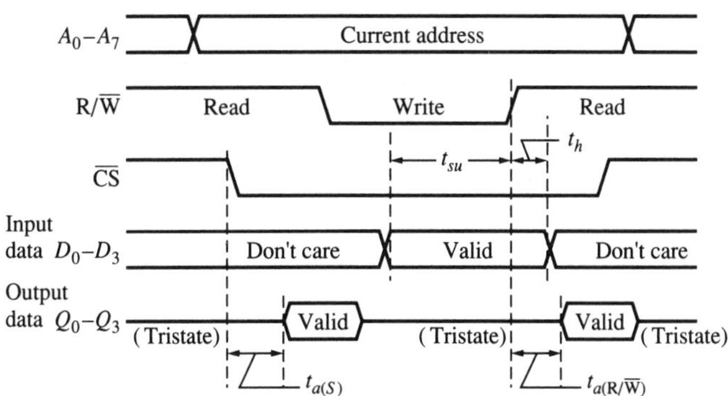

$t_{a(S)}$: Access time, $\overline{\text{CS}}$ to Data out
t_{su}: Setup time
$t_{a(R/\overline{W})}$: Access time, R/$\overline{\text{W}}$ to Data out
t_h: Hold time

Read-write cycle

read-write head A mechanical device capable of taking information from a magnetic storage medium (such as a disk or tape) and converting it to electrical signals understood by a computer and doing the opposite; that is, taking electrical signals representing computer information and converting them to magnetic patterns stored on a magnetic material (such as a disk or tape).

read-write memory Digital memory that can easily have its bit pattern changed. See RAM.

real number In complex number notation, real numbers are represented by the horizontal axis of the complex plane. A real number is any positive or negative number. See imaginary number.

real power The actual power delivered by the ac source to a reactive load. Real power represents the actual power consumed by the load and is measured in watts. Real power can be found

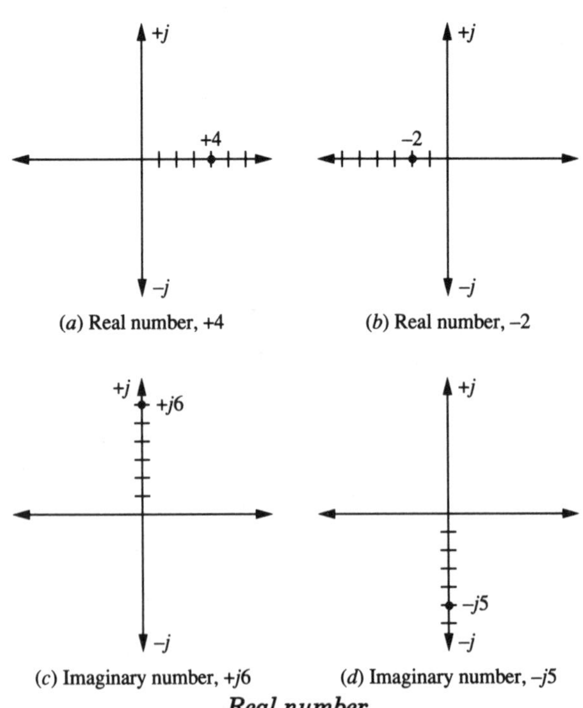

(*a*) Real number, +4
(*b*) Real number, −2
(*c*) Imaginary number, +*j*6
(*d*) Imaginary number, −*j*5

Real number

by the product of the apparent power and the power factor.

real-time clock A clock that represents the actual passage of time. Usually used in a computer to measure real time between selected events.

receiver In electronic communications, the device used to capture a radio wave, remove the intelligence, and reproduce it in a usable form. That part of a communications system that converts transmitted electrical energy into an understandable and/or useful form. Examples of receivers are an FM radio or television set and the headset of a telephone.

reciprocal The reciprocal of a number is that number divided into 1. The reciprocal of a number can be taken by exchanging the numerator and the denominator. Thus, the reciprocal of 2/3 is 3/2.

reciprocity theorem The concept that for single-source networks, the current in any branch will equal the current through the branch in which the source was originally located if the source is placed in the branch in which the current was originally measured.

recorder Any electronic instrument that makes a permanent record of electrical events that represent some form of information.

recording format The method used to store data magnetically.

record/playback head The electromechanical element used to place (record) a magnetic pattern on a recording surface and to read (play back) the same information at a later time.

recovery The time required for an electronic device to ready itself to receive or process another signal.

rectangular form In representing complex numbers, a representation of a point on the complex plane by stating the magnitude of the real part (the horizontal axis) and the magnitude of the imaginary part (the vertical axis). For example, the expression $12 + 4j$ is the rectangular form of a point of a complex plane that is 12 units in the horizontal (real) direction and 4 units in the vertical (imaginary) direction.

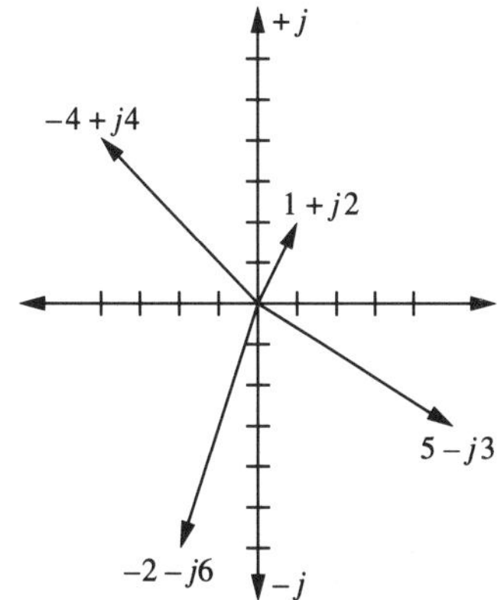

Rectangular form

rectangular waveguide A hollow enclosure with four perpendicular sides for the purpose

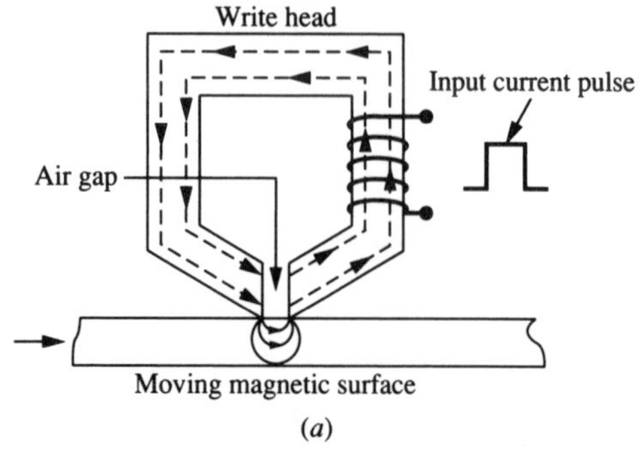

(*a*)

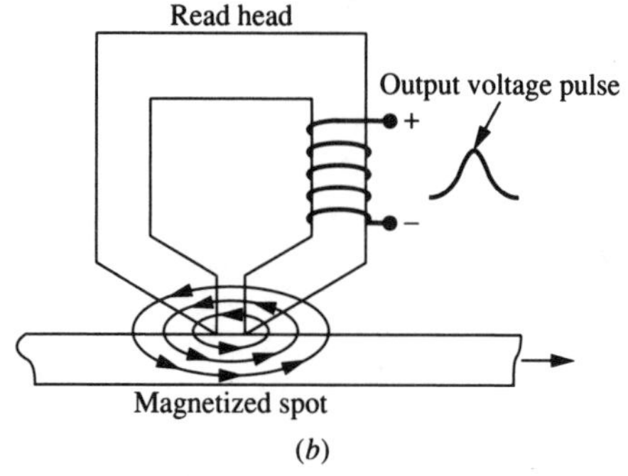

(*b*)

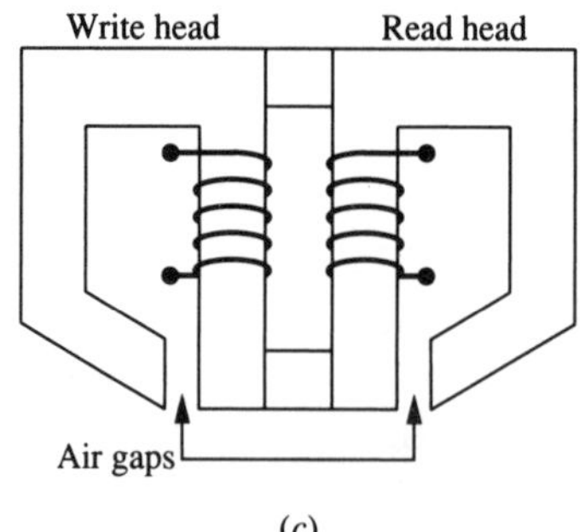

(*c*)

Record/playback head

of transmitting microwave energies.

rectification The process of taking an ac wave and converting it into pulsating dc. An ac wave has current flow in first one direction and then the other. Rectification causes the current to flow in just one direction or not at all, resulting in a pulsating dc.

rectifier A circuit or device that forces current to flow in only

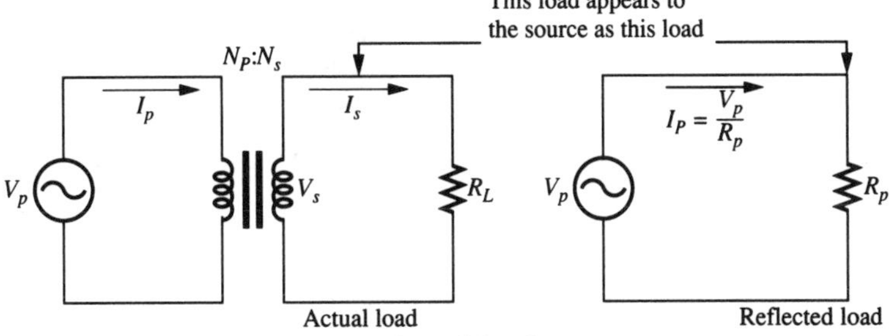

Reflected load

one direction. An electronic circuit that converts ac into pulsating dc.

rectilinear Being bounded by or moving in a straight line.

red gun In a three-color television tube, the electron gun that strikes the red phosphor on the screen.

redundant group In Karnaugh mapping, a group of cells that are already used by other groups used in a logic simplification process.

reed relay An on/off electrical device consisting of two flat metal conductors sealed in an airtight container and surrounded by a coil of wire. When current is passed through the wire, the resulting magnetic field affects one or both of the flat metal conductors, causing them to open if they are normally closed or close if they are normally open. Thus, if a permanent magnet is brought next to them, the metal contacts will open if they were normally closed or close if they were normally open.

reflected load The actual load as seen by the source looking into the primary of the transformer. The reflected load of a transformer is equal to the square of the ratio of the primary to the secondary times the value of the secondary load.

reflected power In transmission lines, the ratio of the reflected power to the incident power. A measure of how much the transmission line is unmatched. See unmatched transmission lines.

reflected ray In the study of light, the ray of light reflected from a surface as a result of an incident ray. The angle at which a reflected ray leaves a flat surface is equal to the angle of the incident ray. See incident ray.

reflection coefficient The ratio of the voltage of the reflected wave to the voltage of the incident wave in an unmatched transmission line. A measure of how much the transmission line is not matched. See unmatched transmission line.

refraction The bending of electromagnetic energy (usually light) as it passes from one me-

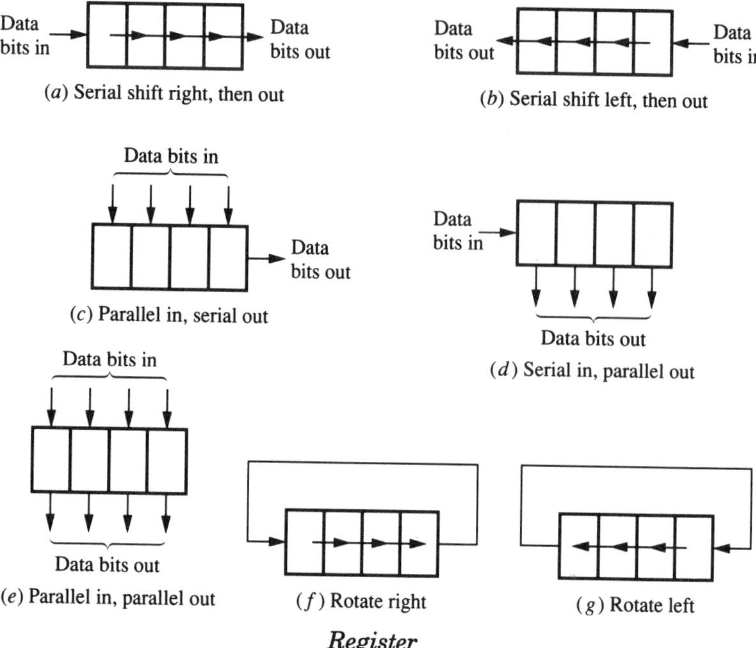

(*a*) Serial shift right, then out
(*b*) Serial shift left, then out
(*c*) Parallel in, serial out
(*d*) Serial in, parallel out
(*e*) Parallel in, parallel out
(*f*) Rotate right
(*g*) Rotate left

Register

dium to another. Refraction is caused by the energy having different velocities in different media.

refresh The process of maintaining the electrical cell charge in a dynamic RAM.

regeneration The process of taking part of an output signal and feeding it back to the input. Also called positive feedback. An oscillator uses regeneration to sustain its oscillations.

register A group of flip-flops treated as a unit. May be used as a storage place for a binary word. How a bit pattern is loaded into a register and copied from a register determines the kind of register. Parallel in registers have their bit patterns loaded all at one time. Parallel out registers have their bit patterns copied all at one time. A serial in register has its bit patterns placed in it one bit at a time. A serial out register has its bit pattern copied out one bit at a time.

register architecture As pertaining to a microprocessor, a diagram of the internal registers of interest to the programmer.

register set In a microprocessor, the internal registers used to store and process data.

regulated power supply A source of dc voltage that maintains a constant voltage output for a wide range of current demands.

	15 … 8	7 … 0	
Data set	AH	AL	Accumulator
	BH	BL	Base
	CH	CL	Count
	DH	DL	Data

	15 … 0	
Pointer & index set	SP	Stack pointer
	BP	Base pointer
	SI	Source index
	DI	Destination index

Register set

regulator An electrical device designed to maintain a constant electrical quantity over a range of operating conditions.

relative addressing In computer programming, a method of obtaining a memory location by starting with a given base value and adding to that base value.

relative permeability Ratio of the permeability of a material to that of free space. Letter symbol is μ_r.

relay An electromechanical device that converts electrical energy into a mechanical ON/OFF action.

rels Measurement of reluctance. See reluctance.

reluctance The opposition that a material has to setting up magnetic flux lines. Reluctance is measured in rels or ampere-turns per weber.

remote shutdown Used in voltage regulators. Voltage regulators using remote shutdown have an extra control lead that will cause them to turn off when a specified voltage is applied to this lead. This feature is useful where power can be quickly removed during a system emergency, such as fire or water immersion of the system.

repair To restore an inoperative system back to its normally designed operating characteristics.

repetitive pulse A periodic waveform consisting of a series of pulses. (See pg. 285 for illustration.)

reproducer The device that converts electrical information into a usable or understandable form. A headset is a reproducer. It converts electrical signals into sound.

residual charge The small charge left on the plates of a

A typical relay.

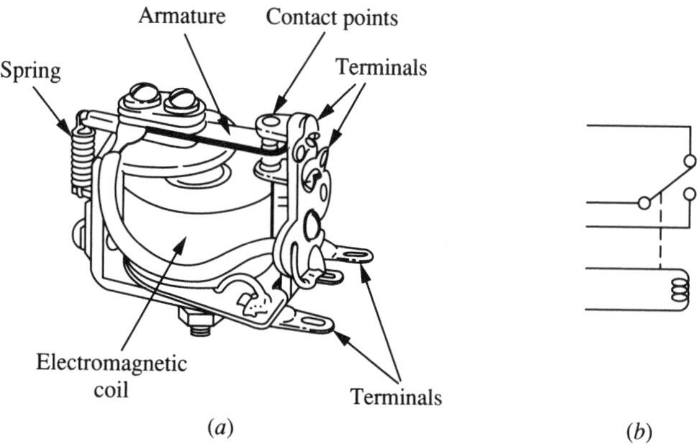

(*a*) (*b*)

Basic structure of a single-pole–double-throw relay.

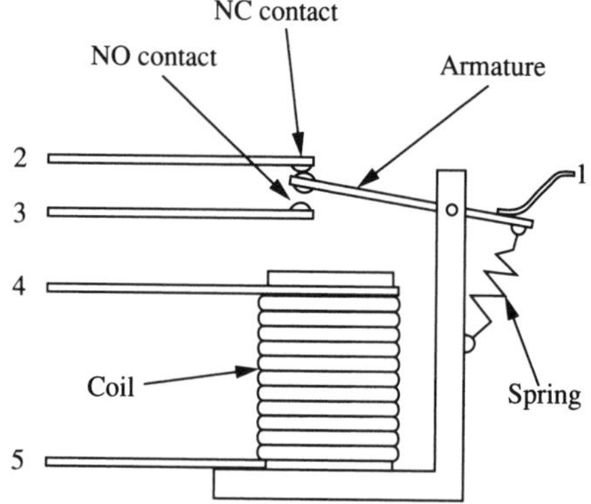

(a) Unenergized: continuity from 1 to 2

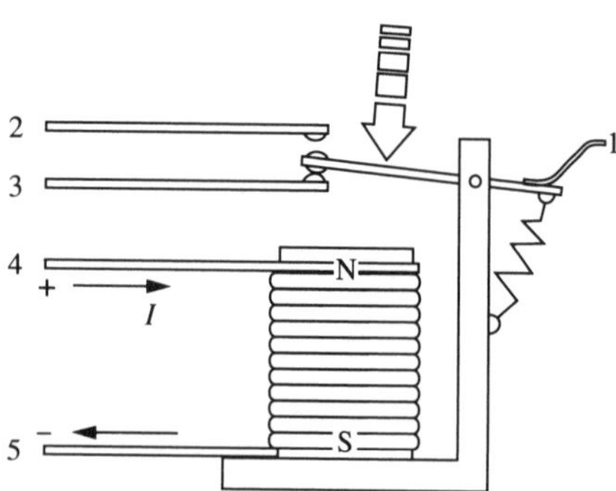

(*b*) Energized: continuity from 1 to 3

Relay

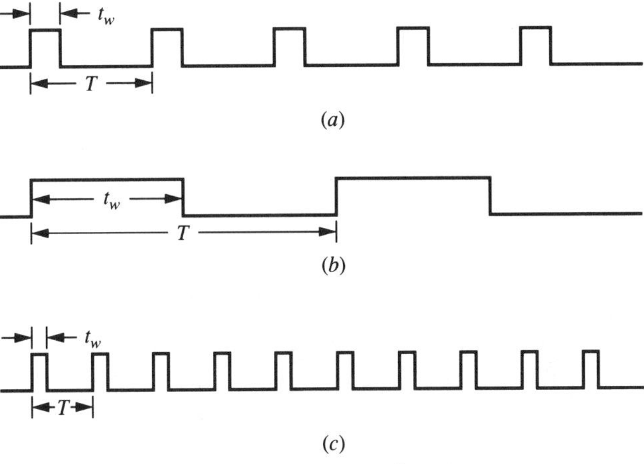

Repetitive pulse

capacitor after it has been discharged.

residual magnetism The magnetic field that remains on an electromagnetic after the exciting current has been removed.

resistance A measure of the opposition to current flow. Resistance is measured in ohms.

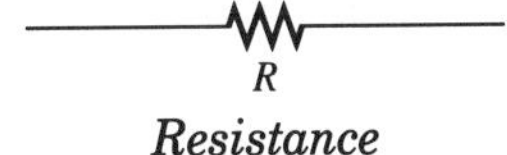

Resistance

resistance-capacitance coupling Electrically connecting two stages together using a resistor and a capacitor so that there is dc isolation between the stages while allowing signals to pass between the stages.

resistance loss The electrical power lost as a result of current flow in a resistor.

resistive device An electrical element that exhibits the property of resistance.

resistive pad A resistive network used to impedance match an antenna. Usually used to match the characteristic impedance of an RF generator to the characteristic impedance of an antenna.

resistor An electrical element that is intended to contain a specified amount of resistance. Resis-

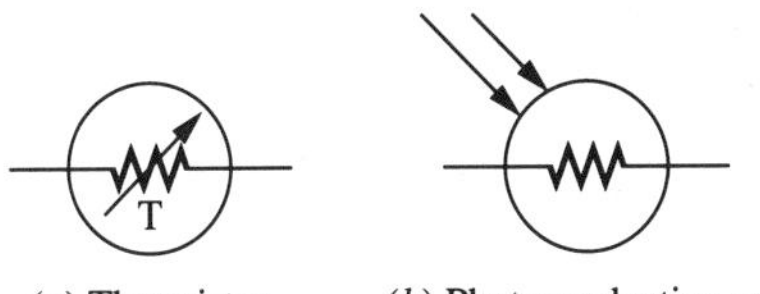

(a) Thermistor (b) Photoconductive cell
Resistive device

Color	Failures (%) during 1000 Hours of Operation
Brown	1.0%
Red	0.1%
Orange	0.01%
Yellow	0.001%

Resistor reliability band

tors are used extensively in electronic circuits.

resistor color code A method of printing the value of a resistor on the body of the resistor through the use of color bands.

resistor networks A combination of several resistors in a single package.

resistor reliability band The fifth band used by some color-coded resistors that indicates the resistor's reliability in a percentage of failure for 1000 hours of use.

resolution 1. A measurement of the smallest possible detectable change. For example, in a three-digit display digital multimeter, the resolution is 1 part in 1000. The ability of an A/D converter to convert accurately small changes in an analog signal to binary data.

resonance The frequency, in an inductive-capacitive circuit, where the inductive reactance is equal to the capacitive reactance. Mathematically expressed as $f_r = 1/(2\pi\sqrt{LC})$, where f_r is the resonant frequency (in hertz), L is the value of the inductor (in henrys), and C is the value of the capacitor (in farads).

resonance curve The graphical representation of the electrical characteristics of a resonant circuit above and below its resonant frequency. The vertical axis usually represents impedance, while the horizontal axis represents frequency.

resonant cavity A type of waveguide for microwave frequencies that consists of a hollow chamber whose physical dimensions determine its resonant frequency. The resonant frequency can be changed by adjusting a mechanical plunger that in turn affects the physical dimensions of the cavity. Usually used for the alignment and maintenance of communications equipment.

resonant circuit An electrical circuit containing capacitance and inductance operated at a frequency where the capacitive and inductive reactances are equal. A parallel resonant circuit will have maximum impedance at resonance, while a series resonant circuit will have minimum impedance at resonance.

resonant frequency See resonance.

resonant line A transmission line operating at a frequency where the reactances of its distributed inductances and capacitances are equal.

resonate To cause to bring to resonance. A capacitive and inductive circuit will resonate at the frequency where its capacitive and inductive reactances are equal.

response curve The graphical representation of the frequency characteristics of an electrical circuit. The vertical axis is usually measured in decibels and the horizontal axis in hertz.

restivity The constant of proportionately between a materials resistance and its physical dimensions. Letter symbol is *p*.

retentivity The property of a material to retain a residual induction after the removal of the magnetizing force that caused a saturation induction of the material.

retrieve To get specific information from a computer or computer system.

retriggerable one shot A monostable multivibrator that can be retriggered during its unstable state.

reverse bias To cause a device not to conduct. For example, to reverse bias the pn junction of a diode, a negative voltage is applied to the p-material and a positive voltage to the n-material.

reverse breakdown In a pn junction, when the external reverse-biased voltage is increased to the point where conduction occurs across the junction. Most diodes are not designed to withstand reverse or avalanche breakdown. The zener diode is especially made to utilize reverse breakdown.

reverse current The small current that flows in a reverse-biased pn junction. Reverse current is caused by minority carriers. See forward current.

reverse direction The direction of current flow when a pn

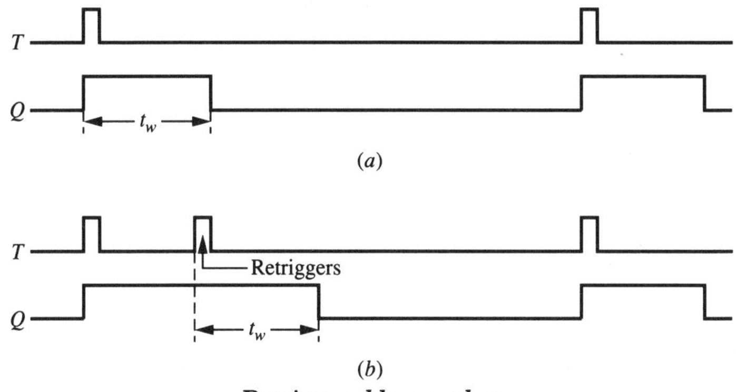

Retriggerable one-shot

junction is reverse biased. See reverse current.

reverse leakage current In semiconductors, the small current that flows due to minority carries when the pn junction is reverse biased. Reverse leakage current is sensitive to temperature changes in the device. Ideally, the reverse leakage current is zero.

RF Abbreviation for radio frequency. See radio frequencies.

RF amplifier An electrical circuit designed to increase the strength of radio frequency waves. Typically, an RF amplifier is tuned to a small range of frequencies. This small range of frequencies is amplified, while those frequencies outside this small range are not amplified.

RFC Abbreviation for radio frequency choke. See radio frequency choke.

RF choke Abbreviation for radio frequency choke. See radio frequency choke.

RF component That portion of a radio wave that consists of only the radio frequency carrier and not any of the lower modulating frequencies.

RF generator Abbreviation for radio frequency generator. See radio frequency generator.

RF oscillator Abbreviation for radio frequency oscillator. See radio frequency oscillator.

RF probe A specially designed device used for extracting electrical energy from radio frequency circuits for the purpose of measuring them on an electrical instrument.

RF transformer Abbreviation for radio frequency transformer. See radio frequency transformer.

RF voltmeter An electronic instrument used to measure the amount of voltage of radio frequencies. RF voltmeters are usually used in the repair and maintenance of communications equipment.

rheostat A variable resistance with two active terminals. The resistance of a rheostat can be changed by turning its wiper arm.

ribbon cable A group of insulated wires connected together in a parallel fashion. Ribbon cables are usually terminated by multiple-pin connectors.

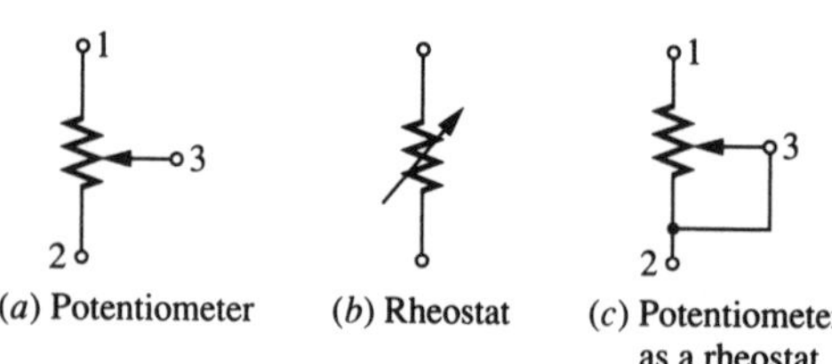

(*a*) Potentiometer (*b*) Rheostat (*c*) Potentiometer as a rheostat

Rheostat

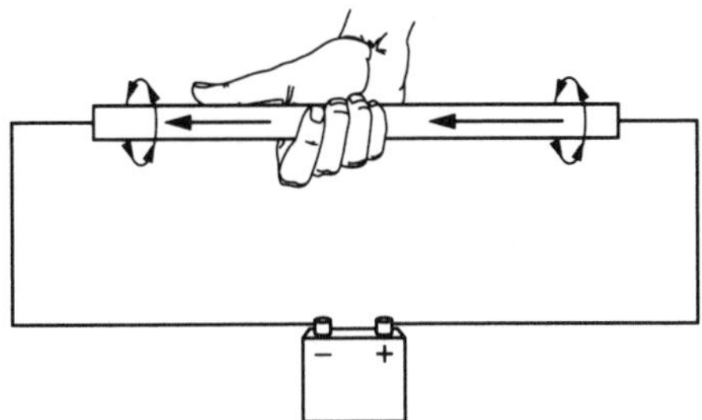

Right-hand rule

ridged waveguide A waveguide made with an indentation along its horizontal axis. A ridged waveguide is designed to accommodate a lower frequency than that dictated by its outside dimension. Used where space is at a premium.

right-hand rule Used as an aid in remembering the direction of the lines of force in a current carrying conductor when conventional current flow ("+" to "−") is used.

right triangle A triangle with one angle equal to 90 degrees.

ring counter 1. A form of digital counter where a given bit pattern continually circulates through the counter. 2. A group of interconnected flip-flops where the last one is connected back to the first one to form a closed loop (like a ring). Only one flip-flop is TRUE, where this TRUE state is transferred from one flip-flop to the next at each clock pulse.

ringing In a pulse waveform, amplitude changes that follow overshooting. They are usually a series of dampened sine waves. Any series of dampened sine waves caused by a circuit.

ripple Undesirable fluctuations in the dc output voltage of a dc power source.

ripple counter A counter where each flip-flop, with the ex-

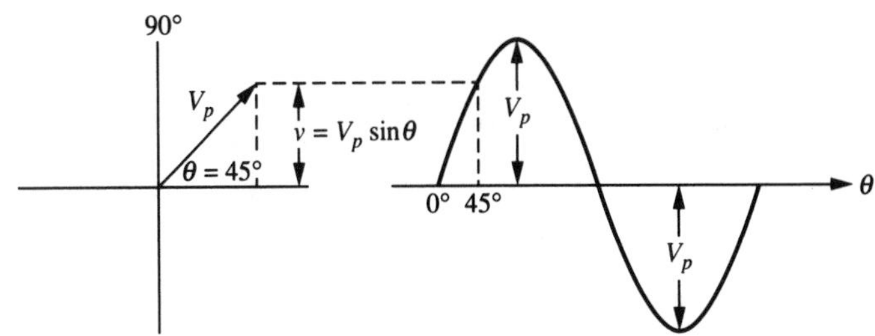

Right triangle

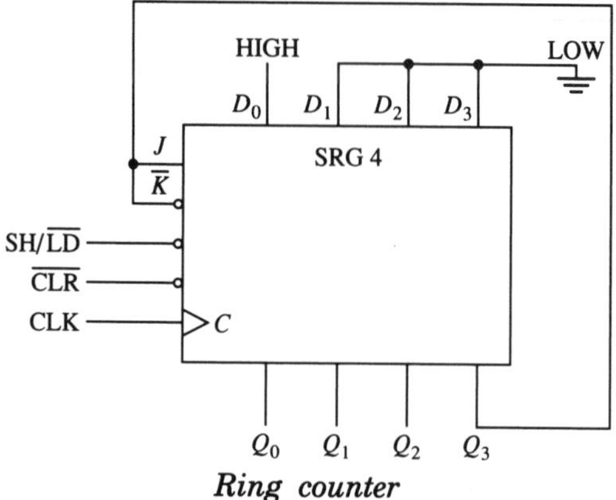

Ring counter

ception of the first, is clocked by the preceding flip-flop.

ripple factor An indication of the effectiveness of a filter. For a power supply filter, the ripple factor is defined as $r = V_r/V_{dc}$, where r is the ripple factor, V_r is the rms ripple voltage, and V_{dc} is the dc (average) value of the output filter voltage.

rise time The amount of time it takes for a pulse to go from 10% of its maximum value to 90% of that value.

RL Abbreviation for resistor-inductor network.

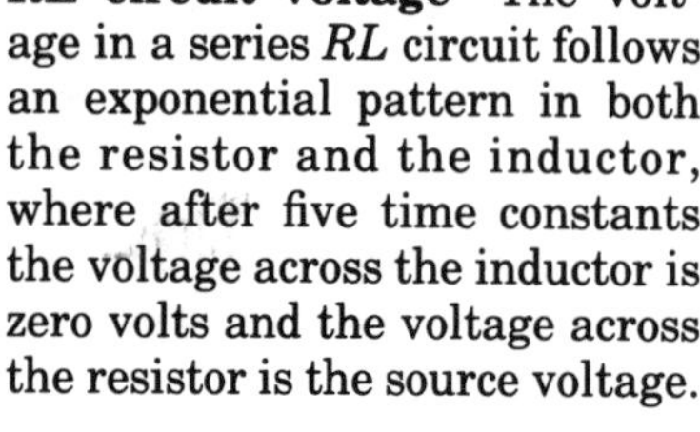

***RL* circuit voltage** The voltage in a series *RL* circuit follows an exponential pattern in both the resistor and the inductor, where after five time constants the voltage across the inductor is zero volts and the voltage across the resistor is the source voltage.

***RL* differentiator** An *RL* circuit that approximates the mathematical function of differentiation. A simple *RL* differentiator consists of a series connection of a resistor and a capacitor with the output taken across the resistor.

***RL* differentiator pulse response** The effects on a pulse

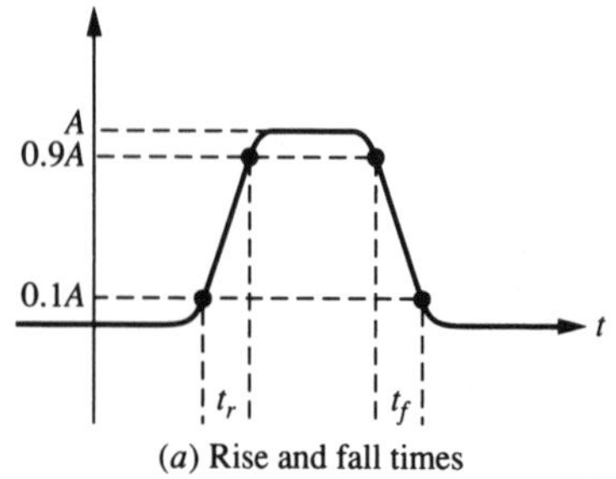

(*a*) Rise and fall times

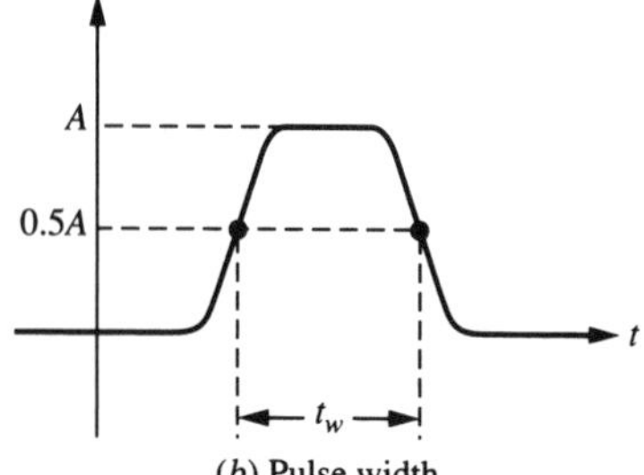

(*b*) Pulse width

Rise time

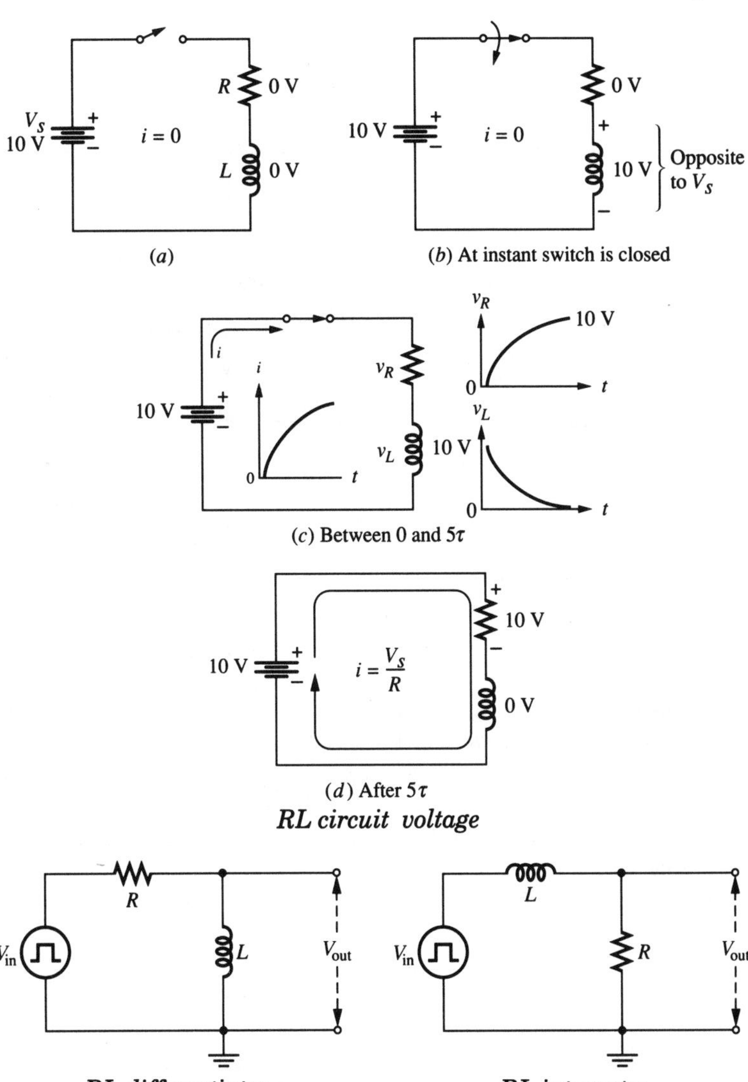

RL circuit voltage

RL differentiator

RL integrator

waveform due to the electrical actions of an *RL* differentiator.

***RL* differentiator time constant** The time constants involved in an *RL* differentiator circuit. Equal to the value of the resistor and the inductor.

***RL* integrator** An *RL* circuit is a circuit that approximates the mathematical function of integration. A simple *RL* integrator consists of a series connection of a resistor and an inductor with the output taken across the inductor.

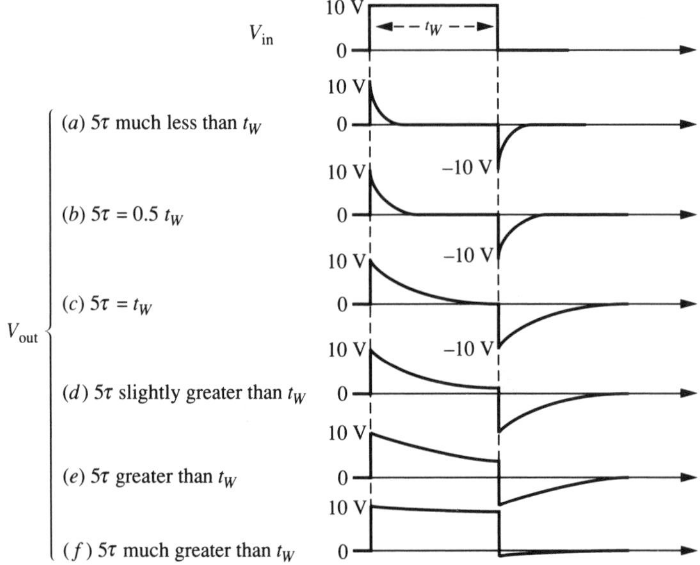

RL differentiator time constant

***RL* integrator pulse response** The effects on a pulse waveform due to the electrical actions of an *RL* integrator.

***RL* integrator time constant** The time constants involved in an *RL* integrator circuit. Equal to the value of the resistor and the inductor.

***RL* lag network** An electrical circuit consisting of an inductor in series with a resistor where the output voltage is taken from across the resistor. In this ar-

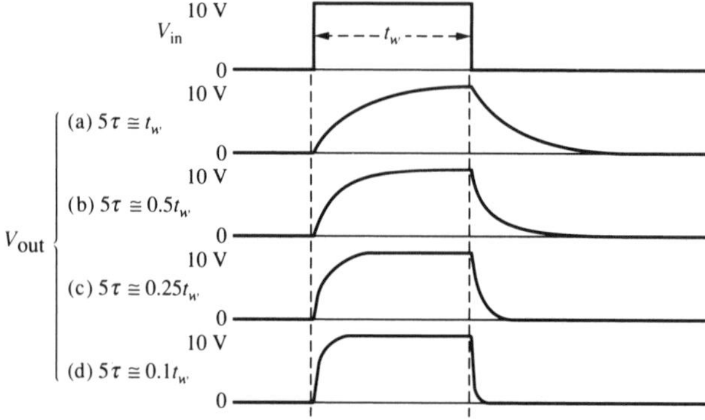

RL integrator pulse response

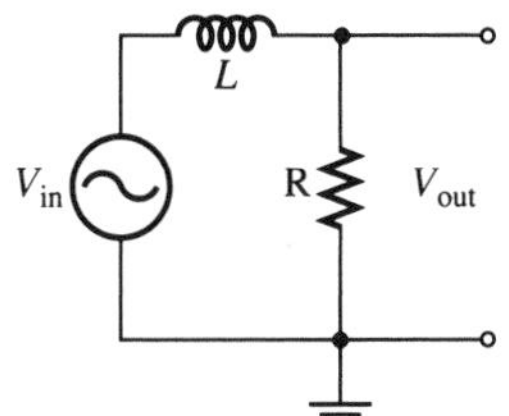

(*a*) A basic *RL* lag network

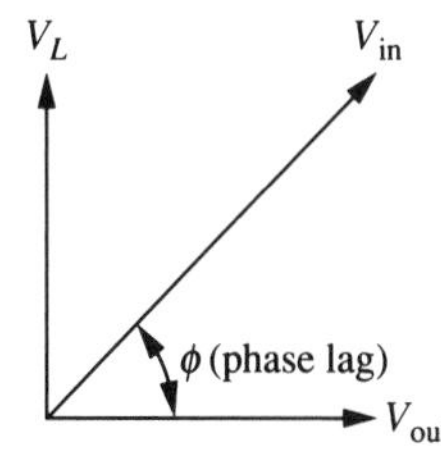

(*b*) Phasor voltage diagram showing phase lag between V_{in} and V_{out}

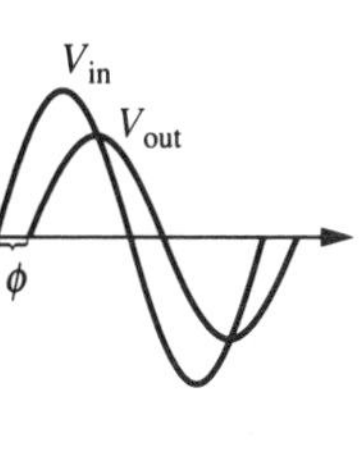

(*c*) Input and output waveforms

RL lag network

rangement, the output voltage will lag the input voltage by an amount determined by the frequency of the source and the values of the inductor and resistor.

***RL* lead network** An electrical circuit consisting of an inductor in series with a resistor where the output voltage is taken from across the inductor. In this arrangement, the output voltage will lead the input voltage by an amount determined by the frequency of the source and the values of the inductor and resistor.

***RL* network** An electrical circuit consisting of resistors and inductors. *RL* networks respond to frequency changes and are usually used as high-pass or low-pass filters. *RL* networks are also found in timing circuits where the values and value ratios of the inductors and resistors determine the frequency and duty cycle of the resultant timing waveform.

rms Abbreviation for root mean square. See root mean square.

robot A reprogrammable multifunctional mechanical manipulator.

robotics The study of systems dealing with robots. See robot.

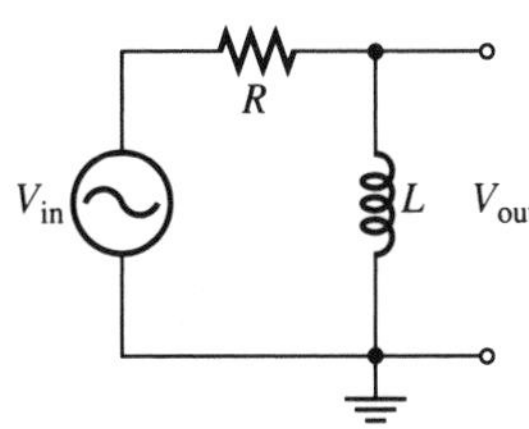

(*a*) A basic *RL* lead network

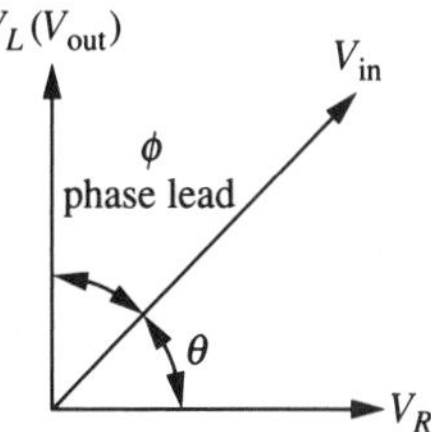

(*b*) Phasor voltage diagram showing phase lead between V_{in} and V_{out}

RL lead network

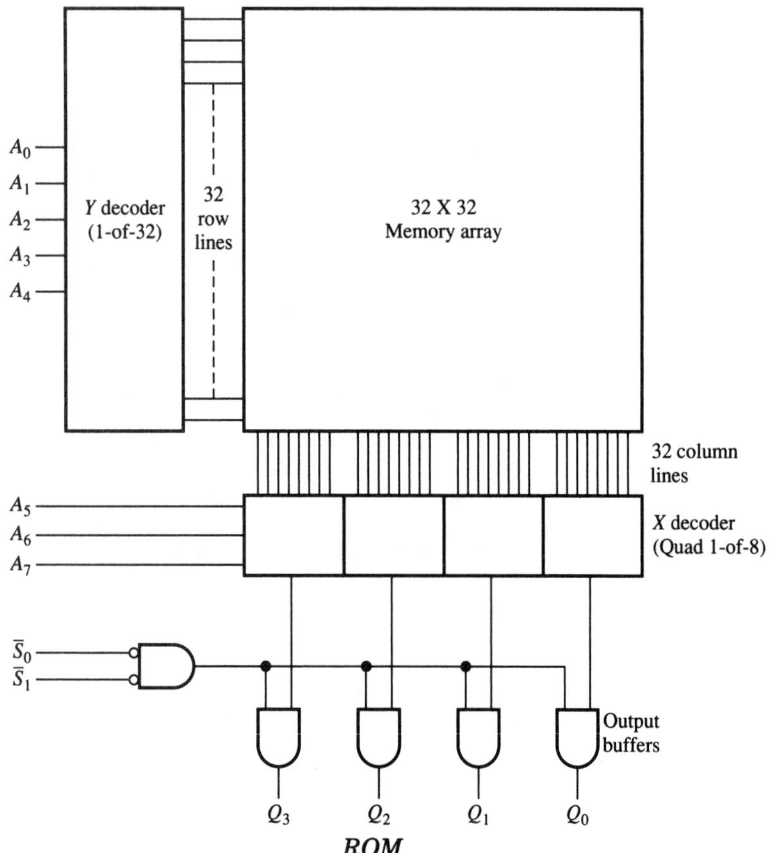

ROM

ROM Abbreviation for read-only memory. See read-only memory.

ROM family The grouping of integrated circuit technologies for read only memories.

root mean square The measure of the heating effect of a sine wave. Equal to 0.707 of the peak value. Also known as the effective value.

rotary switch A mechanical on/off device that is activated by rotation. A rotary switch may have several ON/OFF positions capable of connecting more than one circuit at any one time.

rotational motion Circular movement. The movement of a wheel is rotational motion.

rotator A motor-driven electromechanical device that is used to turn antennas for optimizing signal transmission and reception.

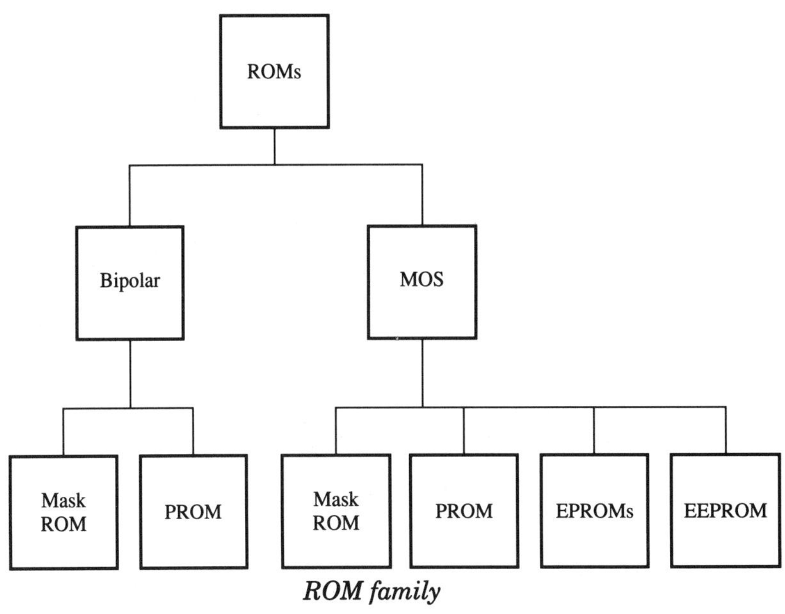

ROM family

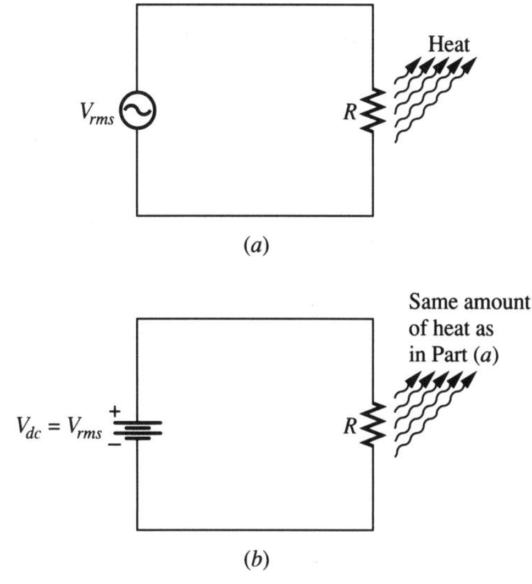

Root mean square

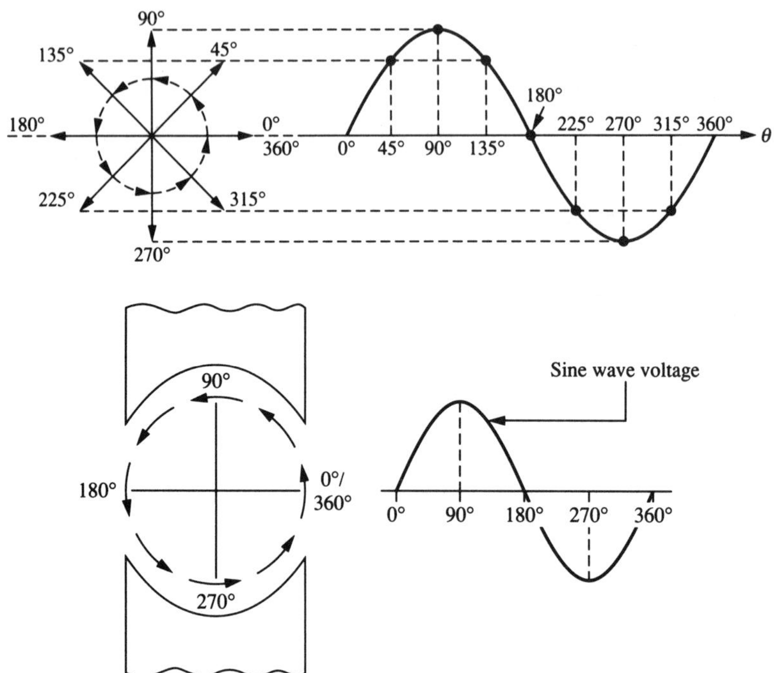

Rotational motion

rotor plates The movable plate of a variable capacitor.

R-S flip-flop A bistable circuit with two complementary outputs (Q and NOT Q) and two inputs designated R and S (for RESET and SET). The R and S inputs are not allowed to be active at the same time. If both the R and S inputs are not active, then the output condition will not change. An active R input will cause the NOT Q output to become TRUE and an active S input will cause the Q output to become TRUE.

RS-232 A standard, as defined by the Electronics Industries Association, for transferring digital information between digital and communications equipment.

run time The amount of time it takes to execute a single continuous computer program.

A typical EIA-232-D/RS-232-C interface.

DTE

Protective ground
Transmitted data (TD)
Received data (RD)
Request to send (RTS)
Clear to send (CTS)
Data set ready (DSR)
Signal ground
Data terminal ready (DTR)

DCE

A full EIA-232-D/RS-232-C interface.

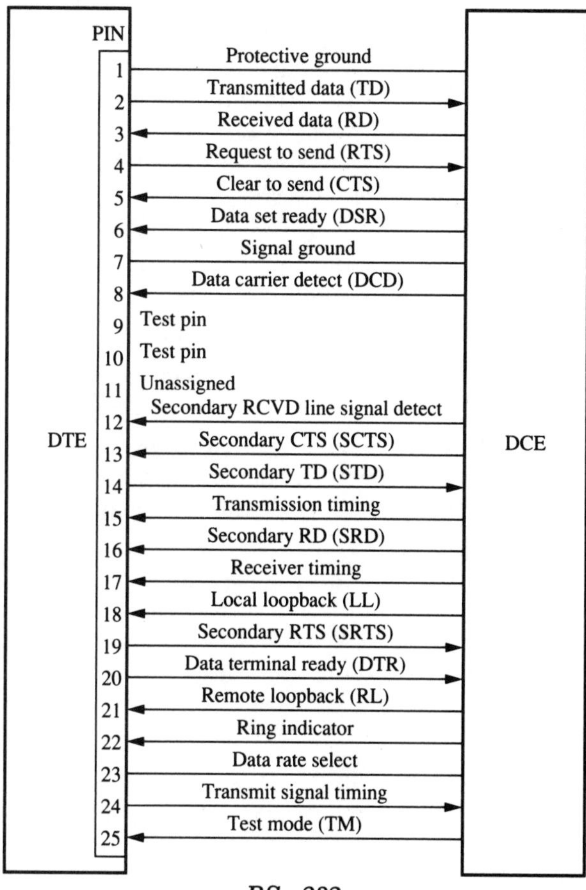

RS - 232

S

s Letter symbol for second. See second.

S Letter symbol for siemen. See siemen.

safety The awareness of protection of human life of self and others followed by the protection of equipment used by humans.

safety factor The amount by which equipment may be operated over its design limits without damage to the equipment.

sample and hold In an analog-to-digital conversion scheme, the value of the analog input signal is sampled, and this sampled value "remembered" (held) by the circuit until the resulting digital output is constructed to the desired accuracy.

sampling The process of taking a periodic sample of the waveform to be transmitted and transmitting the samples. An example of this is pulse amplitude modulation, where the pulse is a sample and its amplitude represents the amplitude of the information to be transmitted.

sampling frequency The frequency at which information is sampled. See sampling theorem.

sampling theorem If a pulse-modulated system is given by $f_s = 2f_{N(\max)}$ where f_s is the minimum sampling frequency to ensure that the samples contain all of the information of the original signal and $f_{N(\max)}$ is the maximum frequency of the modulating signal.

SAP Abbreviation for second audio program. See second audio program.

saturable reactor A magnetic core device whose reactance can be varied by changing the saturation of its core.

saturation 1. The condition in a current controlling device such as a vacuum tube, transistor, or FET when the controlled current is at its maximum and is limited only by external circuit elements. See cutoff. 2. A circuit condition where an increase of the input signal no longer produces an increase in the output signal. As an example, when an amplifier is saturated, it means that any

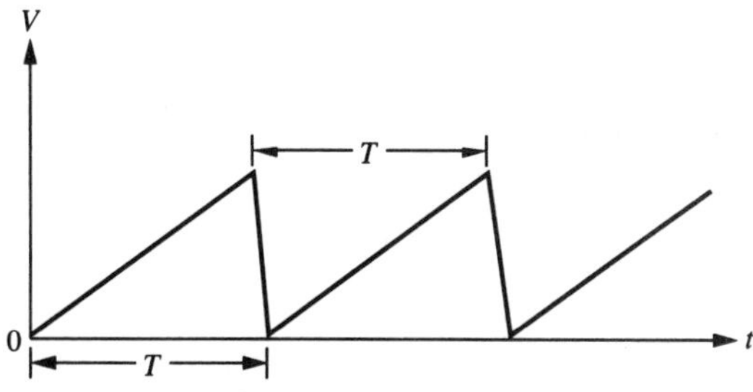

Sawtooth waveform

increase in amplitude of the input signal will not cause a corresponding change on the output signal of the saturated amplifier.

sawtooth waveform A special case of a triangular waveform having two ramps where one ramp is of a much longer duration than the other.

scattering loss In fiber optics, this is a signal loss due to the slight imperfections contained within the optical fiber. With improved manufacturing methods, this loss is usually very small.

schematic A drawing of electrical components and their connections done in an agreed-to symbolic form. Schematics are used as a convenient way of representing electrical circuits.

Schmitt trigger A circuit that changes state abruptly when the input signal crosses a specified triggering level.

Schottky diode A solid-state device used in high-frequency and fast-switching applications. Also known as hot carrier diodes. Constructed as an n-material metal junction and uses only majority carriers.

scientific notation The use of powers of 10 that makes it easier to express large and small numbers and to do calculations with the use of such numbers.

scope Oscilloscope. See oscilloscope.

SCR Abbreviation for silicon controlled rectifier. See silicon controlled rectifier.

second The unit of measurement for time in the SI, MKS, CGS, and English system of measurements.

$1{,}000{,}000 = 10^6$	$0.000001 = 10^{-6}$
$100{,}000 = 10^5$	$0.00001 = 10^{-5}$
$10{,}000 = 10^4$	$0.0001 = 10^{-4}$
$1{,}000 = 10^3$	$0.001 = 10^{-3}$
$100 = 10^2$	$0.01 = 10^{-2}$
$10 = 10^1$	$0.1 = 10^{-1}$
$1 = 10^0$	

Scientific notation

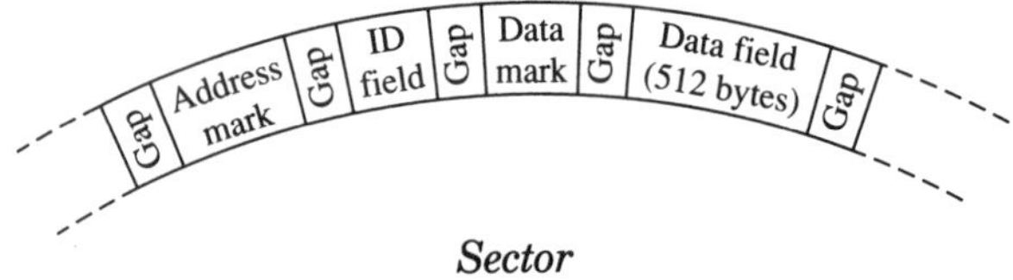

Sector

second audio program In a television signal using the BTSC system, another audio channel transmitted along with regular television information. This second channel can be used to transmit in another language.

second source A manufacture of a device or system that is not the original manufacture.

secondary winding The winding of a transformer from where the output is taken.

sector On a disk, the electronically "slicing up" of the disk in order to organize data into pie-sliced parts of the disk called sectors

segmentation registers In a microprocessor, internal storage for the address of a section of memory called a segment.

selectivity The ability of an electrical device or circuit to select one frequency and reject all others. The greater the selectivity, the narrower the bandwidth. Good selectivity causes one transmitted signal to be easily selected from a group of transmitted signals that have carrier frequencies close to the desired signal. When a receiver receives more than one signal at a time (at different frequencies), it is said to have poor selectivity.

selectivity curves A graphical representation of the selectivity of a circuit.

self-bias Using resistance to cause a voltage drop that will determine the dc operating point of an electronic device such as a vacuum tube or FET. For an FET a resistor is placed in the source

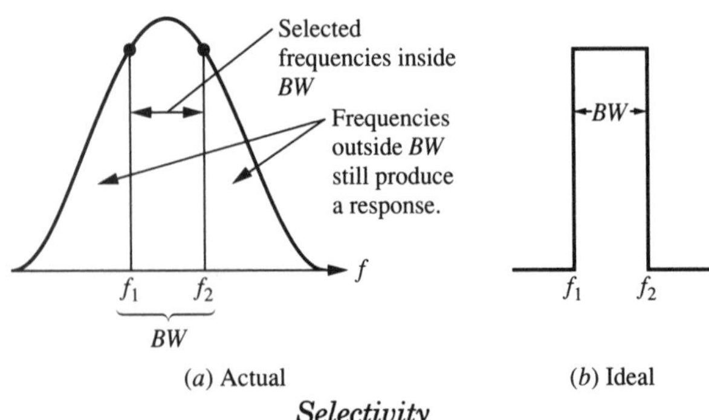

Selectivity

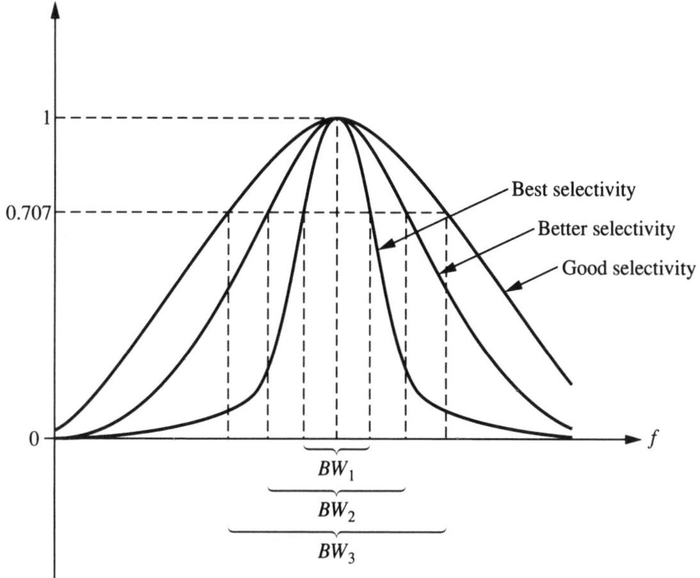

Selectivity curves

circuit while the gate is referenced to ground. Such an arrangement produces a small offset or bias voltage that sets the dc operating point of the device. For a vacuum tube, see cathode bias.

self-inductance The ability of a coil to oppose any change in current is a measure of the self-inductance of the coil. Inductors are measured in henrys (H), after Joseph Henry.

semiconductor Material that is between being a good conductor and a good insulator. Semiconductor material, such as germanium and silicon, is used in the construction of active devices such as transistors, diodes, and integrated circuits.

semiduplex In electronics communication between two stations, where one station is operated in the duplex mode and the other in the simplex mode. This system requires two operating frequencies.

sensitivity The ability of an electrical device or circuit to respond to weak signals. A circuit with good sensitivity will be able to respond to very small signals, while a circuit with poor sensitivity will not be able to respond to a weak signal.

sequential access In digital systems retrieving and storing data in a serial fashion. In this method, the amount of time it takes to retrieve or store informa-

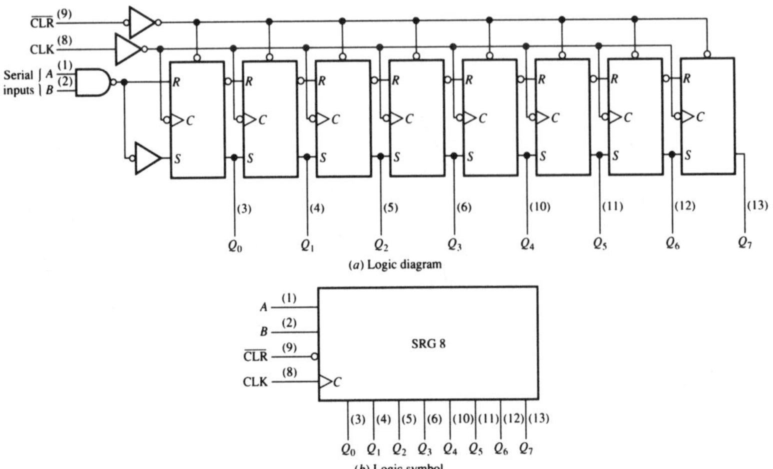

(a) Logic diagram

(b) Logic symbol

Serial in, parallel out

tion depends upon where it is located in memory. See random access.

serge current A worst case condition for the amount of current in an electrical circuit. In a power supply, surge current is the maximum current that can occur if the unit is turned on when the applied ac voltage is at its peak value. This can be a large value because on a capacitor input filter, the capacitor is not yet fully charged and presents a low impedance to the peak voltage.

serial in A term used to describe the entering of data 1 bit at a time. An example is a serial-in register where data are entered into the register 1 bit each clock pulse. See shift register.

serial out A term used to describe the transferring of data 1 bit at a time. An example is a serial-out register where data are copied from the register 1 bit each clock pulse. See shift register.

serial in, parallel out A digital register that inputs data serially, 1 bit at a time, and outputs

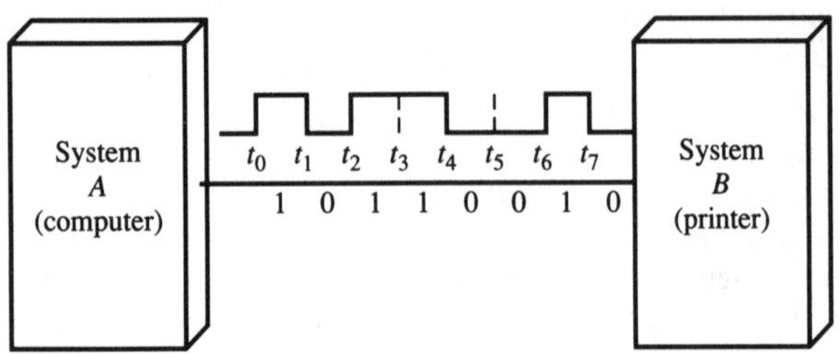

Serial transfer of bits from A to B. Interval t_0 to t_1 is first. Computer to printer is an example.

Serial transfer

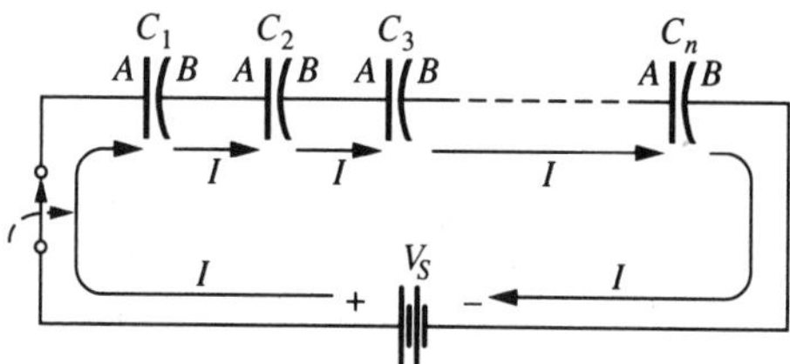

(*a*) Charging current is same for each capacitor, $I = Q/t$.

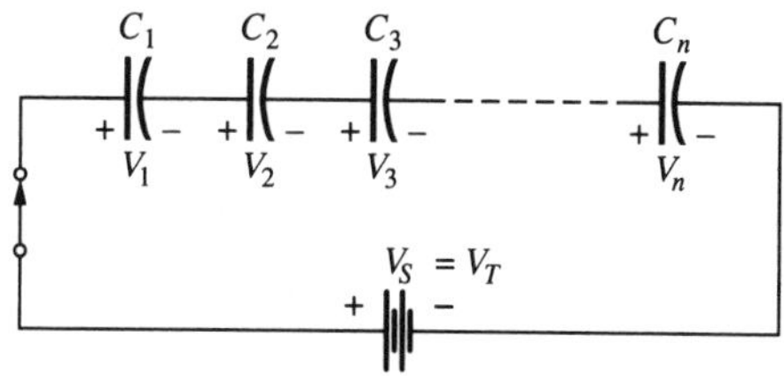

(*b*) All capacitors store same amount of charge and $V = Q/C$.

Series capacitors

data in parallel, all bits at the same time.

serial transfer In a digital system, the process of sending 1 bit at a time from one point to another. When transferring digital information in serial form, the bits are sent on a single line 1 bit at a time.

series capacitors Capacitors that are connected in series. The total capacitance of capacitors connected in series is less than the value of the smallest capacitance in the series connection.

series circuit A circuit arrangement where all the circuit elements are connected in such a fashion that there is only one path for current flow; as a result, the current in a series circuit is everywhere the same. See parallel circuit.

series parallel A combination of electrical components consisting of both series and parallel branches.

series-parallel circuit A circuit arrangement where some of the components are connected in series and some of the components are connected in parallel.

series *RC* circuit An electrical circuit consisting of a capacitor in series with a resistor. (See pg. 305 for illustration.)

series *RC* circuit phase The response of a series circuit consisting of a resistor and a capacitor to different frequencies in terms of the phase relationship between the applied signal voltage and the circuit current. (See pg. 305 for illustration.)

series regulator An electrical circuit that attempts to maintain

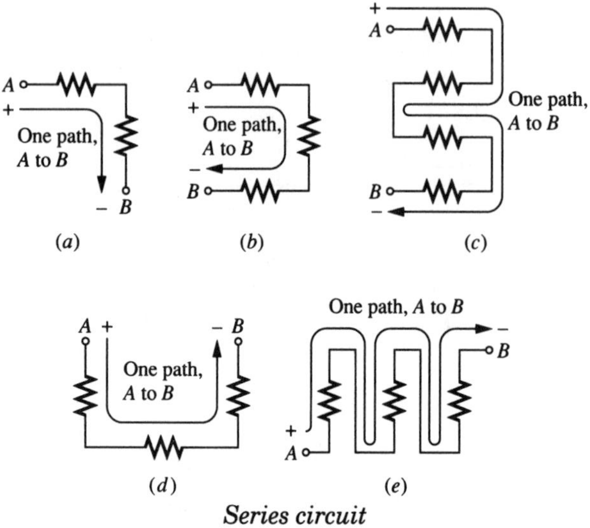

Series circuit

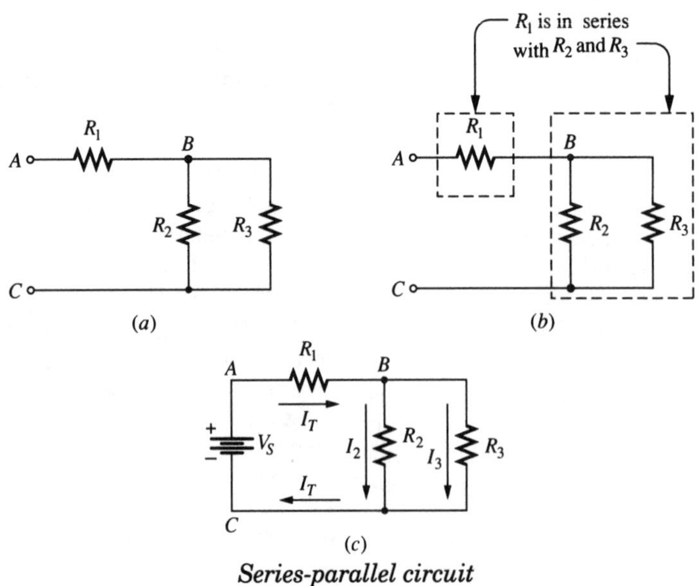

Series-parallel circuit

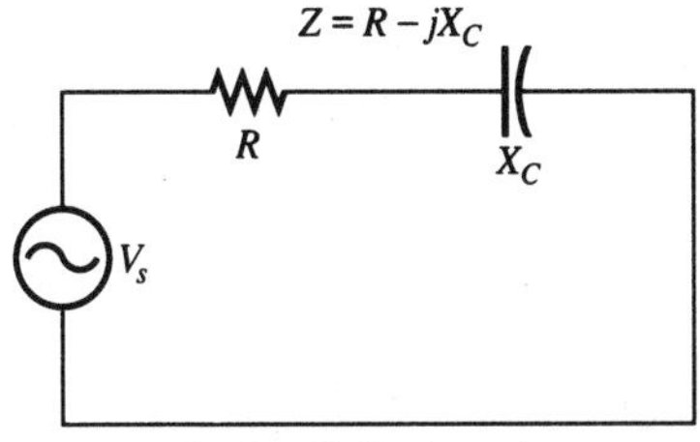

Series RC circuit

a constant output voltage by using a regulating device in series with the voltage source. Series regulators are used with power supplies to help maintain a constant output voltage under a wide range of load conditions.

series resonance The condition in an electrical circuit where capacitive reactance is subtracted from the effects of inductive reactance. At resonance, the impedance of a series resonant circuit is at a minimum; equal only to the resistance of the circuit.

series *RL* circuit An electrical circuit consisting of an inductor in series with a resistor.

series *RLC* circuit An electrical circuit consisting of a series connection of a resistor, inductor, and capacitor. (See pg. 307 for illustration.)

series *RLC* impedance The response of a series circuit consisting of a resistor, inductor, and capacitor to different frequencies in terms of its oposition to current flow (impedance). (See pg. 307 for illustration.)

series *RLC* phase The phase relationships in a series *RLC* circuit. The phase relationships depend upon the source frequency and values of the resistor, inductor, and capacitor. (See pg. 308 for illustration.)

series winding In an electrical motor or generator, a field winding that carries the same current as the armature; this is accom-

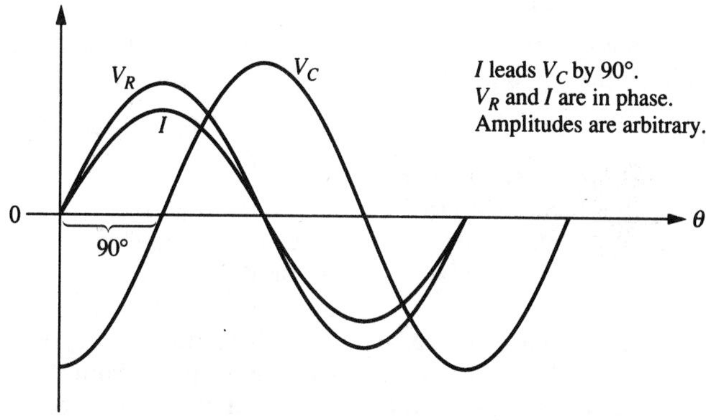

Series RC circuit phase

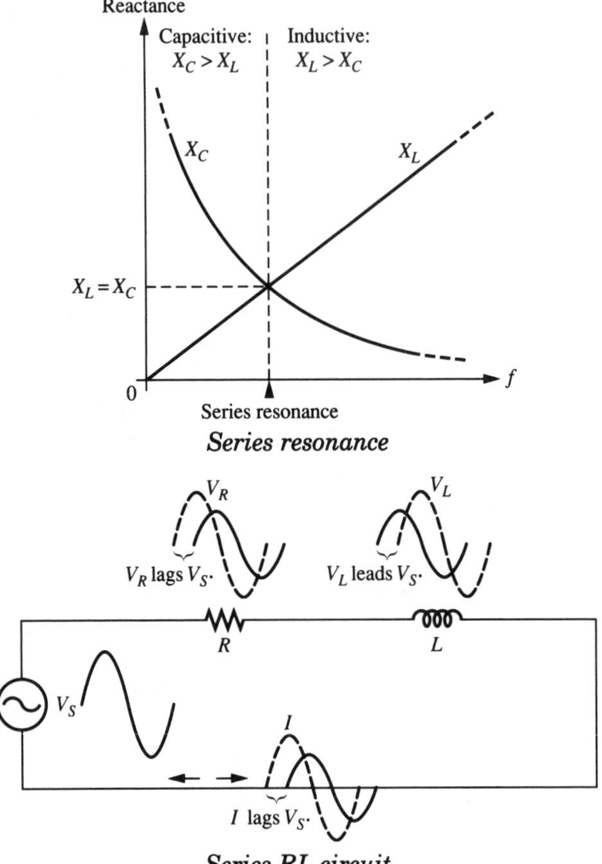

Series resonance

Series RL circuit

plished by having the armature and field windings connected in series.

servomotor An electrical motor whose direction of rotation and speed of rotation are controlled by electrical signals.

servo system An automatic control system where the variable conditions are electrically measured and automatically adjusted to accommodate a given set of conditions.

set-reset flip-flop See SET-RESET latch.

SET-RESET latch A type of bistable storage device. A SET-RESET latch has two complementary outputs with a SET and RESET input. This type of latch can be formed from two cross-coupled NAND gates or two cross-coupled NOR gates. A digi-

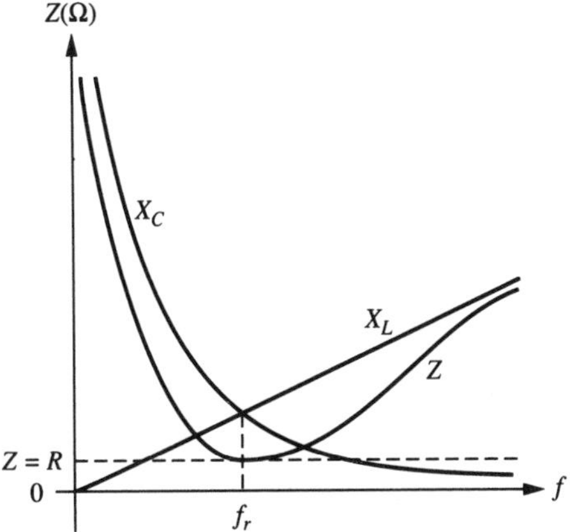

Series RLC impedance

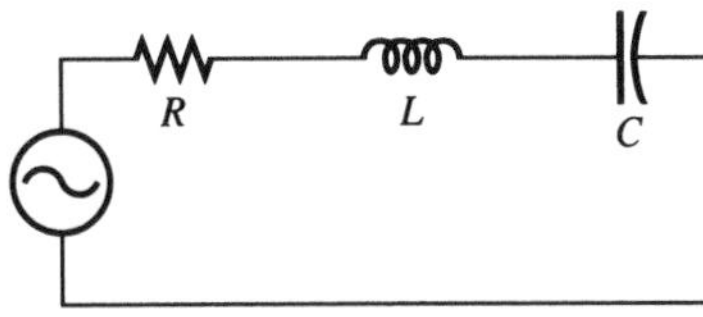

Series RLC circuit

tal circuit capable of storing a two-state logic condition usually in the form of 0 volts or +5 volts. Consists of two inputs, one called the SET, the other the RESET, and two complementary outputs. The SET input is used to set the latch and the RESET input is used to reset the latch.

settling time In a pulse waveform, the time it takes for the pulse to reach its maximum amplitude, not counting overshoot. See overshoot.

setup time The minimum amount of time it takes for the logic level to be maintained constantly on the inputs of a digital device (such as a flip-flop), prior to the triggering edge of the clock, in order for the input levels to affect reliably the condition of the device. (See pg. 310 for illustration.)

seven segment A device consisting of seven unique parts, such as a seven-segment display. See seven-segment display. (See pg. 310 for illustration.)

seven-segment display 1. An electrical device containing seven separately controlled light-emitting diodes (or other such light-emitting devices) arranged as the figure 8. Used to display numbers from 0 to 9 and letters from A to F. Thus the device is capable of displaying values in number systems from the base 2 (binary) to the base 16 (hexadecimal). (See pg. 310 for illustration.)

sexadecimal See hexadecimal.

shelf life The amount of time an item can be stored and still

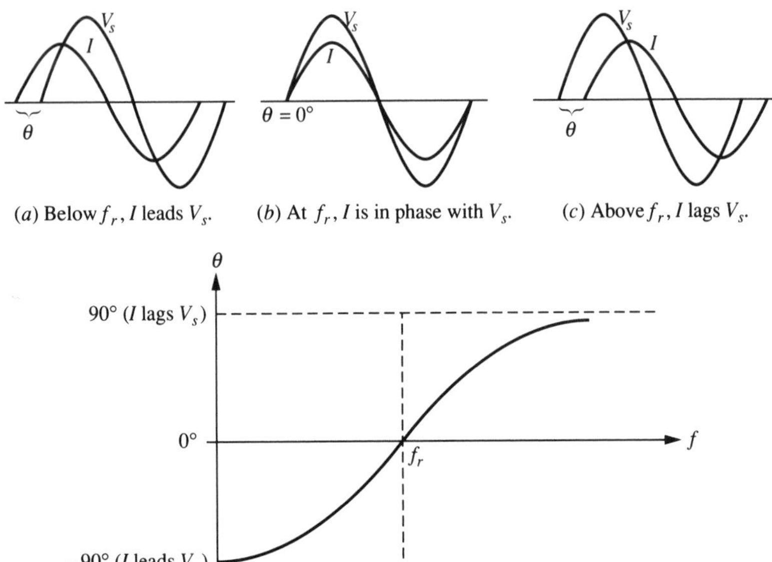

(a) Below f_r, I leads V_s. (b) At f_r, I is in phase with V_s. (c) Above f_r, I lags V_s.

(d) Phase angle versus frequency

Series RLC phase

Two versions of SET-RESET latches.

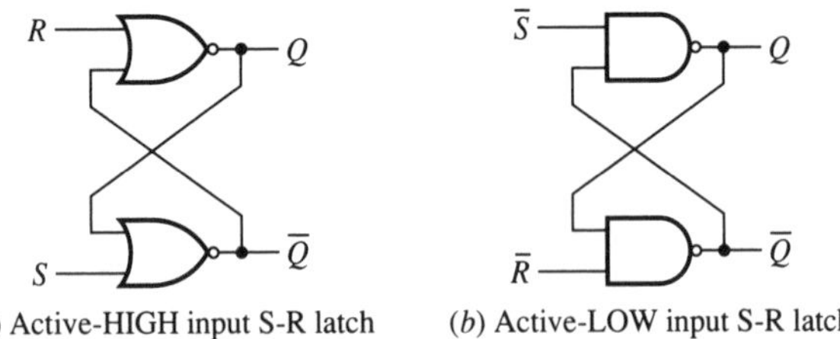

(a) Active-HIGH input S-R latch (b) Active-LOW input S-R latch

SET-RESET latch

maintain its required operational characteristics.

shell In the Bohr model of the atom, the distance orbiting electrons are from the nucleus. Electrons are allowed to be at only discrete distances from the nucleus. These discrete distances, called shells, represent the energy levels of the electrons. The farther away from the nucleus, the higher the energy of the electron. See Bohr model.

The four modes of basic latch operation.

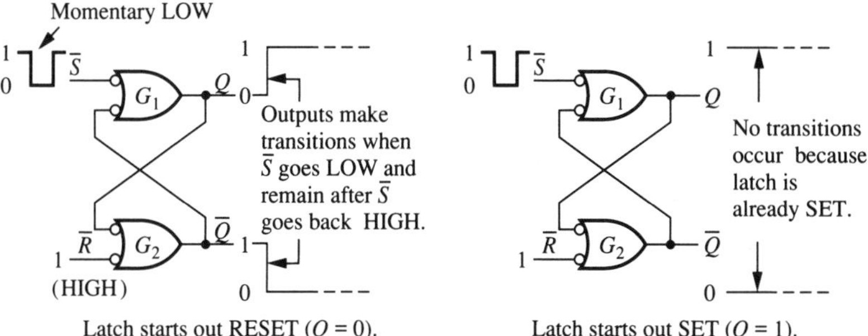

(*a*) Two possibilities for the SET operation

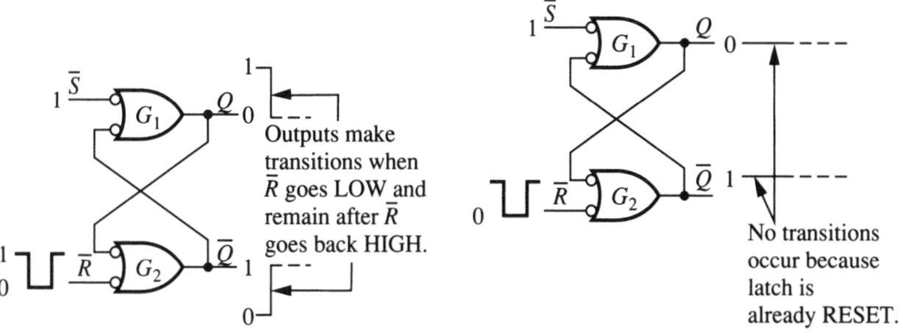

(*b*) Two possibilities for the RESET operation

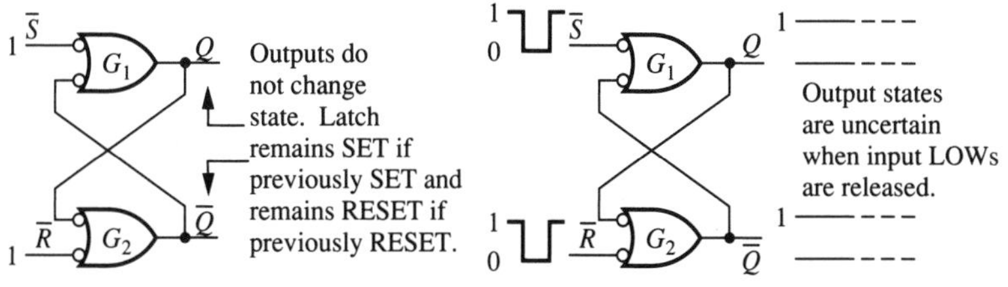

(*c*) No-change condition

(*d*) Invalid condition

SET-RESET latch (con't)

shell

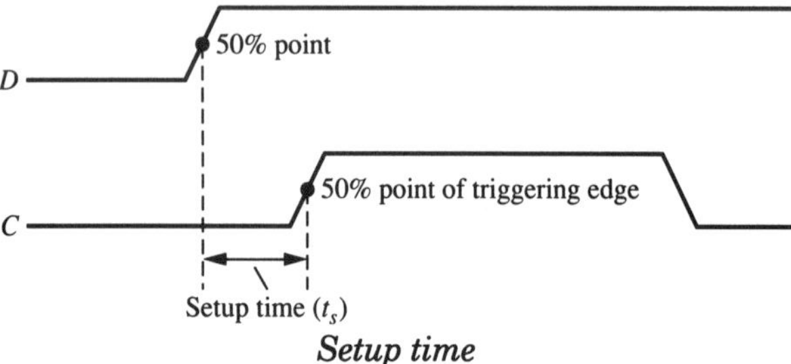

Setup time

Display of decimal digits with a seven-segment device.

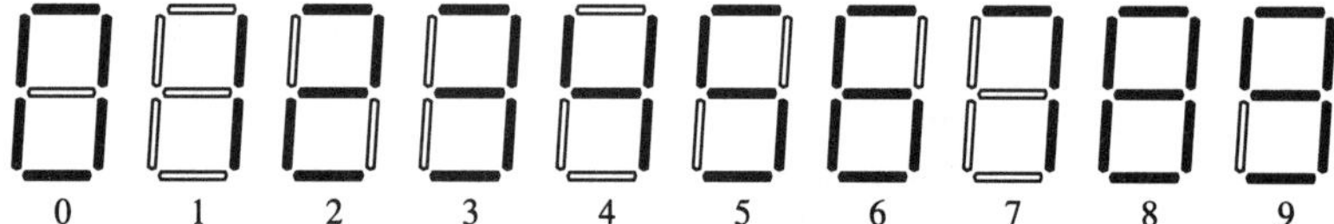

Seven-segment display format.

Digit	Segments Activated
0	*a, b, c, d, e, f*
1	*b, c*
2	*a, b, g, e, d*
3	*a, b, c, d, g*
4	*b, c, f, g*
5	*a, c, d, f, g*
6	*a, c, d, e, f, g*
7	*a, b, c*
8	*a, b, c, d, e, f, g*
9	*a, b, c, d, f, g*

Seven segment

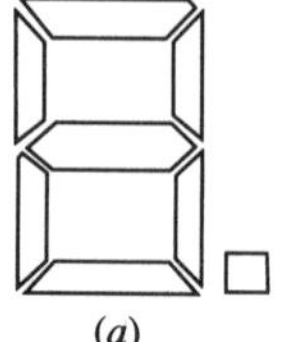

(*a*)

(*b*)

Seven-segment display

shf Abbreviation for super high frequency.

shielded cable An electrical conductor surrounded by an insulator where the insulator in turn is surrounded by a flexible conductive jacket called the shield. The shield is normally at ground potential and serves the purpose of eliminating radiation to or from the center electrical conductor.

shift register 1. A series of digital storage elements (such as

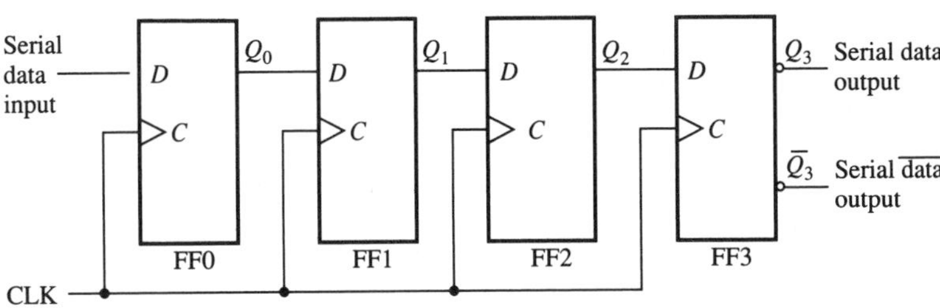

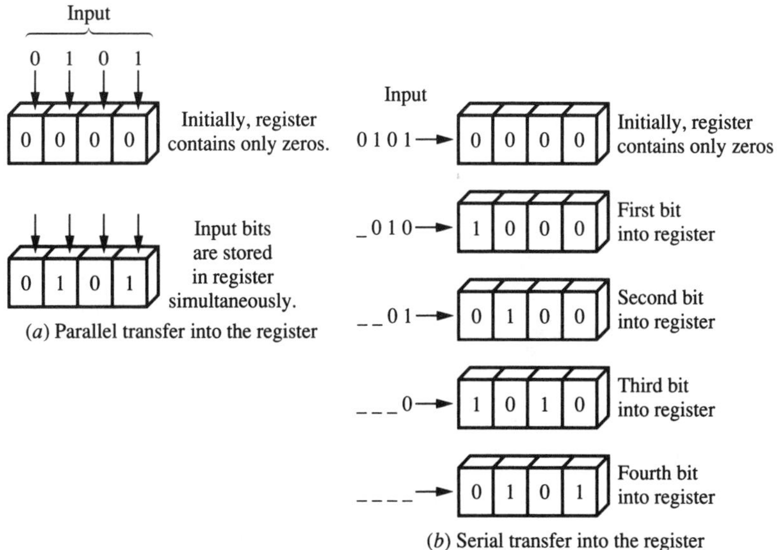

(*a*) Parallel transfer into the register

(*b*) Serial transfer into the register

Shift register

flip-flops) connected serially so as to accept 1 bit at a time. A shift register accepts data serially 1 bit at a time. The stored information is produced on its output 1 bit at a time. 2. A register where the data are copied in and transferred out serially one bit at a time. See serial out, serial in.

short A direct electrical connection between two otherwise separate mechanical points. An ideal short has 0 ohms' resistance. An electrical condition offering 0 ohms' resistance to the power or signal source. Typically, shorts in power equipment are very dangerous because they can cause electrical fires due to high currents. This is especially true if the unit is not fused. See open.

short circuit See short.

short-circuit current The maximum amount of current an electrical power source is capable of delivering.

shorted transmission line A transmission line that is terminated by a short. If the length of the line is an exact multiple of a quarter wavelength of the transmitted wave, then the line will appear as an open to the source.

shunt regulators An electrical circuit that attempts to maintain a constant output voltage by using a regulating device in parallel (shunt) with the voltage source. Shunt regulators are used with power supplies to help maintain a constant output voltage under a wide range of load conditions.

shunt resistance Resistance in parallel.

sideband The range of harmonic frequencies caused by the modulation of a carrier wave with another lower-frequency wave that represents the information to be transmitted. An upper and

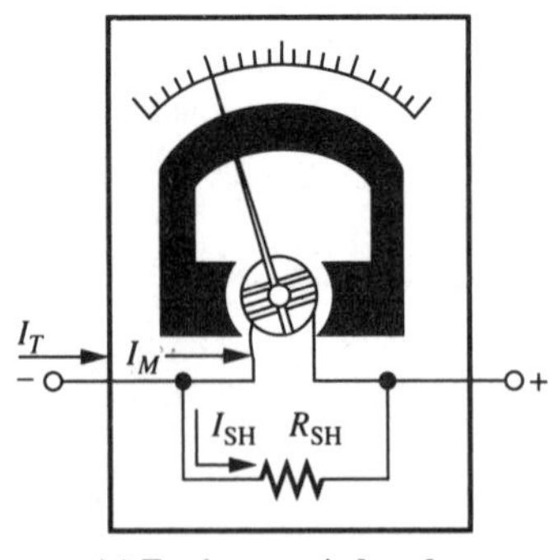

(*a*) Total current is $I_M + I_{SH}$.

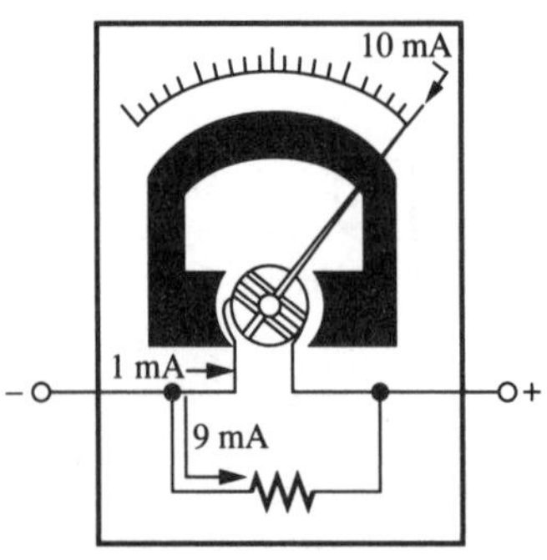

(*b*) Meter indicates 10 mA (1 mA + 9 mA).

Shunt resistance

a lower sideband are produced from the process of modulation. The upper sideband contains the harmonics that are higher in frequency than the carrier; the lower sideband represents the frequencies that are lower in than the carrier.

siemen 1. The unit of measurement for conductance. See conductance. 2. The reciprocal of resistance.

sign bit In complementary binary arithmetic, the leftmost bit represents the sign of the binary number. If the leftmost bit is a 1, then the number is negative; if a 0, then the number is positive.

signal The variations of an electrical quantity (such as voltage, phase, frequency, current, or power) in time that represent usable information.

signal generator 1. An electrical instrument used for the creation of various waveforms (signals). Signal generators are used extensively in troubleshooting all types of electronic systems. 2. Electrical test equipment that provides a calibrated frequency variable over a large range; where the output can be varied with one or more types of signal modulation available.

signal injection A method of troubleshooting an electronic system by electrically putting a signal into each stage, one stage at a time, and observing the output of the system. For example, the signal injection method for troubleshooting a communications receiver would require the use of a modulated RF generator. First, the signal would be injected in the IF amplifier and then the RF amplifier and so on. When an output signal was no longer observed, the defective stage has been found.

signal strength The amplitude of an electromagnetic transmission at a given point usually measured in volts or watts or decibels.

signal strength meter An electrical instrument used for measuring the amount of received signal. Usually calibrated in decibels or arbitrary "S" (for strength) units.

signal-to-noise ratio The ratio of the signal strength to the noise strength. Mathematically expressed as SNR = S/N, where SNR is the signal-to-noise-ratio, S is the signal strength, and N is the noise strength. Generally speaking, the larger this ratio, the better.

signal tracer An electrical piece of test equipment used for detecting the presence of a signal in different locations in a circuit.

signal tracing A method of troubleshooting an electronic system by observing the output signal of each stage. As an example, the signal tracing method for troubleshooting a communications receiver would involve the use of an oscilloscope and the observation of the signal at the output of the RF amplifier, the LO, the mixer, the IF, the detector, and finally, the audio amplifier. A missing output signal indicates the defective stage.

signature analysis A method of troubleshooting digital systems that is based on a comparison of measured bit patterns (called signatures) and documented signatures at various test points (called nodes). When troubleshooting, measured signatures are compared against these documented ones. If a comparison fails, it's an indication that there is a problem at that node, though it does not say what the problem may be.

signature analyzer A digital troubleshooting instrument. See signature analysis.

silicon An element used in the manufacturing of semiconductor devices. Silicon has an atomic weight of 14.

silicon-controlled rectifier A three-terminal unidirectional device whose conduction is controlled by a gate. When the gate is ON, current flows in the device; when the gate is OFF, current will not flow.

simplex mode A mode of transmission in which data are sent in only one direction from transmitter to receiver.

simulcast To transmit information over more than one type of broadcast station at the same time. Any program broadcast in such a manner.

sine A trigonometric function that represents the values of the ratio of the opposite side to the hypotenuse of a right triangle as the opposite angle goes from 0 to 90 degrees.

sine wave A graph of the sine function. A sine wave is used to represent graphically the values (over time) of alternating current (ac).

sine wave measurement The act of taking readings of voltage, current, phase, frequency, or other parameters of the sine wave. Sine wave measurements are usually performed through the use of the oscilloscope.

single-ended transmission The simplest method of connecting two systems together. A single, usually shielded, conductor is used, and the shield is grounded.

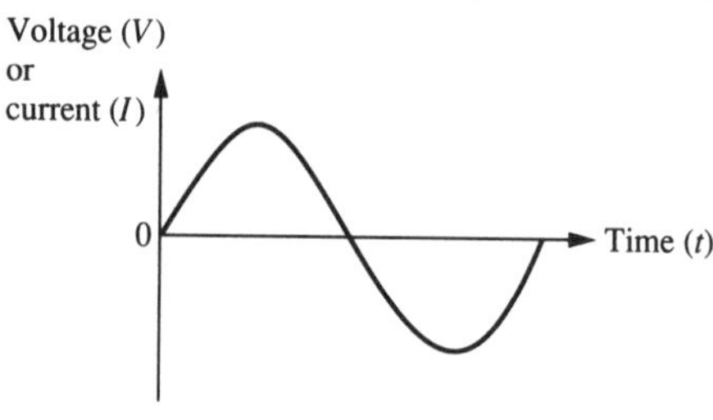

Sine wave

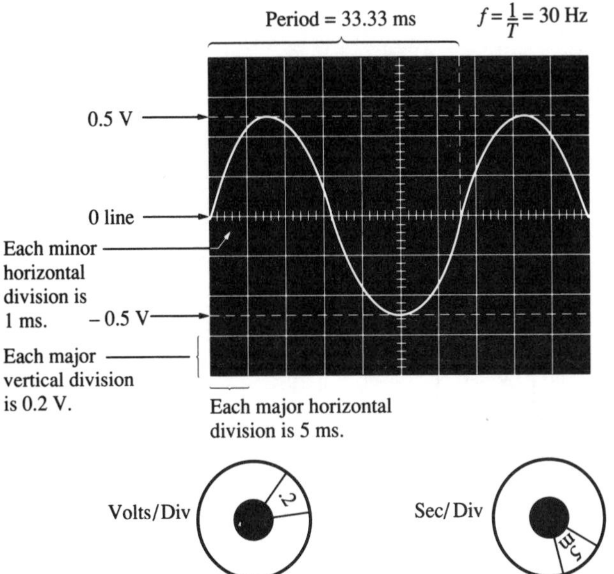

Measurement of the sine wave output for its minimum amplitude and minimum frequency

Sine wave measurement

This system usually suffers from noise pickup.

single in-line package An arrangement of electrical components that allows for easy insertion in a printed circuit board. Packages are characterized by a single row of external connecting terminals that are inserted into holes in the printed circuit board.

single-pole, double-throw switch A single mechanical switch consisting of three contacts. Connection is made from one contact to either of the other two. Thus, at any one time, there is one open and one closed position in the switch.

single-pole, single-throw switch A single mechanical switch having either an open or a closed position.

single shot See monostable.

single sideband An electrical term used in reference to the transmission and reception of only one sideband of the radio signal. Single-sideband transmission is much more efficient than normal transmission (where the carrier as well as both sidebands are transmitted). Since there is no information in the carrier and the information contained in both sidebands is identical, only one sideband (either the upper sideband or the lower) need be transmitted. The disadvantage is that the receiver must have the carrier reinserted at the receiver.

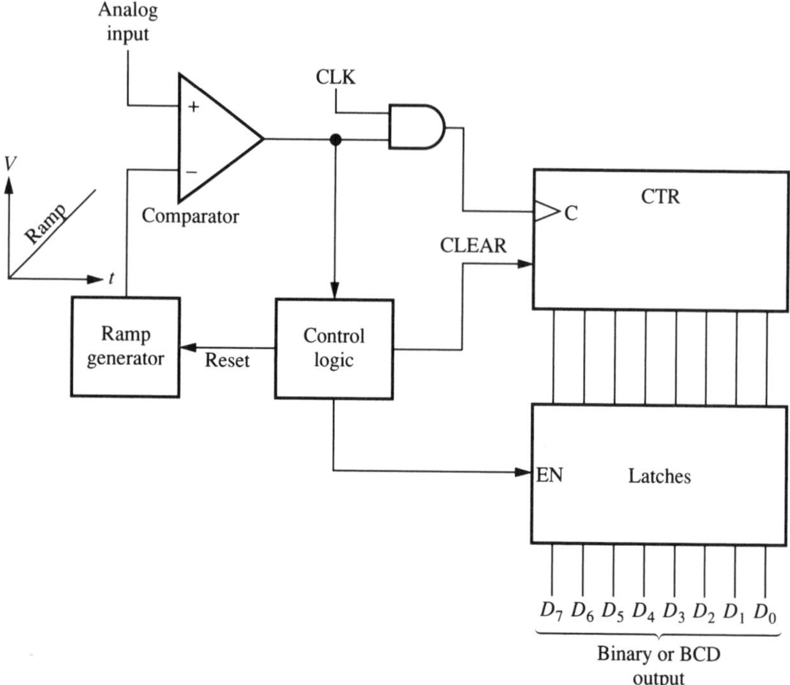

Single-slope A/D converter

single-slope A/D converter An analog-to-digital converter that uses a single ramp function as a part of the conversion process from an analog quantity to a digital code.

single-throw switch A mechanical device used for making or breaking electrical contact where only one set of contacts are needed to be moved to create the contact.

SIP Abbreviation for single inline package. See single in-line package.

SI system Abbreviation for Systéme International d'Units. The system of units adopted by the IEEE in 1965.

skin effect The phenomenon where most of the current transmission in a conductor is done at the surface of the conductor when the currents involved are high frequencies. The higher the frequency of transmission, the greater this effect. Caused by the equivalent series inductance present in all conductors.

skip distance The distance between the source of a transmitted radio wave and its return back to earth resulting from its reflection from the ionosphere. See sky wave.

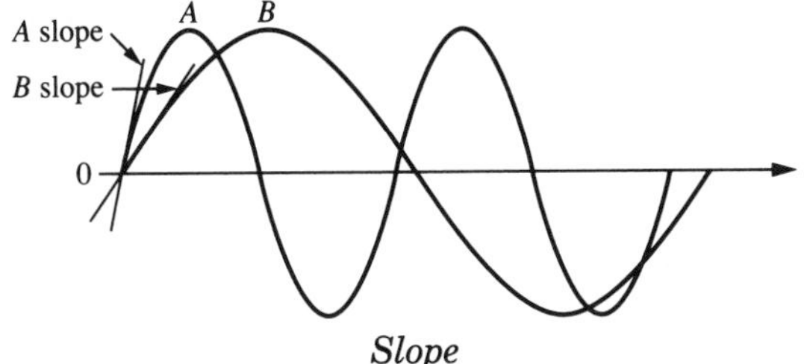

Slope

sky wave The electromagnetic radiation from an antenna highly dependent on the characteristics of the ionosphere. Sky waves occur at frequencies between 300 kHz and 30 MHz. The characteristics of the ionosphere can cause sky waves to be reflected back to the earth.

slide switch A mechanical device used for making or breaking electrical contact; it is activated by moving a control in a parallel manner across the switch housing.

slope The rate of change of a curve at a particular point on the curve.

slope detector A circuit that converts the frequency changes of a carrier into amplitude changes. A slope detector uses a parallel resonant circuit that is tuned to the carrier frequency being detected. The circuit takes advantage of the fact that the amount of signal developed across a parallel *LC* circuit is proportional to the frequency of the incoming signal. Thus, an FM signal will cause different voltage amplitudes across the circuit. These different voltage amplitudes will represent the modulating signal.

SLSI Abbreviation for super-large-scale integration (represents about 100,000 transistors per chip).

slug A unit of measurement for mass in the English system. A slug is equal to 14.6 kilograms in the SI or MKS system.

smart terminal A computer terminal that can do more than just display and send information from and to a main computer. Most smart terminals can perform some processing of the data contained in it, such as word processing.

Smith chart A special polar diagram used in the solution of transmission line and waveguide problems. Most of these types of problems are now more easily and accurately solved on the computer and programmable calculators. Named after its inventor.

smoke detector An electrical device capable of detecting byproducts of combustion, such as smoke.

socket A mechanical device, usually electrically connected to a printed circuit board for the purpose of making a good mechanical and electrical connection to an electronic device. An example is an IC socket, which is normally connected to a printed circuit board and into which an IC chip is inserted.

soft sector A method of keeping track of how data are stored on a computer disk by electrical means rather than physical means. See hard sector.

software The nontangible part of a processing system that determines the process performed by the hardware.

software maintenance The task of keeping software working properly by making necessary programming changes. This may come about because of changes in the computer system or changes in the requirements of the software or the finding of software programming errors once the software is in use.

solar cell 1. An electronic device capable of converting sunlight directly into electricity. 2. A voltage source based on the principle of photovoltaic action where the process of light energy is converted directly into electrical energy.

solder A metal that is easily melted with a heating device called a soldering iron to make a permanent and reliable electrical connection at room temperature.

soldering gun An electrical heating device used for melting solder that has an on/off switch that is operated by the index finger with a pistol grip.

soldering iron An electrical heating device used for melting solder.

solenoid An electromechanical device that consists of a movable iron rod housed in a coil of wire. When electrical current flows through the coil of wire, it creates a magnetic field that causes the movable rod to move. Thus, the solenoid converts electrical energy into mechanical motion.

solid angle The steradian is the unit of measurement of a solid angle. Used in laser technology

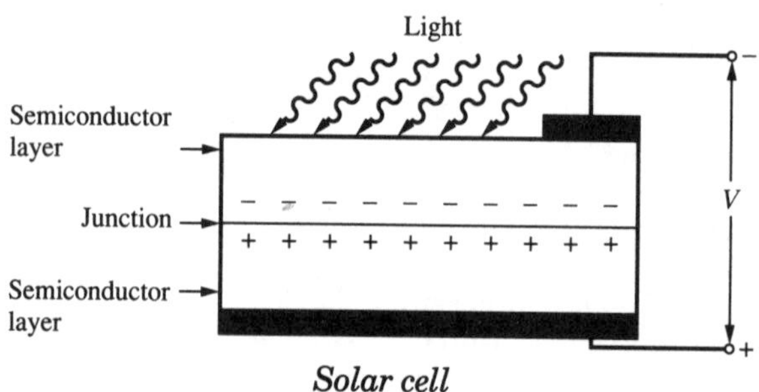

Solar cell

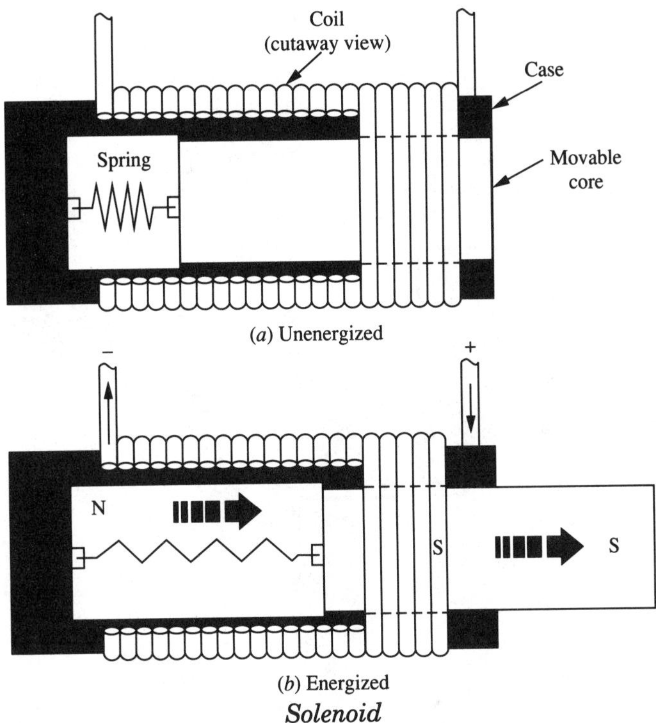

(*a*) Unenergized

(*b*) Energized

Solenoid

to calculate laser radiance. The equation for determining a solid angle is $w = A/r^2$ where w = solid angle in steradians, A = area of the curved surface, r = radius of the sphere.

solid state Pertaining to electrical components that are constructed from semiconductor materials.

S-100 bus A digital bus system designed to operate around the 8080A microprocessor. It used 100 lines and is one of the oldest standard bus arrangements. There were 38 control lines, 16 address lines, 8 data output lines, 8 data input lines, 4 power supply lines, 8 interrupt lines, and 2 grounds. The remaining 16 lines were spare lines.

sound Alternations of air pressure at such a repetition rate that it is detectable by the ear.

source One of the connections of a field-effect transistor. Voltage applied to the gate controls the current between the source and the drain.

source code That part of a computer program as originally entered by the programmer (the source).

source conversion The mathematical changing of a voltage

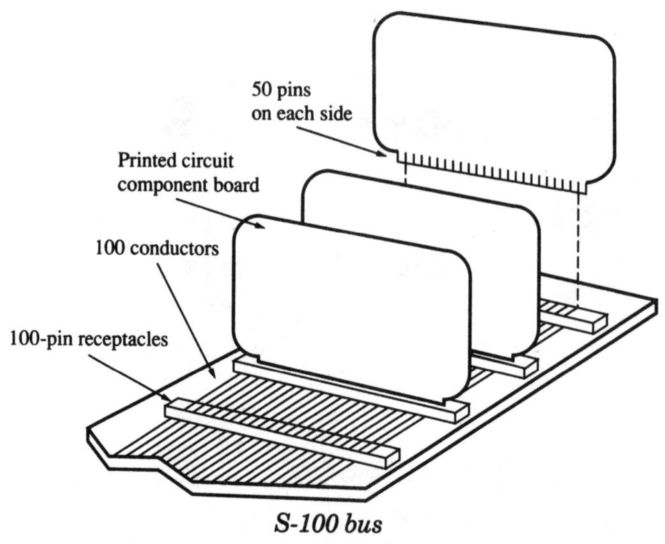

S-100 bus

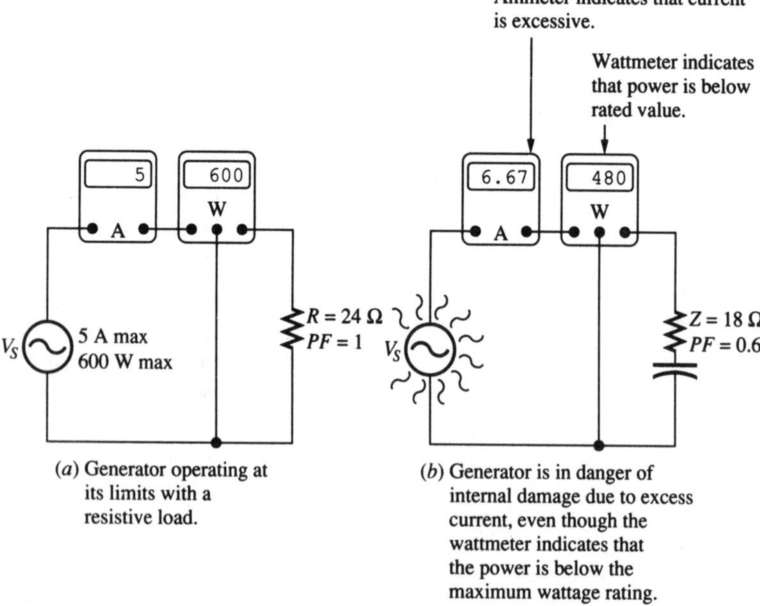

(*a*) Generator operating at its limits with a resistive load.

(*b*) Generator is in danger of internal damage due to excess current, even though the wattmeter indicates that the power is below the maximum wattage rating.

Source wattage rating

source to a current source or a current source to a voltage source. Such source conversions are used as an aid in the mathematical analysis of electrical circuits.

source follower See common drain.

source power rating The maximum safe power that can be delivered by the source, usually expressed in watts. For an ac source where reactive loads may be used, the source power rating should be given in VA (volt-amps) rather than in watts.

space attenuation The loss of energy of a signal as it is sent from one point to another through space. Usually measured in decibels.

spark A short brilliant burst of electrical energy between two electrodes.

SPDT Abbreviation for single-pole, double-throw switch. See single-pole, double-throw switch.

speaker An electromechanical device that converts electrical signals into sound waves. A speaker is usually operated by an audio-frequency amplifier. See audio amplifier. An ideal speaker will amplify the full range of audio frequencies with equal efficiency.

specifications Statements concerning the description of a part or a system in a concise and technical way. Specifications are usually published by the manufacturer of the part or system.

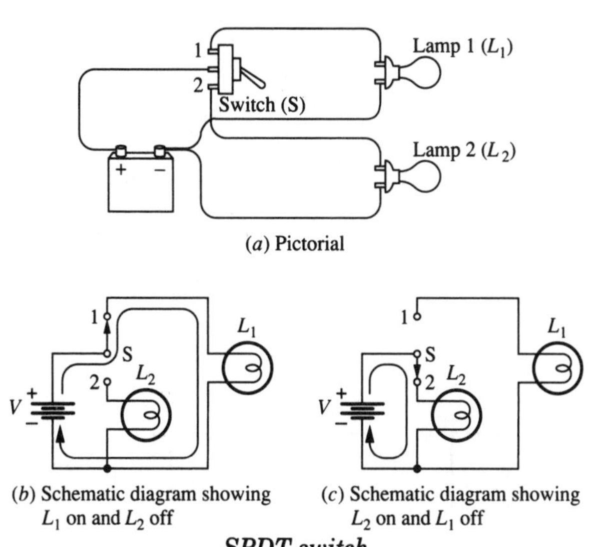

(*b*) Schematic diagram showing L_1 on and L_2 off

(*c*) Schematic diagram showing L_2 on and L_1 off

SPDT switch

Permanent magnet

Coil

Flexible cone

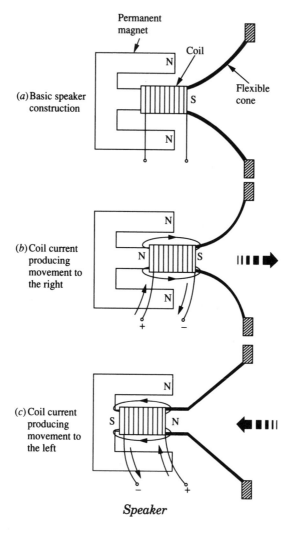

(*a*) Basic speaker construction

(*b*) Coil current producing movement to the right

(*c*) Coil current producing movement to the left

Speaker

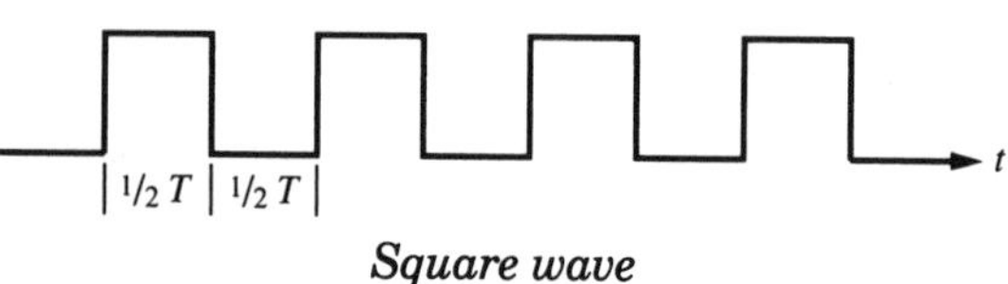

Square wave

spectral Having to do with or as a function of a wavelength. A spectral quality is evaluated for a single wavelength.

spectrum A continuous range of frequencies. Also the frequency components that make up a complex waveform.

spectrum analysis The study of the frequencies present in a given electrical signal. For example, a spectrum analysis of a pure square wave would show that it consisted of an infinite number of odd harmonics of the fundamental frequency.

spectrum analyzer An instrument that analyzes the frequencies contained in a complex waveform.

SPST Abbreviation for single-pole, single-throw switch. See single-pole, single-throw switch.

square wave A periodic waveform with flat tops and bottoms in the shape of squares. The vertical axis is usually measured in volts and the horizontal axis measured in time. An ideal square wave is where the "tops" (called the ON time) is equal in time to the "bottoms" (called the OFF time).

S-R latch Abbreviation for SET-RESET latch. See SET-RESET latch.

SSB Abbreviation for single sideband. See single sideband.

st Letter symbol for steradian. See steradian.

stack A location in memory usually different from the memory locations used by the main program. The address of the stack is kept in an internal microprocessor register called the stack pointer.

stack pointer A counter or register that contains the address of a particular place in computer memory called the stack. A stack pointer may point to the top or to the bottom of the stack.

stage An amplifier with its associated circuit. A transistor requires external components such as resistors in order to operate as an amplifier. The transistor

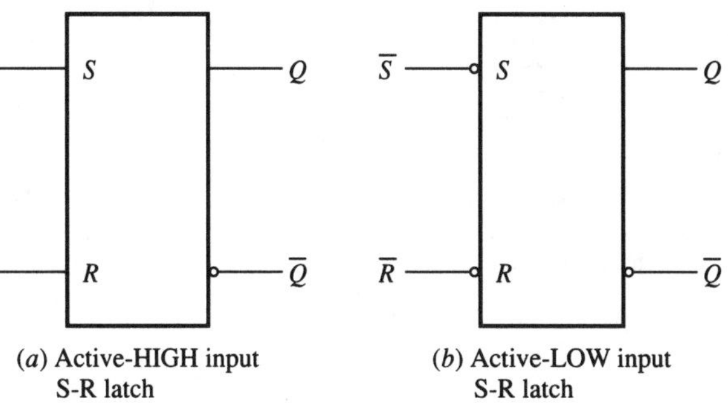

(*a*) Active-HIGH input S-R latch

(*b*) Active-LOW input S-R latch

S-R latch

Illustration of the pushing of data onto a RAM stack.

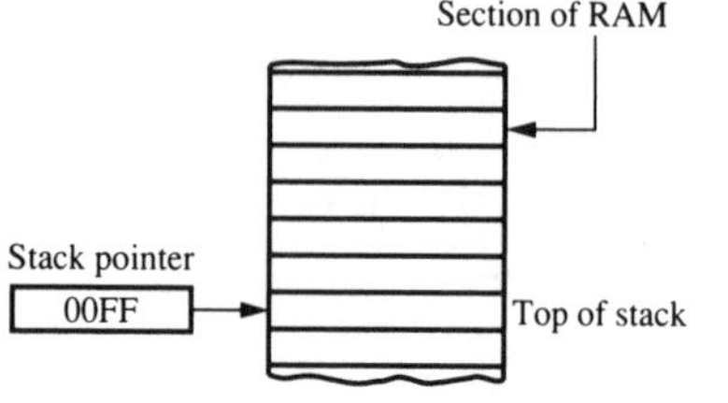

(*a*) Stack pointer addresses top of stack.

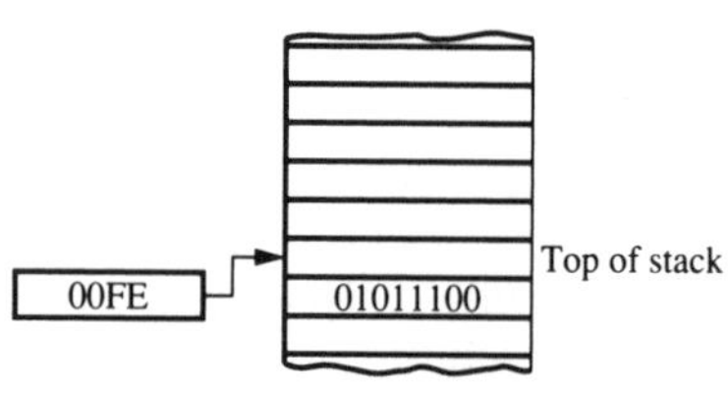

(*b*) First byte "pushed" onto stack; stack pointer decremented.

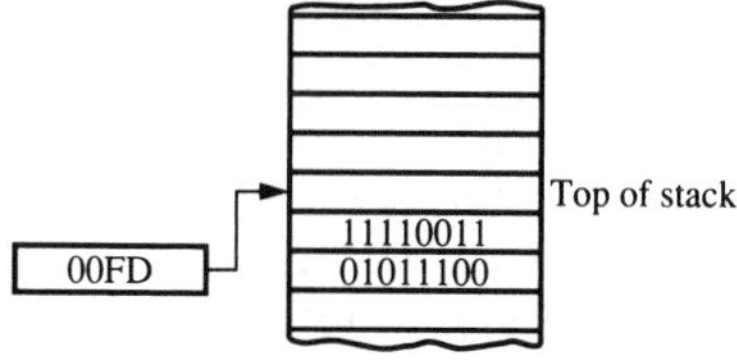

(*c*) Second byte "pushed" onto stack; stack pointer decremented.

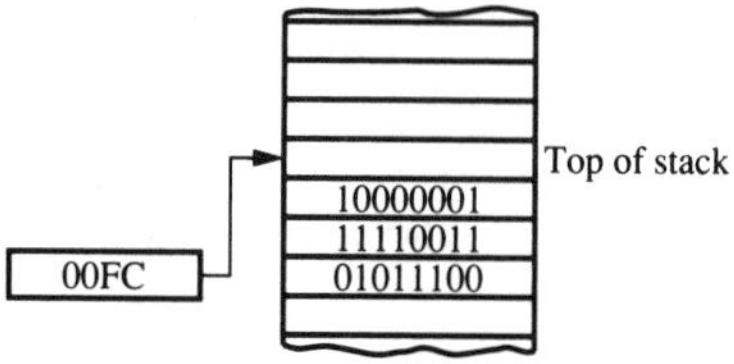

(*d*) Third byte "pushed" onto stack; stack pointer decremented.

Illustration of the pulling of data out of the stack.

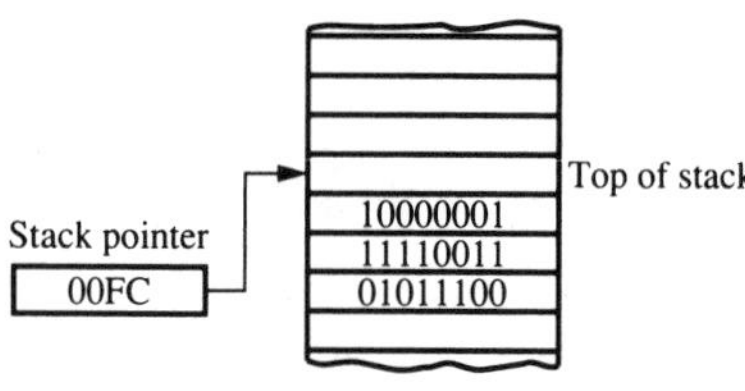

(*a*) Three bytes stored in stack.

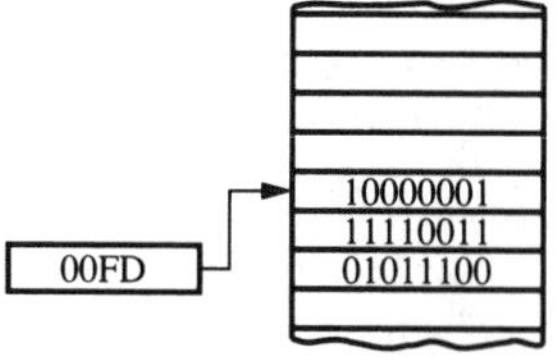

(*b*) Stack pointer incremented; last byte in read out.

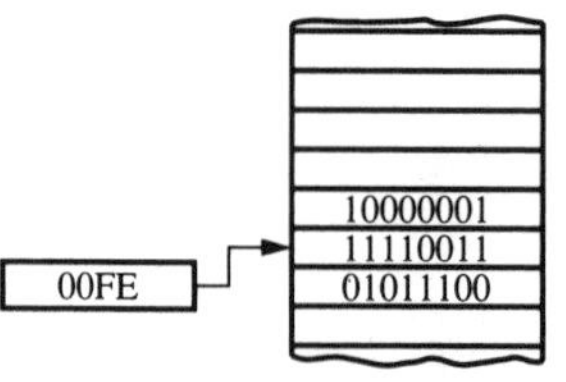

(*c*) Stack pointer incremented; second byte in read out.

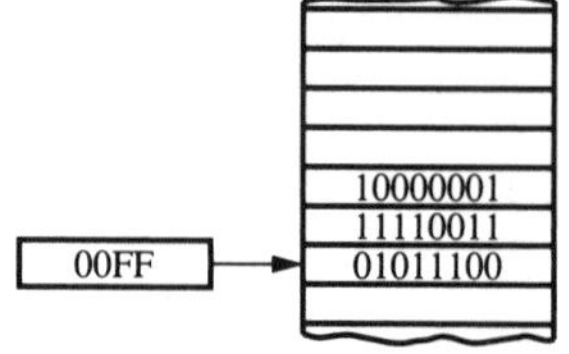

(*d*) Stack pointer incremented; first byte in read out.

Stack

along with these necessary components is called a stage of amplification. Thus a two-stage amplifier would consist of two transistors and their associated external components.

staggered tuning A method of making a wideband amplifier in a multistage amplifier by tuning each amplifier to a different center frequency, each of which is within the desired band of frequencies to be amplified.

staircase An electrical signal having the shape of an ascending or descending staircase. A series of decreasing or increasing levels of an electrical current or voltage.

stairstep-ramp A/D converter An analog-to-digital converter that uses a binary counter as a part of the conversion process from an analog signal to a digital code. As a part of this process, a signal that appears as a stairstep (on a ramp) is produced.

stand-alone debugger A separate program used as an aid for finding programming bugs in another source program. A stand-alone debugger is available for most versions of Pascal and C as well as other programming languages.

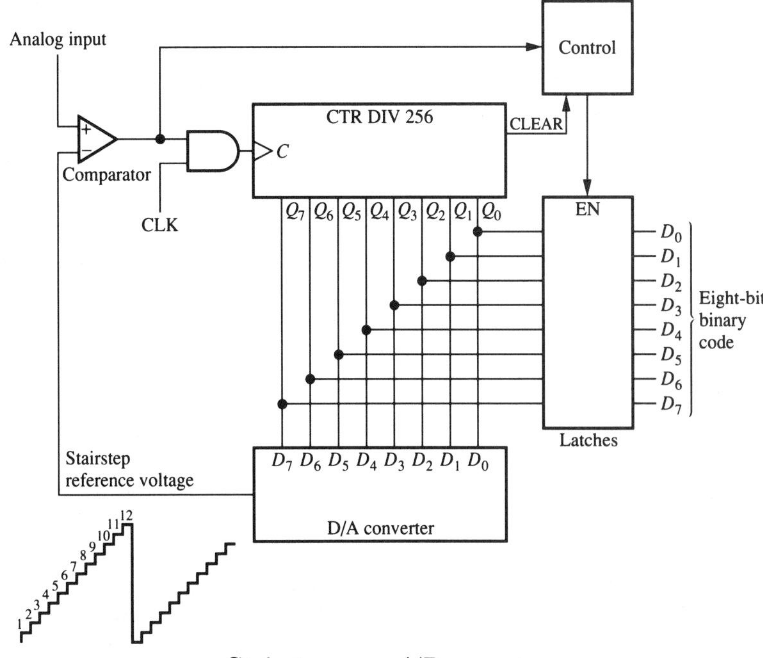

Stairstep-ramp A/D converter

standard An agreed-to concept or value established by physical laws, agreements, or customs.

standing waves The wave pattern along a transmission line produced when electromagnetic energy is transmitted along the line. Consists of an incident wave and a reflected wave; the phase relations of these two waves is what produces the standing wave. The optimum condition for transmitting power to a load over a transmission line occurs when there are no standing waves.

standing wave ratio A ratio, used in transmission lines, of the maximum value of the current or voltage along the line to the minimum value of the current or voltage along the line.

state diagram A drawing that shows a sequence of events accomplished by a sequential circuit. Referred to as a state machine.

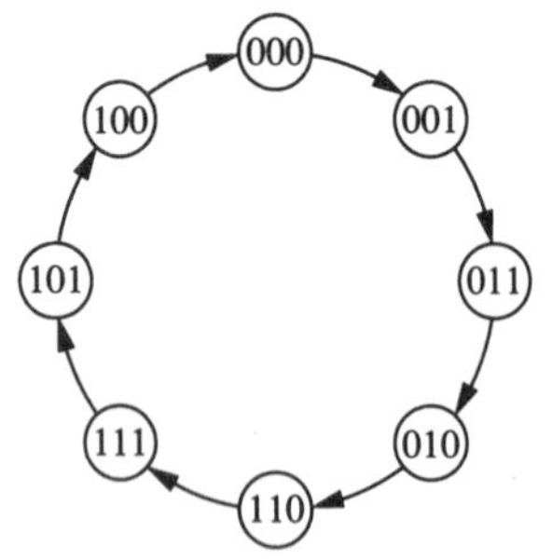

State diagram

state machine See state diagram.

static electricity Electrical charge caused by an excess of deficiency of electrons that are in a state of equilibrium.

static RAM Random access memory (read-write memory) that is either bipolar or MOS latches. Static RAM will store a bit pattern and maintain it as long as power is applied. See dynamic RAM.

status flag In a microprocessor, a 1-bit indicator used to reflect a certain condition after an arithmetic or logic operation by the ALU.

steady state A condition where no circuit values are changing. Descriptive of a stable, nonchanging condition.

step-down transformer A transformer with fewer windings on its secondary than on its primary. A step-down transformer will produce a smaller voltage on the secondary over that of the primary with a corresponding increase in secondary current.

step-index fiber A glass fiber with a uniform index of refraction. See graded-index fiber.

step recovery diode A solid-state device with a very fast turn-off time. The diode uses a graded doping technique. Used in high-frequency switching applications.

step-up transformer A transformer with more windings on its secondary then on its primary. A step-up transformer will produce a larger voltage on the secondary

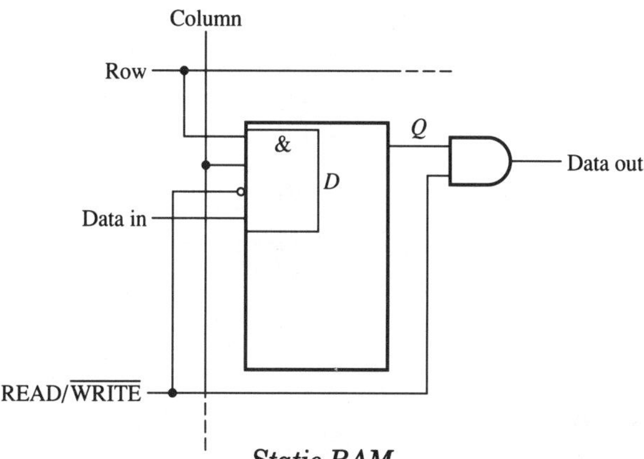

Static RAM

over that of the primary with a corresponding reduction in secondary current.

steradian Unit of measurement of a solid angle. A solid angle of one steradian is equal to a surface area of the square of the radius.

stereo Common usage implies an audio amplifier system that has at least two separate sound channels to give the listener the illusion of musical depth.

storage battery An electrochemical source of electricity made by combining two or more storage cells in series. See storage cell.

storage cell An electrochemical source of electrical energy that can be charged whereby electrical energy is converted into chemical energy, and later, the stored chemical energy can again be converted back to electrical energy.

stored program concept The idea of storing instructions along with data in computer memory.

stray capacitance Capacitance produced by the proximity of conductors. Stray capacitance exists between the leads of resistors and other similar devices such as between the anode and cathode of a diode. See capacitance.

strobing A signal that causes a predictable result. Strobing is used in decoders to prevent the effects of glitches.

subatomic Smaller than an atom. An electron can be considered a subatomic particle.

subcarrier A carrier frequency transmitted with a main carrier frequency. Subcarriers are used in FM stereo to help detect the left and right stereo channels.

subroutine That part of a computer program that can be treated as if an independent set

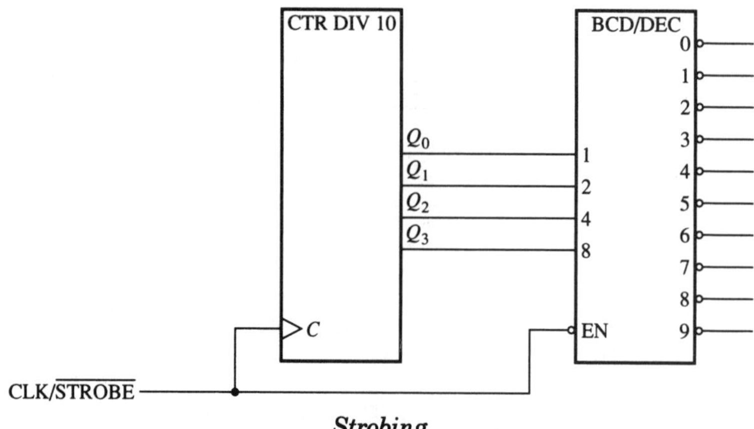

Strobing

of instructions and referred to as such by the rest of program.

subscript A notation used to indicate a different item from a list of similar items, where the notation is made in small print below and following the name of the item. An example is R_1, R_2, and R_3, which represents three different resistors.

substitution theorem A concept that the voltage across and current in any branch of a dc bilateral network that is known can be replaced by any combination of elements that will maintain the same voltage across and current in the chosen branch.

successive approximation A/D converter One of the most popular types of analog-to-digital converters. Uses a successive approximation register and an A/D converter. This system is very fast because it uses digital successive approximation techniques to determine quickly the value of the input analog quantity.

Successive-approximation A/D converter.

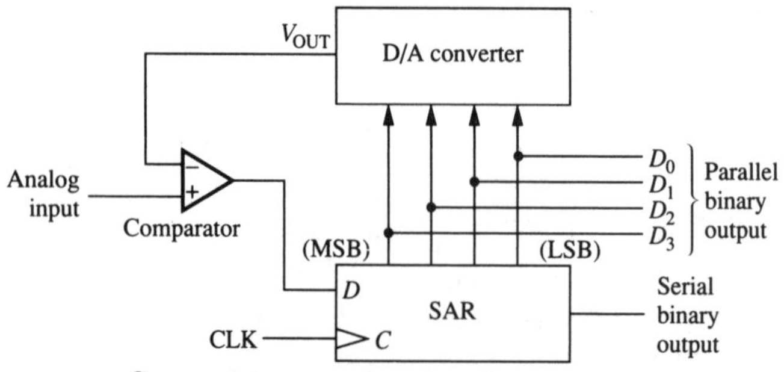

Successive approximation A/D converter

Successive-approximation conversion process.

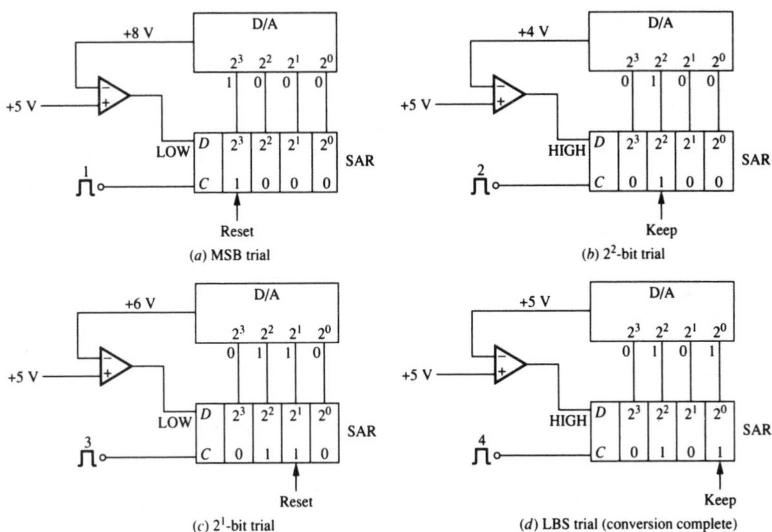

Successive approximation A/D converter (con't)

summing network An electrical circuit where the output signal is equivalent to the sum of two or more input signals.

sum of products A Boolean expression consisting of ANDed terms that have been ORed together. As an example, $(A \cdot B) + (C \cdot D)$.

supercomputers Generally refers to the largest, fastest, and most expensive of the mainframe computers. See mainframe computer.

superconductivity The phenomenon exhibiting a decrease in resistance in certain materials as the temperature of the material is lowered. At a low enough temperature, the resistance of the material becomes zero, making it superconductive.

superconductor A material that possesses no resistance. Su-

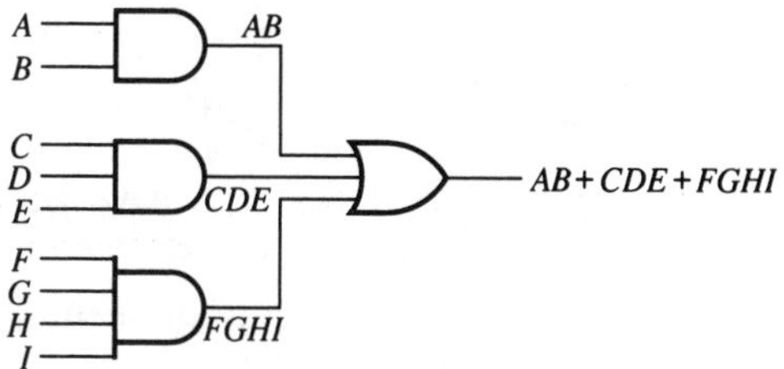

Sum of products

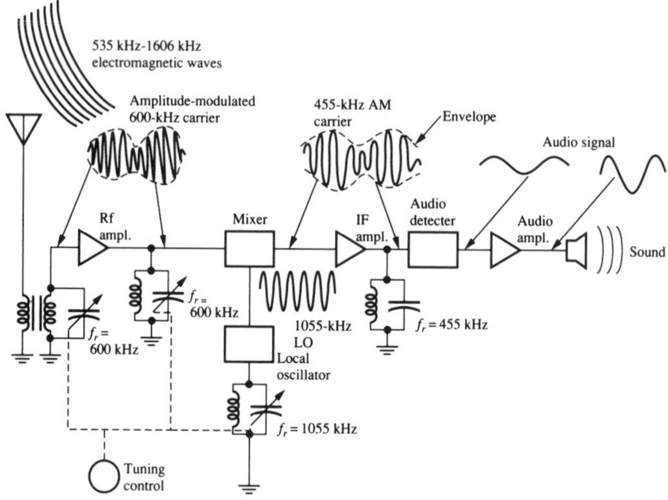

Superherterodyne AM receiver

perconductors are achievable in certain materials near absolute zero.

superheterodyne A communications receiver using heterodyning to receive radio signals. Such a receiver will contain a local oscillator, mixer, and one or more stages of IF amplifiers. See heterodyne.

superheterodyne AM receiver An AM receiver using the superheterodyne principle.

superposition theorem A concept that permits considering the effects of each source independently in a manner where the resulting voltage and/or current is the algebraic sum of the currents and/or voltages developed by each source independently.

suppressed-carrier modulation A type of modulation where only the resulting sidebands are desired and not the original carrier. Results in greater transmission efficiency since the carrier does not contain any information and would require two-thirds of the total transmission power if it were transmitted.

surface wave See ground wave.

surge A sudden increase in current or voltage in a circuit.

susceptance The reciprocal of reactance, measured in siemens. See capacitive susceptance or inductive susceptance.

sweep generator An electrical instrument capable of supplying a continuous range of fre-

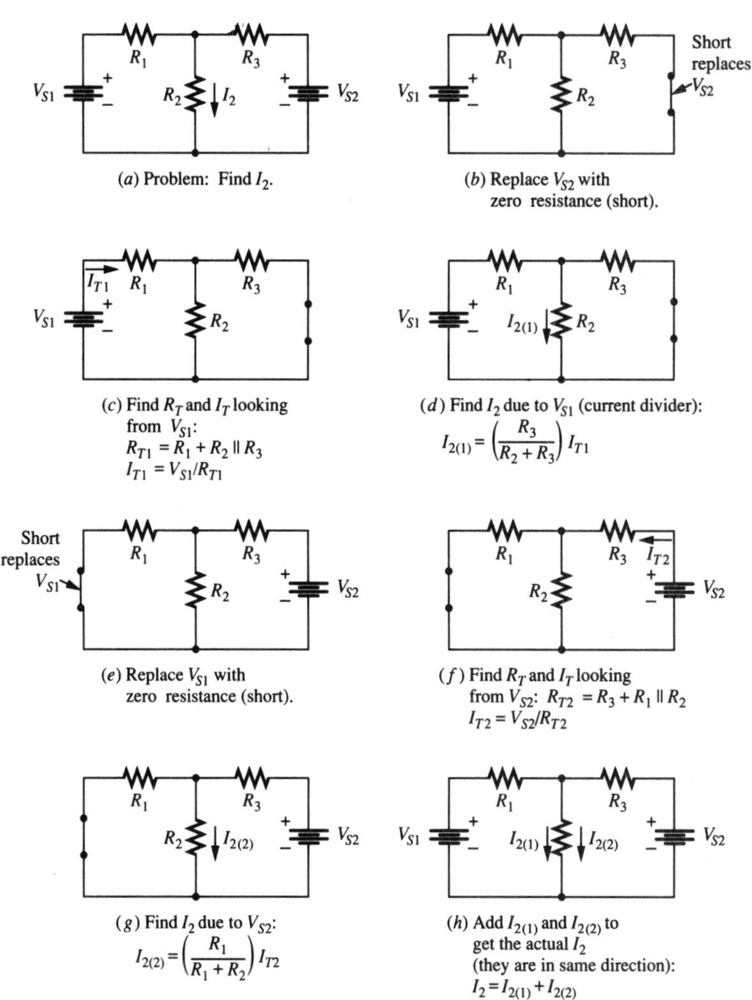

(a) Problem: Find I_2.

(b) Replace V_{S2} with zero resistance (short).

(c) Find R_T and I_T looking from V_{S1}:
$R_{T1} = R_1 + R_2 \parallel R_3$
$I_{T1} = V_{S1}/R_{T1}$

(d) Find I_2 due to V_{S1} (current divider):
$I_{2(1)} = \left(\frac{R_3}{R_2 + R_3}\right) I_{T1}$

(e) Replace V_{S1} with zero resistance (short).

(f) Find R_T and I_T looking from V_{S2}: $R_{T2} = R_3 + R_1 \parallel R_2$
$I_{T2} = V_{S2}/R_{T2}$

(g) Find I_2 due to V_{S2}:
$I_{2(2)} = \left(\frac{R_1}{R_1 + R_2}\right) I_{T2}$

(h) Add $I_{2(1)}$ and $I_{2(2)}$ to get the actual I_2 (they are in same direction):
$I_2 = I_{2(1)} + I_{2(2)}$

Superposition theorum

quencies for the purpose of testing the response of electronic systems to these ranges of frequencies. Commonly used in the alignment and maintenance of communications equipment. See marker generator.

switch An electrical or mechanical device capable of interrupting the flow of current. An example of a mechanical switch would be the simple on/off wall switch used in the home. An electrical switch example is a transis-

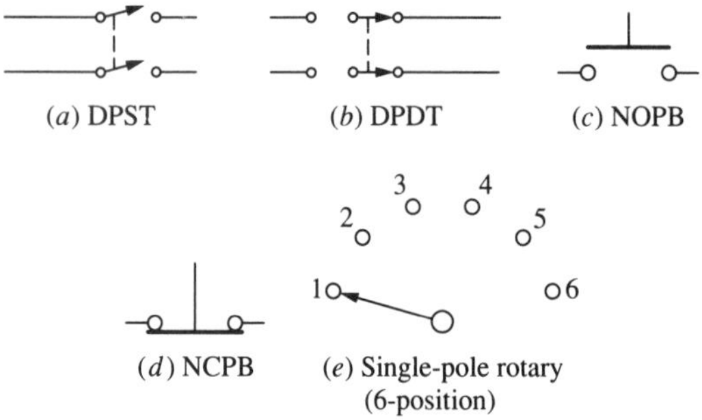

(*a*) DPST (*b*) DPDT (*c*) NOPB

(*d*) NCPB (*e*) Single-pole rotary (6-position)

Switch

tor circuit where a small signal causes the transistor to allow or block current flow. An electromechanical switch is a relay where electrical energy causes the opening or closing of a mechanical switch.

switch equivalent AND A method of representing an AND gate through the use of electrical switches. In this circuit, all switches must be closed (representing a TRUE) before the circuit is active.

switch equivalent OR A method of representing an OR gate through the use of electrical switches. In this circuit, closing any one switch (representing a TRUE) will cause the circuit to be active.

swr Abbreviation for standing wave ratio. See standing wave ratio.

symmetry As used in digital electronics refers to the amount of pulse ON time versus the amount of pulse OFF time. A symmetrical pulse has the same amount of ON time as it does OFF time.

synchronous In electronics a waveform that has some timing relationship with another waveform. See asynchronous.

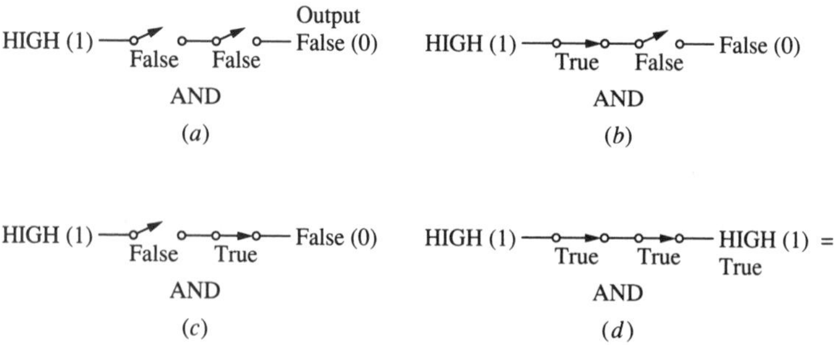

HIGH (1) False False Output False (0)
AND
(*a*)

HIGH (1) True False False (0)
AND
(*b*)

HIGH (1) False True False (0)
AND
(*c*)

HIGH (1) True True HIGH (1) = True
AND
(*d*)

Switch-equivalent AND

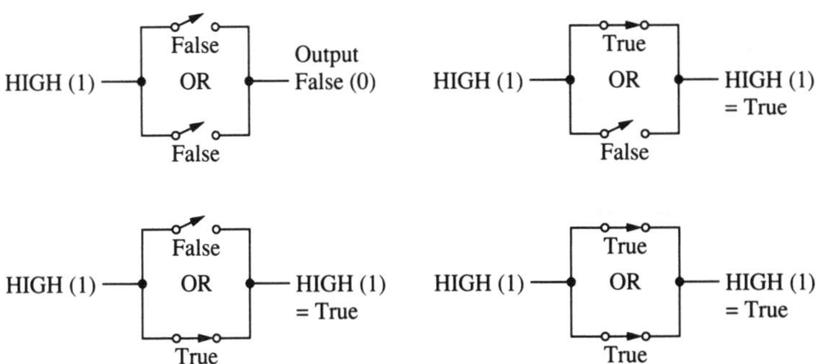

Switch-equivalent OR

synchronous counter A digital counter where all the flip-flops are triggered at the same time by the clock pulse.

sync pulse Abbreviation for synchronized pulses. Pulses used to cause certain digital events to happen at a predetermined timing sequence.

sync separator In television systems a circuit that separates the horizontal and vertical sync pulses from the received signal.

syntax The grammar of a programming language. The rules that apply to the correct arrangement of symbols and groups of symbols for a particular program-

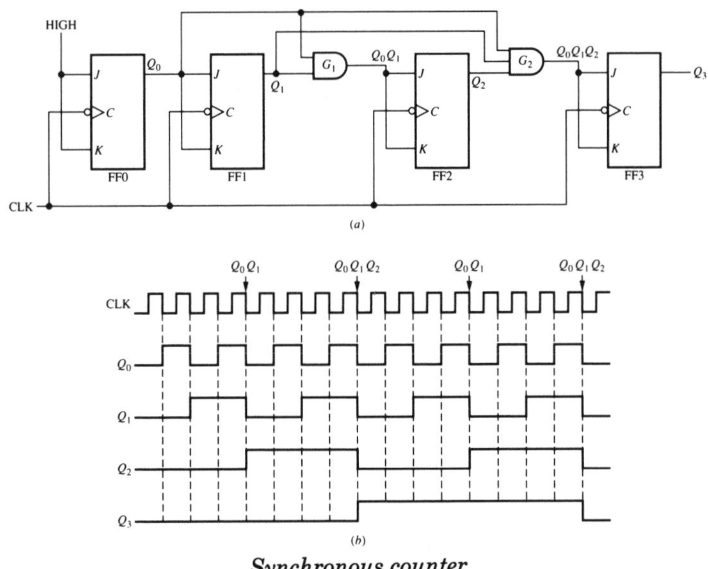

Synchronous counter

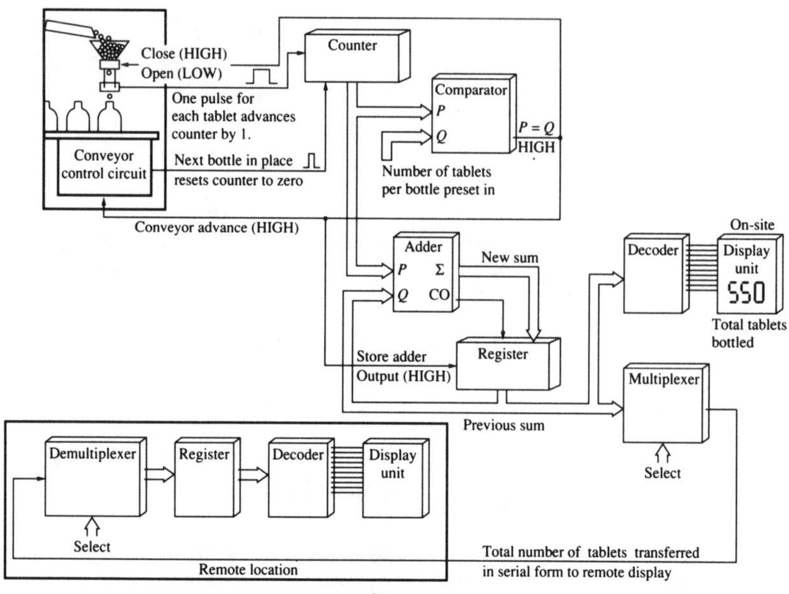

System

ming language, in order to evoke a predictable response from the computer.

system In digital systems, a group of individual functional units that are interconnected in such a manner as to produce a larger functional unit. As an example, a digital system may be used to control an automated processing plant that fills bottles with tablets or manufactures automobile engines.

synthesizer In electronics, emulating something electrically that makes it appear as the real thing being emulated. See voice synthesizer.

t

table A collection of related items each of which is identified by a label.

table lookup A technique used in computer programming where data are stored in a table and the contents of the table are scanned and selected during the execution of the program.

tamper switch An electromechanical device that is installed in a system to detect any unauthorized entry into the system. Usually, the tamper switch itself is constructed in such a fashion as to detect unauthorized removal of itself.

tandem See cascade.

tank circuit An electrical circuit capable of storing dynamic electrical energy, in the form of a parallel resonant circuit at resonance.

tantalum capacitor An electrolytic capacitor constructed from a tantalum foil anode. Since electrolytic capacitors are polarized, you must make certain that the positive anode is connected to the most positive portion of the circuit and the negative lead to the most negative portion of the circuit into which it will be fitted.

tape Typically a plastic ribbon coated with a magnetic material for the recording and playback of information.

tape recorder An electromechanical device used for placing and/or receiving information on a magnetically coated plastic ribbon. A tape recorder consists of the necessary mechanical assembly and electrical circuits.

tapered resistor A variable resistor whose change in value depends upon the mechanical position of the wiper arm. As opposed to a tapered resistor.

tapped resistor A wirewound fixed-value resistor with one or more terminals along its length. Tapped resistors are generally used for voltage divider applications.

tapped transformer A transformer that has its primary and/or secondary windings with extra connections (taps).

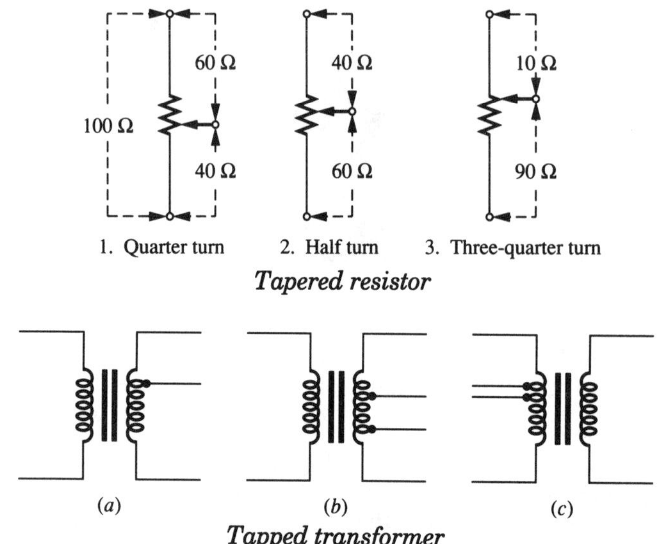

Tapered resistor

(a) (b) (c)

Tapped transformer

tapped winding In an inductor, a coil winding with one or more terminals along its length. Sometimes referred to as an autotransformer.

T-connection See Y-connection.

TDM Abbreviation for time division multiplexing. See time division multiplexing.

TE mode Abbreviation for transverse electric. The mode of operation of a waveguide when the electric field is perpendicular to the direction of wave propagation.

teach box An electronic device that is used to program instructions into a robot, usually hand held.

technician A person who works directly with the repair, maintenance, testing, and fabrication of mechanical or electrical equipment.

telecommunication A general term that applies to the transmission and receipt of information between two remote points.

teleconference Exchanging information between two or more groups of people using audio, visual, and data transmission techniques in real time.

telemetry The process of sensing, measuring, and transmitting information from a remote location to a central place for recording and/or further processing.

telephone An electromechanical device used for the purpose of transmitting and receiving speech between two separate points in real time.

television A communication system that uses audio and visual information.

temperature coefficient A value that indicates a change in an electrical quantity with temperature. Temperature coefficient is usually expressed as a percentage and may be positive or negative. A positive temperature coefficient means that the parameter (such as resistance) of the device increases with an increase in temperature, while a negative temperature coefficient means that the same parameter decreases in value with an increase in temperature.

temporary magnet Any magnetic material that, when magnetized by an outside source of energy, will not retain its magnetism. A temporary magnet is characterized as a material that has a high permeability and a low retentivity.

tera Prefix for the numerical value of 10^{12}, abbreviated T. A value of one million million = 1,000,000,000,000.

terahertz A frequency of one million megahertz or 1^{12} Hz = 1,000,000,000,000 Hz.

terminal In electronics a mechanical device used to attach one or more electrical connections. In computers, a computer station that connects to a larger computer system.

terminal drift The changes in some characteristic of a device, circuit, or system due to changes in ambient temperature.

terminating The completion of a circuit by connecting it to some device.

terminating resistor A resistor that is used to complete a circuit connection.

ternary A number system to the base 3. The ternary number system has three symbols, 0, 1, and 2. Ternary also means to take something in groups of three.

test A procedure used to determine the operation of a device, circuit, or system. A test may also be designed to determine the nature and/or location of a fault in such systems.

text editor In computers, a program that allows the user to modify text material. A text editor will allow the insertion and deletion of letters, words, lines, or paragraphs as well as moving or copying any section of text from one place to the other. Text editors may also come with word searches/replacements as well as other features such as underlining or boldfacing of characters.

text mode In computers, the operation of a computer so that standard alphanumeric characters are displayed on the monitor or printer. See graphics mode.

thermal breakdown A condition where a device is destroyed due to heat caused by its electrical operation.

thermal emf The voltage produced when the junction of two dissimilar metals is heated.

thermal noise Undesirable random signals generated by electronic circuits due to the random interaction between electrical charges in a conductor. This random interaction is directly related to the temperature of the conductor: the greater the temperature, the greater the thermal noise.

thermal relay A mechanical device that will open or close an electrical connection as a result of its temperature.

thermal runaway A condition that results from excessive electrical currents and causes a device to heat up, lowering the resistance of the device and allowing current to increase, which in turn causes the device to heat up even more. This process continues until the device destroys itself.

thermal shutdown In a voltage regulator circuit. The voltage regulator actually turns itself off if the temperature reaches a certain level. Normal output operation is not resumed until the

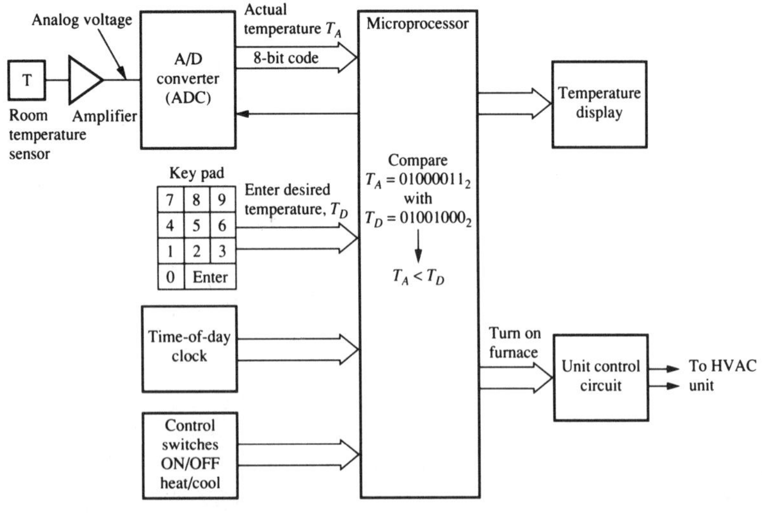

Thermostat

regulator has cooled to a specified temperature.

thermionic emission The emission of charged particles (such as electrons) due to heat or an incandescent body. When a wire is heated in a vacuum, electrons are released from the heated wire. This is a method used by cathode ray tubes to release electrons in its electron gun assembly.

thermistor A variable resistor whose resistance is changed by a change in temperature. Thermistors are sometimes found in power supplies as part of a safety circuit that will shut down the power supply if it overheats.

thermocouple A device that produces the thermocouple effect. See thermocouple effect.

thermocouple effect The effect that produces an electromotive force when there is a difference in temperature between two junctions of dissimilar metals in the same circuit.

thermometer An instrument used for the measurement of temperature. An electronic thermometer may use a temperature sensitive resistor such as a thermistor to help make the measurement.

thermostat 1. A device used to control the temperature settings of an enclosed area. 2. An electronic thermostat that is microprocessor controlled can be used to control the temperature of different areas according to the time of day, where the temperature may be adjusted automatically when the areas are not occupied.

Thevenin equivalent An electrical circuit consisting of a voltage source in series with an impedance. The Thevenin equivalent circuit will model a more complex electrical circuit. Complex electrical circuits may be reduced to their Thevenin equivalents, which are in turn easier to analyze under various loads than their original, more complex circuits.

Thevenin's equivalent circuit A circuit representation of

Thevenin's equivalent circuit.

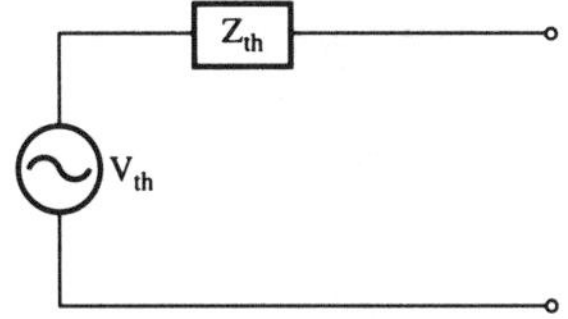

An ac circuit of any complexity can be reduced to a Thevenin equivalent for analysis purposes.

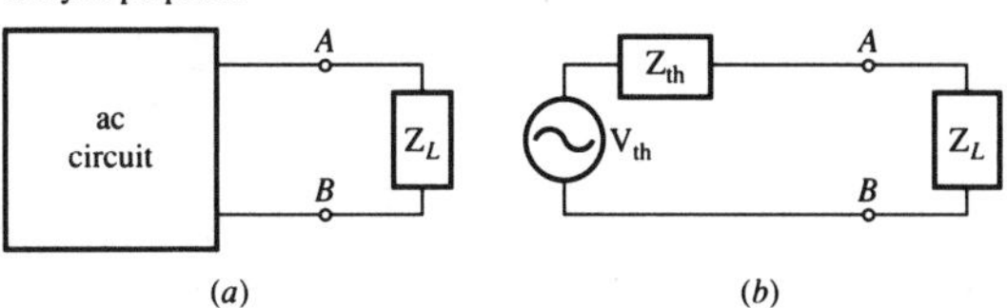

Thevenin's equivalent circuit

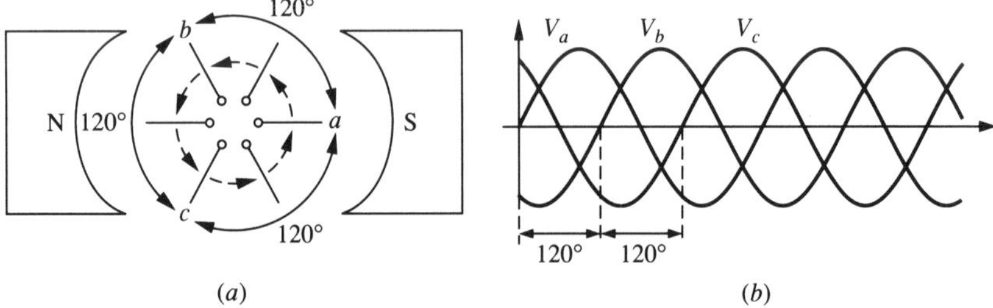

Three-phase generator

a Thevenized circuit. See Thevenin's theorem.

Thevenin's theorem A statement that any two-terminal linear dc network can be represented as a single fixed-voltage source in series with a fixed resistor.

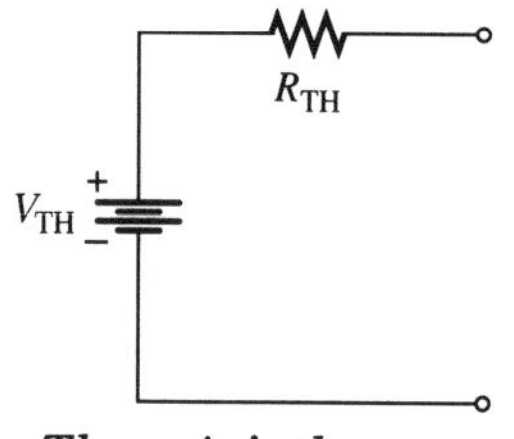

Thevenin's theorem

three phase Having a phase difference of 120 degrees or one-third of a cycle.

three-phase generator An electrical generator that produces three simultaneous multiple sine waves that are separated from each other by 120 degrees.

three-phase inductor motor A specially constructed motor designed to operate from a three-phase electrical system.

three-phase loads An electrical load that is either of a Y- or a delta-connection.

three-phase transformers Various transformer connections designed to operate from a three-phase power system.

three-state logic A logic family that uses devices that have three output conditions: HIGH,

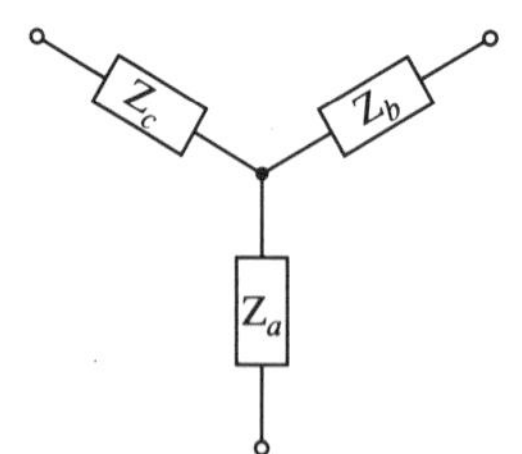

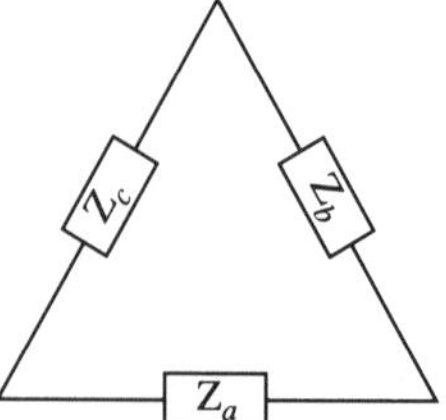

(a) Y-connected load (b) Δ-connected load

Three-phase loads

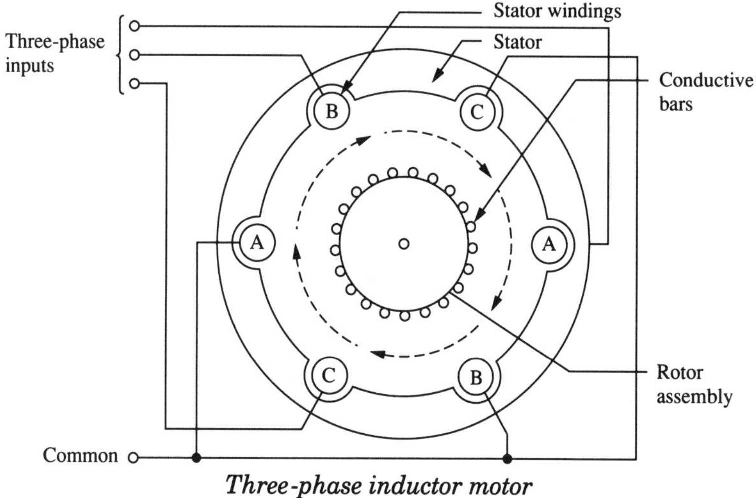

Three-phase inductor motor

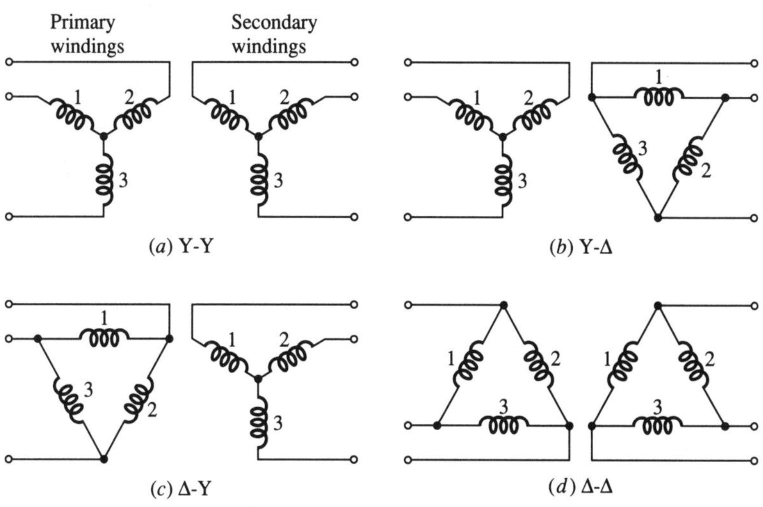

Three-phase transformer

Percentage of final charge after each charging time-constant interval.

Number of Time Constants	% Final Charge
1	63
2	86
3	95
4	98
5	99 (considered 100%)

Percentage of initial charge after each discharging time-constant interval.

Number of Time Constants	% Initial Charge
1	37
2	14
3	5
4	2
5	1 (considered 0)

Time constant

LOW, and OPEN (or high impedance). These devices allow connections of more than one output to a common line or groups of lines called a bus.

threshold current The minimum amount of current required for some kind of electrical action to take place.

thyristor A four-layer semiconductor device that uses internal feedback to produce a latching action. Thyristors can operate only as electrical switches.

time constant A measure of the amount of time it takes for a capacitor to charge up or discharge down to a specified voltage or the amount of time it takes an inductor to build up or collapse its magnetic field resulting in a specified current. One time constant is the amount of time (measured in seconds) required to produce 63% of these processes.

time delay The amount of time that transpires between two events.

time delay circuit A circuit that delays the transmission of an electrical signal for a given amount of time.

time division multiplexing The process of sampling two or more signals for the purpose of transmitting the resultant information over the same channel. Time division multiplexing is used to transmit different kinds of data over a single communications link.

time domain The analysis of electrical signals in regard to changes they make with time. The graphs of electrical waveforms are usually presented in the time domain where the horizontal axis is measured in time (usually seconds) and the vertical axis in volts.

time sharing A method of operating several different systems so that they may communicate with each other over a commonly shared interconnection. This is accomplished by causing only two systems at a time to communicate with each other. This process must be designed to happen quickly enough between all systems so as to not economically justify a more elaborate connection scheme between the systems.

timing diagram A diagram in which pulses (or similar waveforms) are shown in relationship to each other. May consist of a series of pulse waveforms shown in relationship to another series of pulsed waveforms.

tinning The process of soldering the tip of stranded wire before placing it in a circuit or other connection. Tinning helps keep the strands of the wire together.

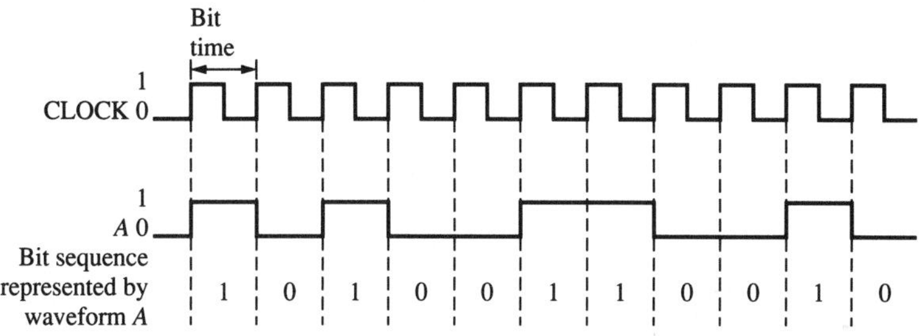

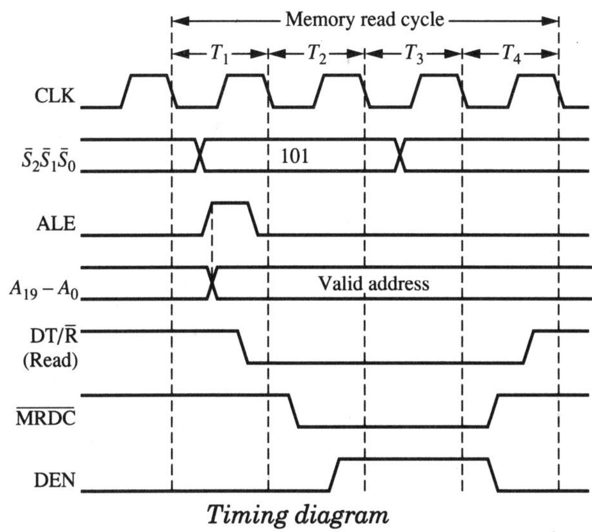

Timing diagram

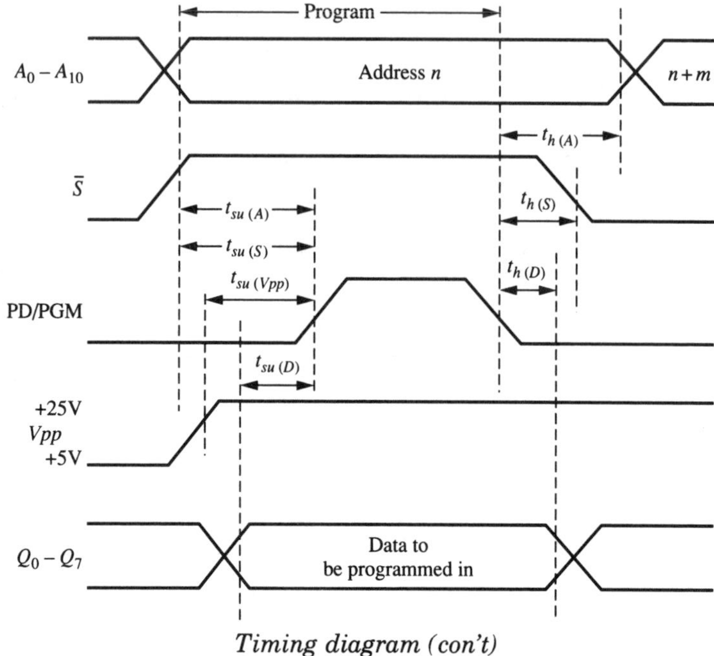

Timing diagram (con't)

tint control In a color television receiver, the adjustment that affects the colors on the screen. This is accomplished by controlling the phase of the 3.58-MHz color oscillator, resulting in a different color being displayed for a given phase difference.

TM mode Abbreviation for transverse magnetic. The mode of operation of a waveguide when the magnetic field is perpendicular to the direction of wave propagation.

T-network An electrical network composed of three branches that schematically form the letter "T."

toggle In an electrical circuit, any action that will cause a circuit to change states abruptly. For example, to toggle a flip-flop means to cause its output logic to change from TRUE to FALSE or from FALSE to TRUE.

tone control An electromechanical device used for modifying the frequency response of an audio amplifier. A tone control is used to emphasize the high frequencies or the low frequencies and is adjusted by the listener for the most pleasing results.

tone decoder A circuit with a predictable output at a specific frequency. Tone decoders are

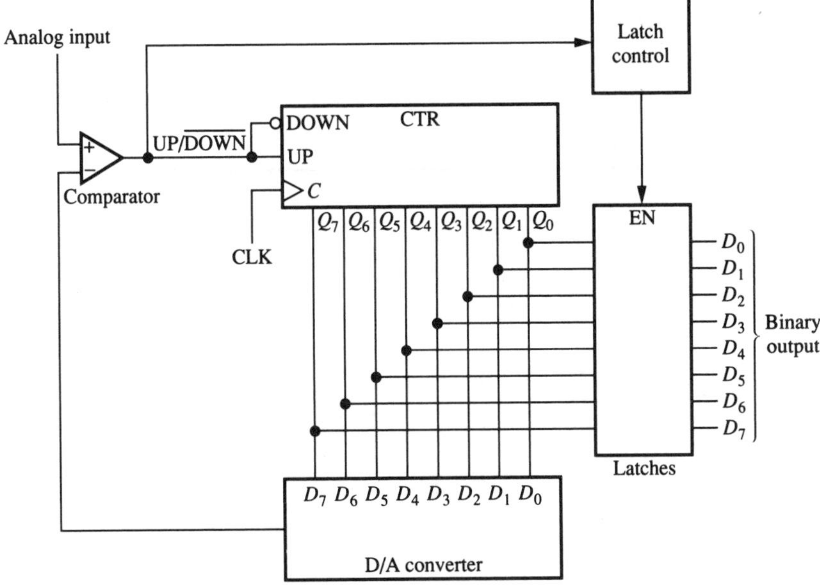

Tracking A/D converter

used in touch-tone telephone systems and other such systems where specific frequencies have special meanings.

touch tone A method of transmitting data by the use of tones. In telephone dialing, touch tone is used to transmit the phone number being dialed. Touch tone for phones actually uses seven frequencies that are combined in such a manner as to produce 12 distinct codes representing the numbers 0 through 9 and the symbols * and #.

tracking A/D converter An electrical circuit that converts an analog signal into a digital code. This type of converter uses a binary up-down counter that keeps "track" of the value of the incoming analog quantity.

trailing edge The ending part of a pulse. For a negative-going pulse, the trailing edge is the transition from a low voltage (usually 0 volts) to a higher voltage (usually +5 volts). For a positive going pulse, the trailing edge is the transition from a higher voltage (usually +5 volts) to a lower voltage (usually 0 volts).

trailing-edge PDM A form of pulse duration modulation where the trailing edge of the pulses varies in accordance with the amplitude of the modulating signal. See pulse duration modulation.

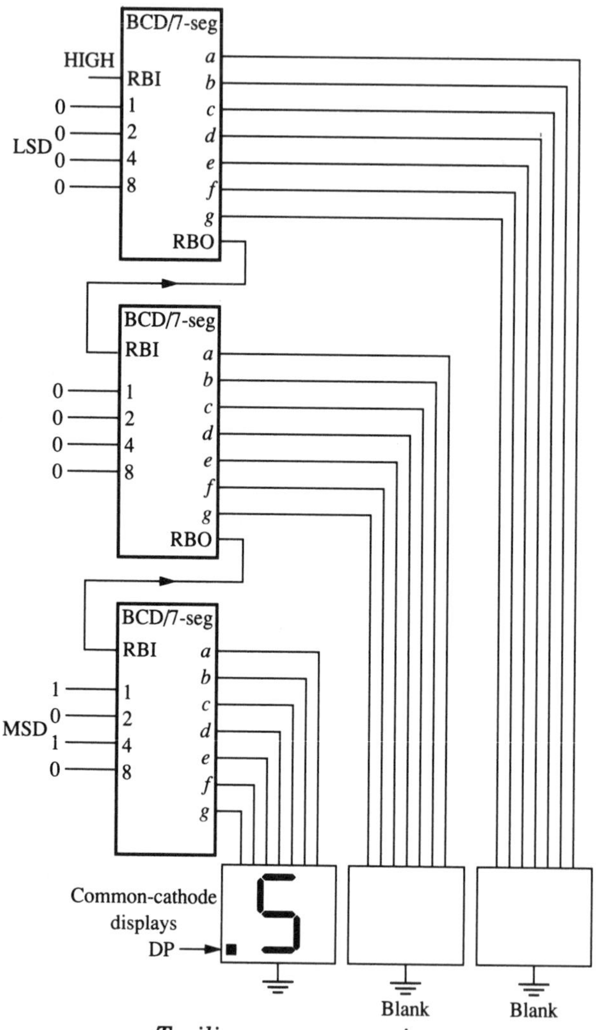

Trailing-zero suppression

trailing-zero suppression A feature sometimes found in seven-segment decoders. Here all the lowest-order digits are blanked if they are a zero. This is done in an attempt to make the display easier to read.

transceiver The combination of a communications transmitter

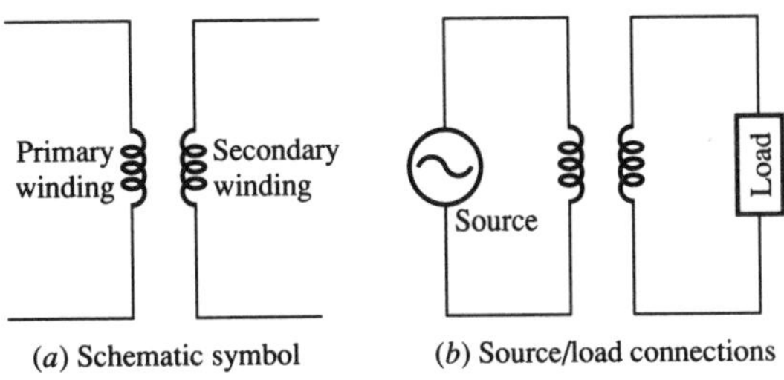

(*a*) Schematic symbol (*b*) Source/load connections

Transformer

and receiver in a common housing. A transceiver is usually a portable unit that employs some circuits in common with both the transmitter and receiver (such as a common power source).

transconductance A measurement of an amplifier that is the ratio of a change in output current for a given change in input voltage. Mathematically, $g_m = dI/dV$.

transducer Any device that converts energy changes from one form to another. As an example, a microphone is a transducer because it converts sound waves to electrical patterns; a loudspeaker is a transducer because it converts electrical patterns into sound waves.

transformer An electrical device that couples electrical energy through a changing magnetic field. A transformer utilizes windings of wire. One set of windings, called the primary, is magnetically coupled via a magnetic core to another set of windings, called the secondary. See autotransformer.

transformer coupling Using a transformer to pass signal variations from one stage to the other while blocking the dc voltages. A method of matching the impedance of one stage to that of the other.

transformer testing A method of testing a transformer by using an ohmmeter to check for winding continuity and shorts between the windings.

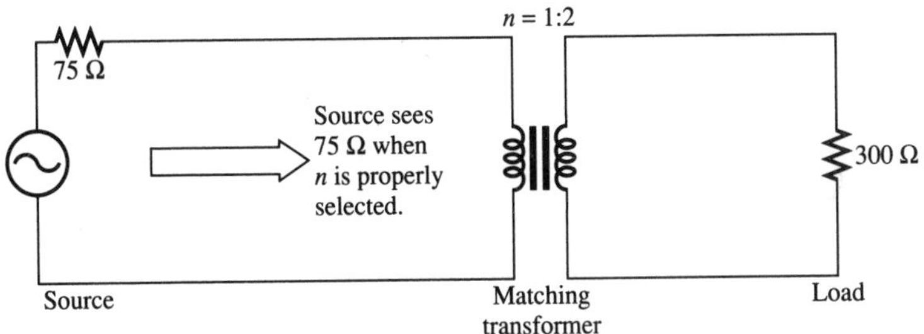

Transformer coupling

Open primary winding.

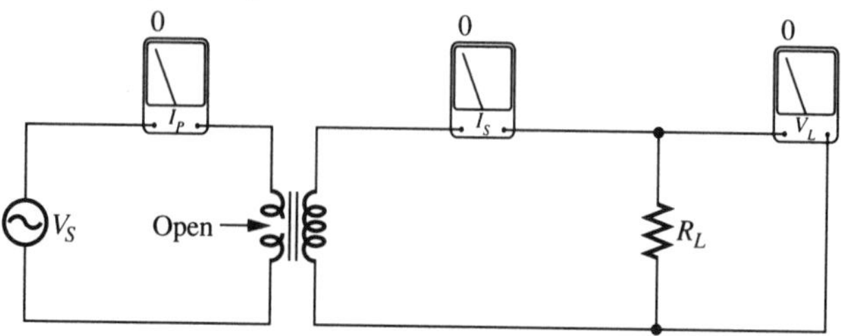

(*a*) Conditions when the primary is open

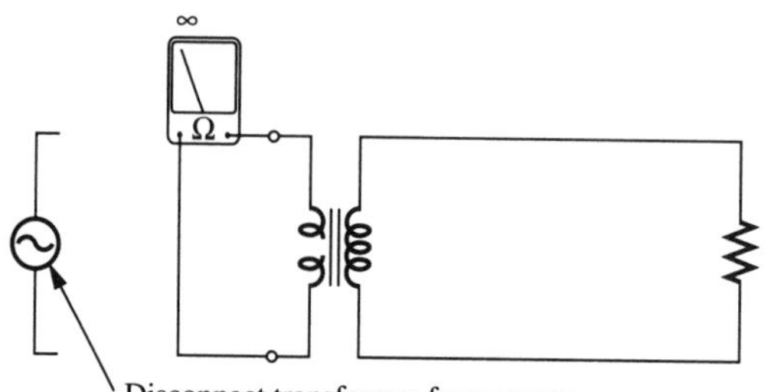

(*b*) Checking the primary with an ohmmeter

Open secondary winding.

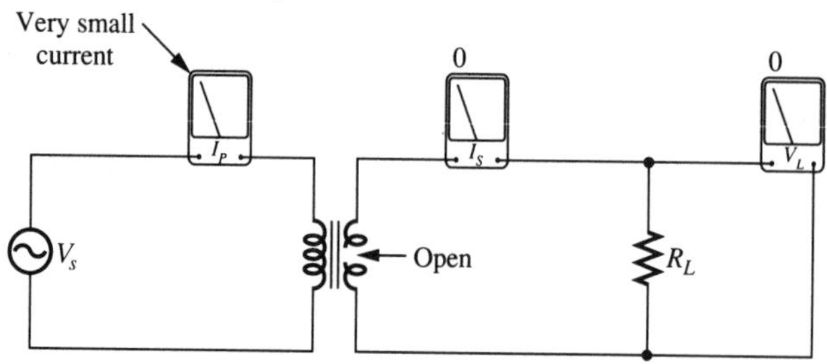

(*a*) Conditions when the secondary is open

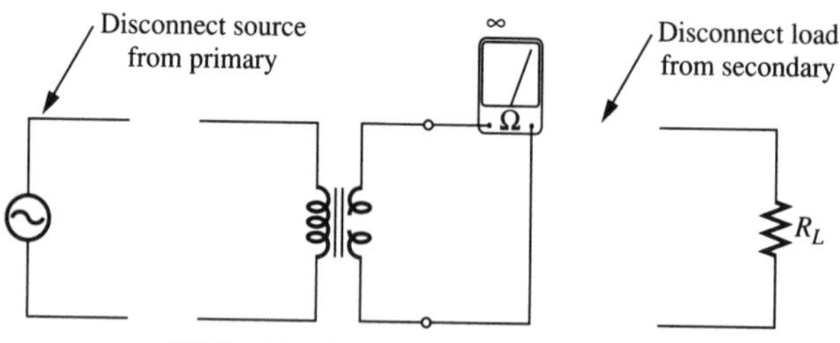

(*b*) Checking the secondary with an ohmmeter

Transformer testing

Shorted secondary winding.

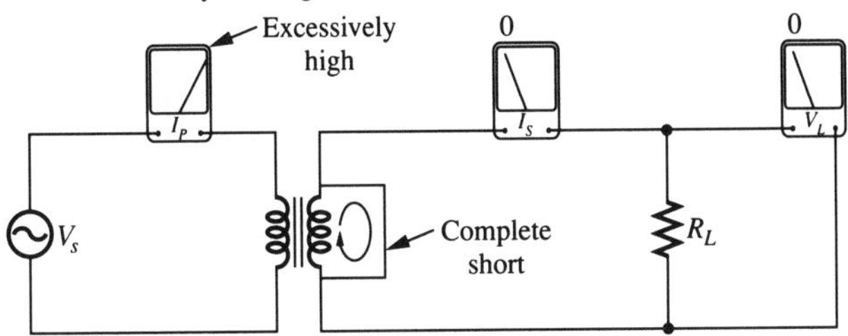

(*a*) Secondary completely shorted

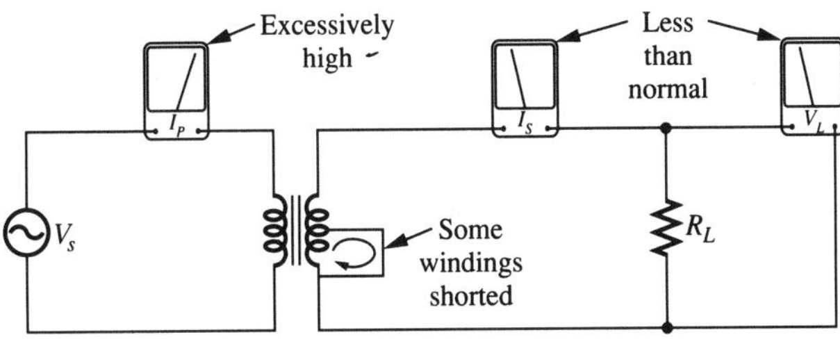

(*a*) Secondary partially shorted

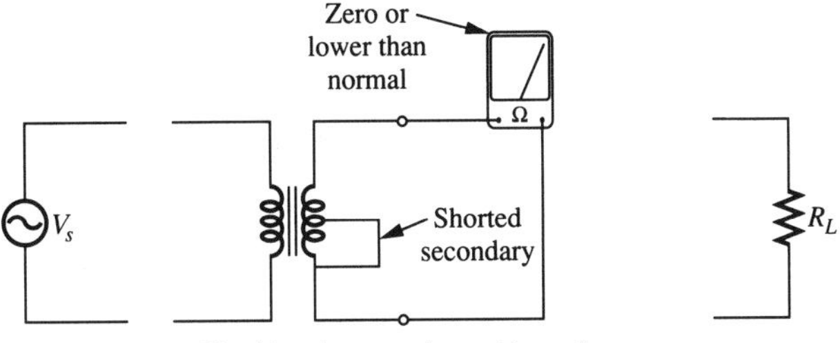

(*c*) Checking the secondary with an ohmmeter

Transformer testing (con't)

transformer winding The direction of the transformer windings with relation to the primary and secondary; determines the relationship of the polarities between the primary and secondary voltages.

transient A momentary change occurring in a circuit or system.

transistor A three-element solid-state current-controlled semiconductor.

transistor-transistor logic Digital logic circuits using transistors fabricated on integrated circuits. TTL consists of a series of logic circuits such as standard TTL, low-power TTL, Schottky TTL, low-power Schottky TTL,

Relative polarities of the voltages are determined by the direction of the winding.

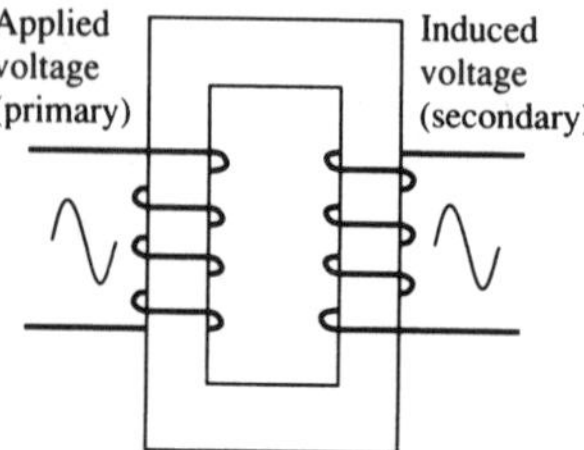

(*a*) The primary and secondary voltages are in phase when the windings are in the same effective direction around the magnetic path.

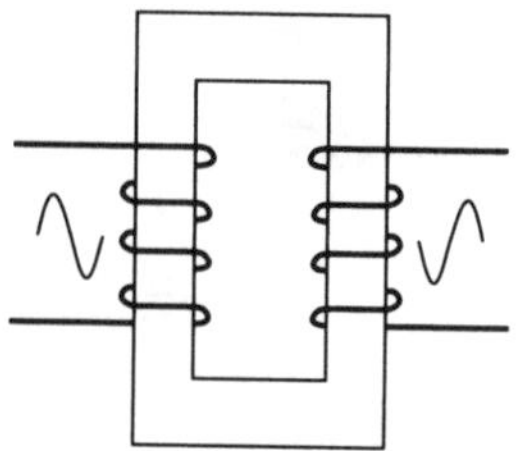

(*b*) The primary and secondary voltages are 180° out of phase when the windings are in the opposite direction.

Phase dots indicate relative polaritites of primary and secondary voltages.

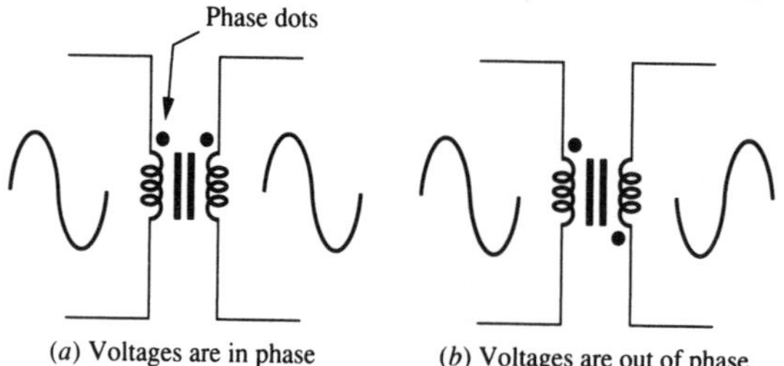

(*a*) Voltages are in phase (*b*) Voltages are out of phase

Transformer winding

advanced low-power Schottky TTL, and other types of characteristics such as delay times.

transition A change from one electrical value to another. As an example, the transition of a clock pulse is from 0 volts to +5 volts.

transmission The sending and receiving of electrical energy or intelligence from one point to another.

transmission line Conductors used to transmit electrical signals. Can be analyzed as containing series inductance and parallel capacitance (between the lines). Transmission lines exhibit specific electrical characteristics that are dependent upon the frequency of transmission.

transmission loss The decrease in power resulting from transmitting electrical energy from one point to another. Usually expressed in decibels.

transmitter Electrical equipment used to send information to another point. A transmitter is used at a radio station to transmit the radio signal.

transmitting The process of conveying information from one point to another. When a radio station is sending information to other radios, the station is said to be transmitting.

transparent In computers, a program or process that is not noticed directly by the user. In materials, any material that allows the passage of light.

transponder A communications repeater unit. Receives an incoming signal, amplifies it, and retransmits it. Communication satellites use transponders to receive signals from one point on the earth and retransmit the signal to another point on the earth, thus assuring reliable communications without the use of mechanical electrical connections such as wires.

treble The higher frequencies of the audio spectrum. In music from middle C (261.63 Hz) upward.

tremolo In electrical musical instruments a subaudio modulation of the audio tone produced by the instrument producing a warbling or fluctuating effect of approxmately seven cycles per second.

trf Abbreviation for tuned radio frequency amplifier. See tuned radio-frequency amplifier.

triac A bidirectional rectifier that consists of two SCRs connected in parallel. Triacs are used as electrically controlled switches for ac loads.

triangular waveform A waveform consisting of a positive-going and a negative-going ramp both the same slope.

Alternating triangular waveform.

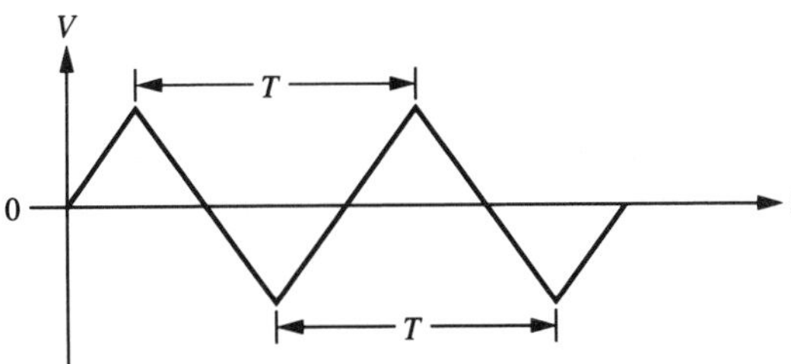

Triangular waveform with a nonzero average value.

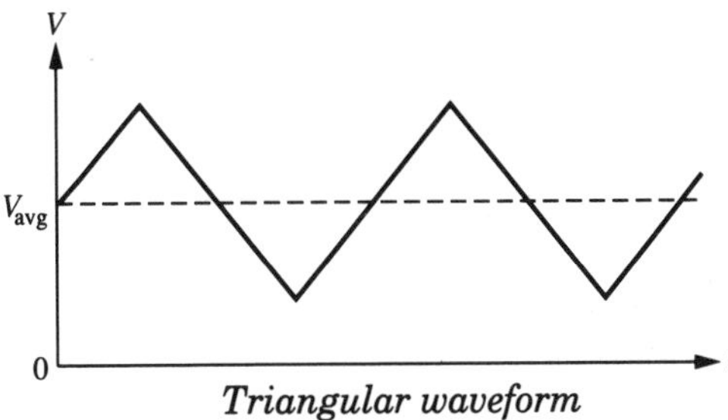

Triangular waveform

trickle charge A small current flow usually used to maintain a standby power source at its peak power capability.

trigger A logic pulse of a short duration used to control a digital process.

trimmer capacitor See trimming capacitor.

trimmer resistor A small variable resistor that allows service adjustments for the purpose of setting it to a required value.

trimming capacitor A small variable capacitor usually associated with another capacitor that contains a service adjustment for the purpose of setting the total capacitance to a required value.

triode A three-element electron tube consisting of a cathode, control grid, and plate. Electrons flow from the cathode (when it is heated by the filament) through the control grid and to the positively charged plate. Small voltage variations on the control grid will cause current variations in the current flow of the triode.

triple nickel (555) An integrated circuit used in timing applications. Called a 555 timer by manufacturers of the device, it can be wired as an oscillator whose frequency and duty cycle are determined by the values of external resistors and capacitors. It also has other applications such as a voltage-controlled oscillator.

tristate See three-state logic.

tristate buffer A digital device that has three states, ON, OFF, and OPEN (or high impedance). In a tristate buffer there is one input, one output, and an ENABLE line. The output is logically the same as the input when the ENABLE line is active. When the ENABLE line is not active, the output appears as an open, regardless of the condition of the input line.

tristate driver A digital device that is similar to a tristate buffer in its operation but is used to increase the power level of the digital pulse in order to operate (drive) other digital circuits. See tristate buffer.

trivalent In semiconductors, meaning three. A trivalent impurity (an element with three valence electrons) is added to pure silicon in order to form an excess of holes.

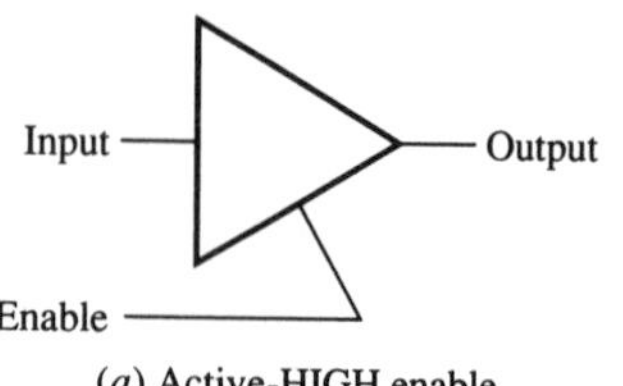

(*a*) Active-HIGH enable

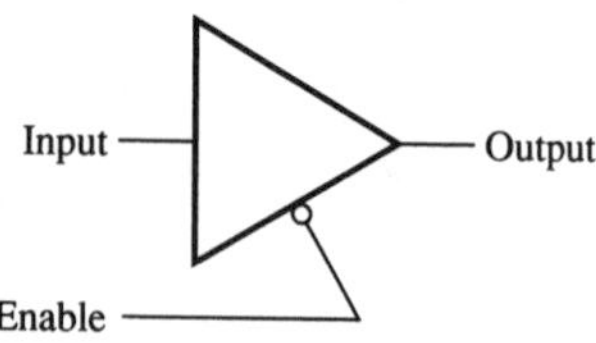

(*b*) Active-LOW enable

Tristate buffer

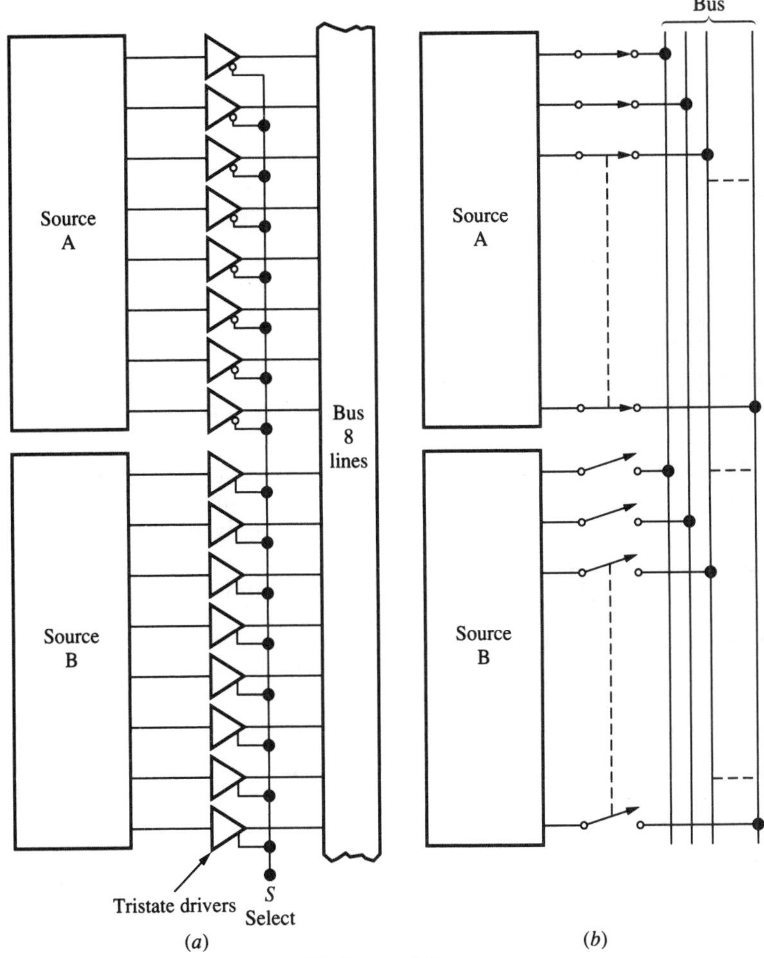

Tristate driver

troubleshooting The act of locating and repairing a fault that causes improper operation of equipment.

true power The actual amount of power delivered by the source to the load. See apparent power.

truncated sequence In a digital counter where the full count of the counter is intentionally made less than its maximum potential count. As an example, in a decade counter that uses four flip-flops, the count is truncated to 10 rather than the maximum potential count of 16.

truth table A listing that shows all possible input combina-

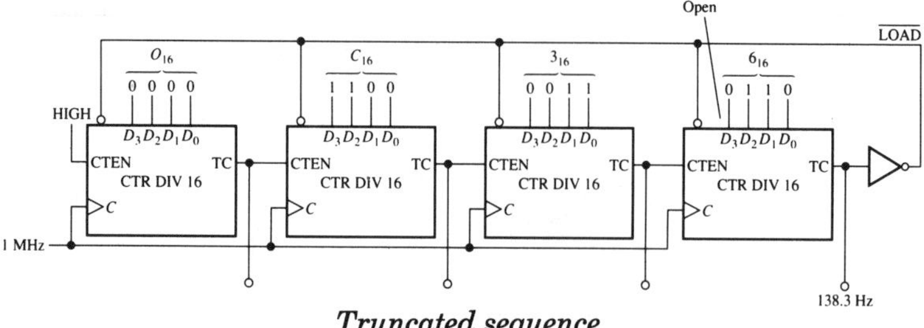

Truncated sequence

tions of two level logic (TRUE and FALSE) and the resultant output for a specific logic condition. Truth tables are used in digital electronics to illustrate all possible conditions of a given logic circuit.

TTL Abbreviation for transistor-transistor logic. See transistor-transistor logic.

TTL AND gate A logical circuit consisting of one or more AND gates, using the TTL method of construction.

TTL OR gate A logical circuit consisting of one or more OR gate, using the TTL method of construction.

tuned circuit A circuit that is selective for only one frequency. A tuned circuit allows a narrow range of frequencies to be selected while rejecting all others.

tuned radio-frequency amplifier An amplifier circuit consisting of resonant circuits. A trf is usually constructed so that it amplifies only a small range of frequencies. Tuned radio frequency amplifiers are sometimes used as the amplifier following the antenna of a communications receiver.

tuner That section of a communications receiver that receives the signal from the antenna and selects one frequency while rejecting all others. The tuner usually consists of a parallel resonant circuit having an inductor and a variable capacitor. The variable capacitor is adjustable,

Technology	CMOS* (silicon-gate)	CMOS* (metal-gate)	TTL Std.	TTL LS	TTL S	TTL ALS	TTL AS
Device series	74 HC	4000B	74	74LS	74S	74ALS	74AS
Power dissipation							
Static	2.5 nW	1 μW	10 mW	2mW	19 mW	1 mW	8.5 mW
@ 100 kHz	0.17 mW	0.1 mW	10 mW	2mW	19 mW	1 mW	8.5 mW
Propagation delay time	8 ns	50 ns	10 ns	10 ns	3 ns	4 ns	1.5 ns
Fan-out (same series)			10	20	20	20	40

*Propagation delay is dependent on V_{CC}. Power dissipation and fan-out are a function of frequency.

TTL

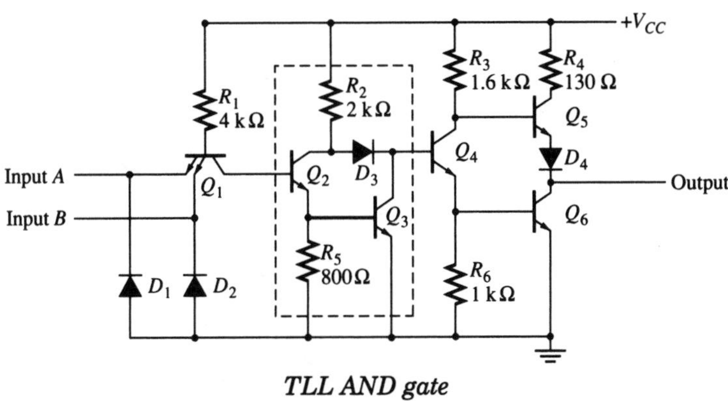

TLL AND gate

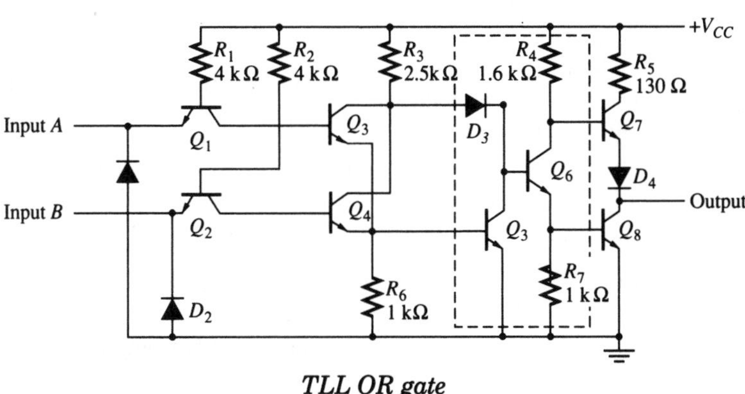

TLL OR gate

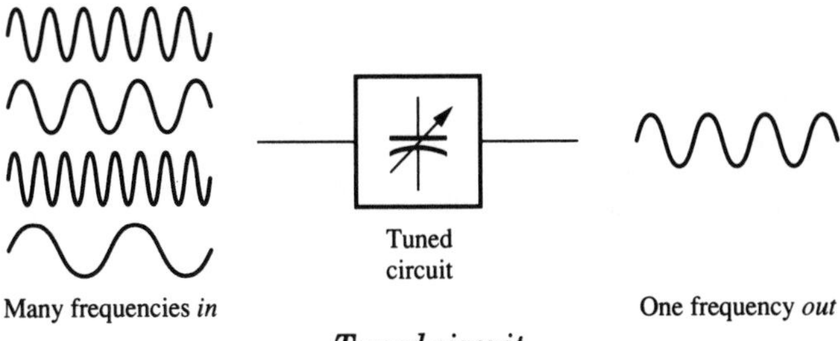

Tuned circuit

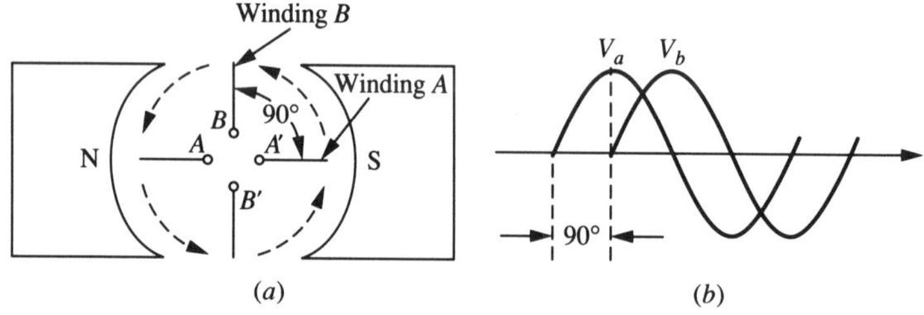

Two-phase generator

so the user may select different frequencies (representing different transmitting stations).

tunnel diode A solid-state device that exhibits a negative resistance that makes it useful for high-frequency oscillators. Constructed with germanium or gallium arsenide where it is more heavily doped in the p- and n-regions than a normal pn-junction diode.

turn 1. One complete loop of wire. 2. To rotate.

turns ratio Used as a measurement of the number of turns in the primary winding of a transformer to the number of turns in the secondary winding. Usually expressed as N_1:N_2, where N_1 is a given number of turns in the primary and N_2 is the equivalent number of turns in the secondary. As an example, 1:2 would mean that the secondary has twice as many turns as the primary of the transformer.

TV Abbreviation for television. See television.

twin-T oscillator A circuit that produces its own signal. Uses a twin-T filter that produces a lead-lag network where the phase shift is zero degrees at resonance. Resulting frequency of sine wave is determined by the values of the resistors and capacitors used in the twin-T network.

twisted pair An electrical cable consisting of two insulated conductors wrapped around each other. The conductors are usually enclosed in a braided wire shield that is in turn enclosed in a plastic jacket.

two-phase Having a phase difference of 90 degrees or one quarter of a cycle.

two-phase generator An electrical generator that produces two simultaneous multiple sine waves. These two waves are separated by a given phase angle.

two-pole, three-phase alternator An alternator constructed with two separate poles, wired in such a manner as to produce three simultaneous multiple sine waves that are separated by given phase angles.

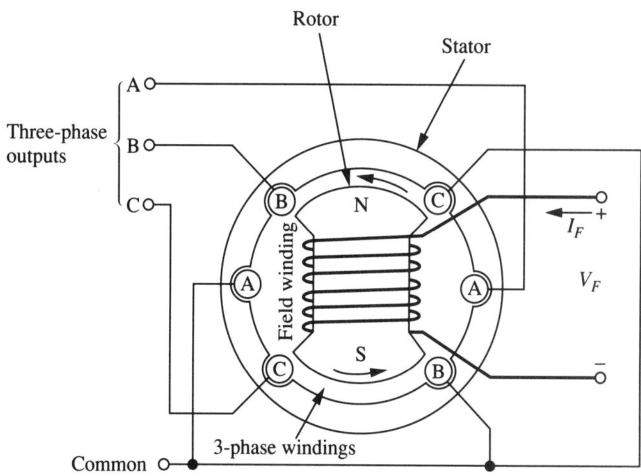

Two-pole, three-phase alternator

two's complement notation Used as an aid in binary subtraction. The two's complement of a binary number is found by taking the complement of each bit of the number and adding 1 to the result.

UHF Ultra high frequency. In television, the term used to indicate channels 14 through 83.

UJT Abbreviation for unijunction transistor. See unijunction transistor.

UL 1. Abbreviation for Underwriter's Laboratories, Inc., a corporation established for setting safety standards on components and equipment. 2. Abbreviation for unit load. See unit load.

ultrasonic Having frequencies above those of audible sound. Applications include ultrasonic cleaning and welding.

ultraviolet Electromagnetic radiation at wavelengths beyond the violet end of the visible spectrum. Ultraviolet radiation lies approximately between 1000 to 4000 Å.

unconditional jump In computer programming, an instruc-

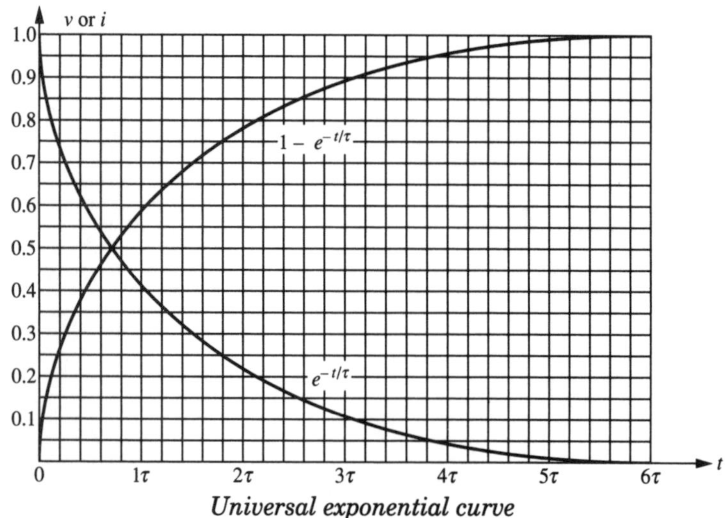

Universal exponential curve

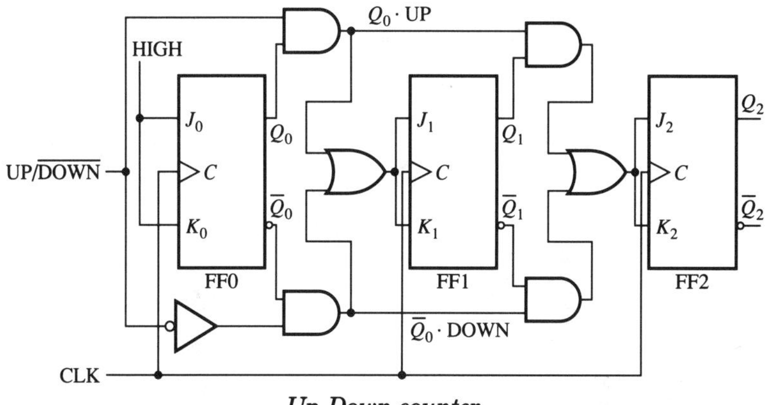

Up-Down counter

tion that causes the program to go to another section of the program. An unconditional jump ignores the normal sequence of having sequential execution of code, one line after the other, and also ignores the result of any computational or logic operation.

unijunction transistor A three-terminal semiconductor with one pn junction. A unijunction transistor exhibits a stable open-circuit negative-resistance property.

unit load Used in digital integrated circuits to indicate the load presented by a single-gate input.

unity gain The condition when the output signal of a circuit has the same electrical characteristics as the input signal, specifically where the output level is the same as the input level.

unity power factor A power factor of 1. The condition of a unity power factor is obtained in an ac circuit only when the voltage and current are exactly in phase.

universal exponential curves The curves that provide a graphic solution of the charge and discharge of capacitors and the rise and fall of the magnetic field of an inductor.

unmatched transmission line A transmission line whose characteristic impedance is not the same as the source impedance.

up-down counter A digital counter that can count in increasing increments (counting up) or in decreasing increments (counting down).

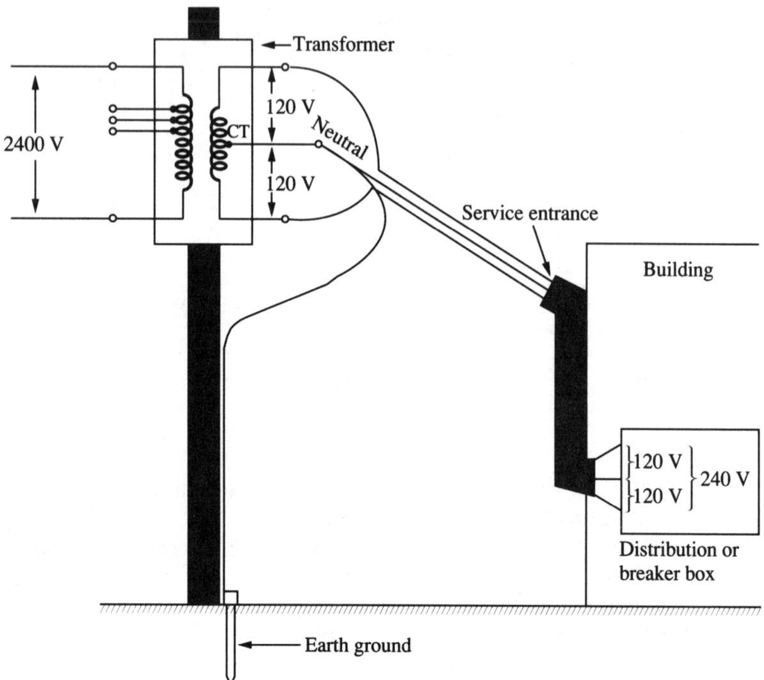

Utility pole transformer

utilities In computers, programs provided to perform routine tasks that are in some way helpful to the programmer or program user.

utility pole transformer A special type of transformer designed for power distribution and usually mounted to a utility power pole.

VAC Abbreviation for ac voltage.

vacuum An enclosed space in which all air and gases have been removed. In an applied usage, vacuum means the removal of enough air and gases to accomplish the job at hand, since the absolute removal of any trace of air or gas is not practical.

valence electrons Electrons in the outer shell of the atom. These electrons are responsible for the chemical characteristics of elements.

valid address In a microprocessor-based system, during processing, the digital patterns on the address bus go through many rapid changes. The address bus can be read (actually used by the system) only when it is not in transition and the data on it are meaningful. It is during this time that the bit pattern on the bus is said to be a valid address.

varactor diode A solid-state device whose capacitance can be controlled by an applied voltage. All diodes exhibit an internal capacitance when reverse biased. The capacitance is proportional to the amount of reverse-biased voltage applied. A varactor diode is intentionally designed to take advantage of the reverse-biased capacitance phenomenon.

variable capacitor A capacitor constructed in such a manner as to have its capacitance changed or varied.

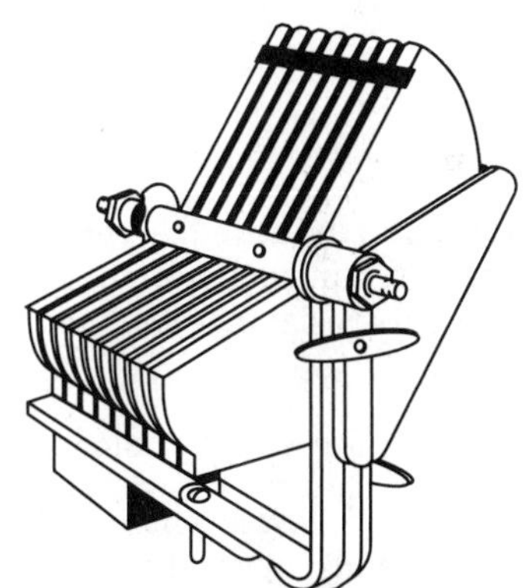

Variable capacitor

variable inductor An inductor constructed in such a manner as to have its inductance varied.

Variable inductor

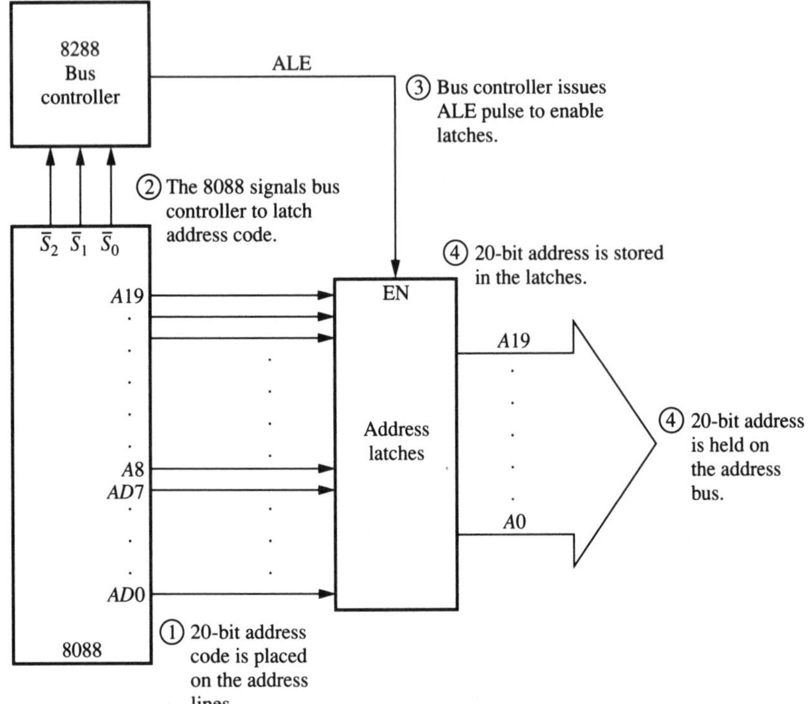
8288
Bus
controller
ALE
③ Bus controller issues ALE pulse to enable latches.
② The 8088 signals bus controller to latch address code.
$\overline{S}_2$ $\overline{S}_1$ $\overline{S}_0$
A19
A8
AD7
AD0
8088
EN
Address latches
④ 20-bit address is stored in the latches.
A19
A0
④ 20-bit address is held on the address bus.
① 20-bit address code is placed on the address lines.

Valid address

variable resistor A resistor constructed in such a manner as to have its resistance changed or varied.

varicap See varactor diode.

varistor A device whose resistance depends upon the voltage applied to it. Varistors are nonlinear devices usually used to suppress high-voltage transients.

V_B Designation for transistor dc base voltage. The dc voltage measured at a transistor's base with respect to ground.

V_{BB} Designation for transistor dc base bias voltage. For a pnp transistor, V_{BB} is negative on the base with respect to the emitter, while for an npn, it is positive on the base with respect to the emitter.

V_C Designation for transistor dc collector voltage. The voltage at a transistor's collector measured with respect to ground.

V_{CB} Designation for transistor dc voltage measured from collector to base.

V_{CC} Designation for transistor dc supply voltage. For a pnp transistor, V_{cc} is a negative voltage; for an npn transistor, it is a positive voltage.

VCCS Abbreviation for voltage-controlled current source. See voltage-controlled current source.

V_{CE} Designation for transistor dc voltage measured from collector to emitter.

VCO Abbreviation for voltage-controlled oscillator. See voltage-controlled oscillator.

VCR Abbreviation for videocassette recorder.

VCVS Abbreviation for voltage-controlled voltage source. See voltage-controlled voltage source.

VDC Abbreviation for dc voltage.

V_E Designation for transistor dc emitter voltage. The dc voltage measured at a transistor's emitter with respect to ground.

V_{EB} Designation for transistor dc voltage measured from emitter to base.

V_{EE} Designation for transistor dc emitter voltage. Any dc voltage applied to the emitter from an external source.

vector Representing both magnitude and direction in graphical form.

vector diagram Showing the relationship of two or more vectors. See vector.

vertical Perpendicular to the horizon. In the direction of gravity. A vertical line is drawn on paper from top to bottom. See horizontal.

vertical deflection The movement of an electron beam, in a CRT, in the vertical direction through the use of an applied sawtooth waveform to the vertical deflection circuits.

vertical polarization Regarding the radiation pattern of an antenna. When the electric field

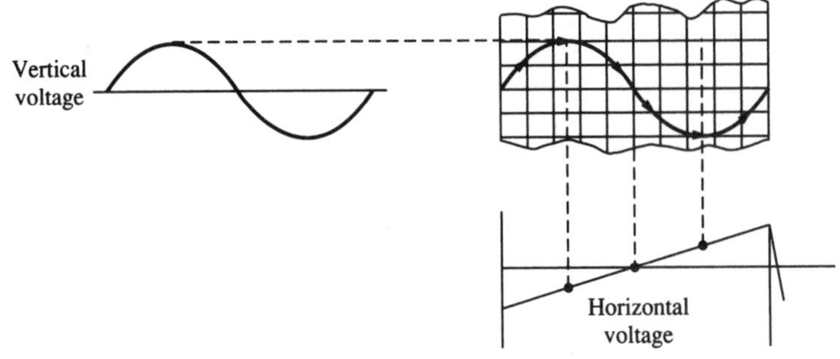

Vertical deflection

is vertical, the antenna is said to be vertically polarized. See horizontal polarization.

VHS Abbreviation for very high speed.

videotex Any of several on-line information and retail services targeted at the home and consumer market. Programs are controlled by personal computers or dedicated terminals in public places.

virtual reality A surrogate or metaphysical environment created by computing and communications systems. A computer-generated counterpart of a physical object, person, place, or thing. See PSpice.

virus In computers, a program that is intentionally made for the purpose of destroying or otherwise modifying other programs to the detriment of the program user whose programs have been unknowingly "infected" by the virus. See worm.

visible spectrum The range of frequencies in the electromagnetic spectrum that can be detected by the human eye.

V_{OH} Abbreviation for voltage output high. Used in specifying the output voltages for digital integrated circuits. See output profile.

voice synthesizer An electrical device that emulates the sound of the human voice. See synthesizer.

V_{OL} Abbreviation for voltage output low. Used in specifying the output voltages for digital integrated circuits. See output profile.

volt The unit of measurement applied to the difference in potential between two points in a circuit. When one joule of energy is required to move one coulomb of charge between two points, the difference in potential is said to be one volt.

voltage The amount of energy it takes to move a specified number of electrons from one point to another. See volt.

voltage-controlled current source A current source whose output value is controlled by the voltage at some other point in the circuit to which it is connected.

voltage controlled oscillator An oscillator whose frequency is determined (controlled) by an independent source of voltage.

voltage-controlled voltage source A voltage source whose output value is controlled by the voltage at some other point in the circuit to which it is connected.

voltage divider A series circuit arrangement where different voltages are produced across the resistors to a common reference due to the voltage drops across each of the resistors.

voltage-divider bias In a transistor amplifier, using two series resistors connected across

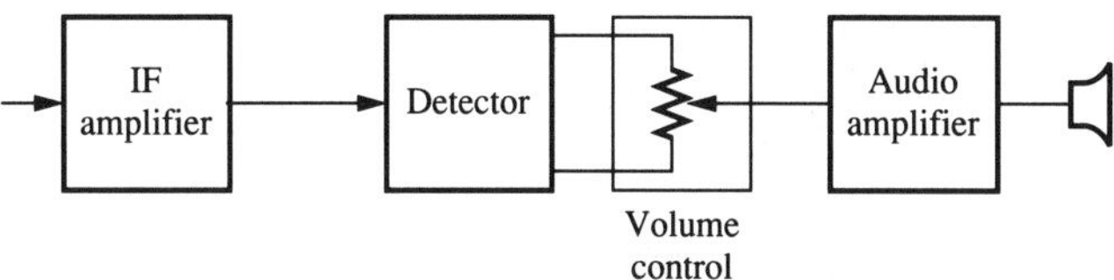

Voltage divider

the power source to provide a smaller voltage to be applied to the base. This bias voltage will cause the transistor to conduct and help determine the dc operating characteristics of the amplifier.

voltage doubler An electrical circuit consisting of diodes and capacitors connected in such a fashion as to produce an output voltage that is double that of an input ac peak voltage.

voltage quadrupler An electrical circuit consisting of diodes and capacitors connected in such a fashion as to produce an output voltage that is four times that of an input ac peak voltage.

voltage regulation The process of utilizing special circuits in an attempt to maintain a constant dc output voltage.

voltage regulator An electronic circuit whose purpose it is to maintain a constant voltage from a voltage source. Voltage regulators are commonly used with power supplies to maintain a constant output voltage over a wide range of load conditions.

voltage source The place in the circuit that produces the electrical energy.

voltage tripler An electrical circuit consisting of diodes and capacitors connected in such a fashion as to produce an output voltage that is triple that of an input ac peak voltage.

voltage-variable capacitor See varactor diode.

voltaic cell A chemical device capable of converting chemical energy into electrical energy.

voltmeter An electrical instrument for measuring voltage.

V_R The maximum reverse dc voltage that can be applied across a diode.

V_{RRM} The maximum reverse-peak voltage that can be applied repetitively across a diode.

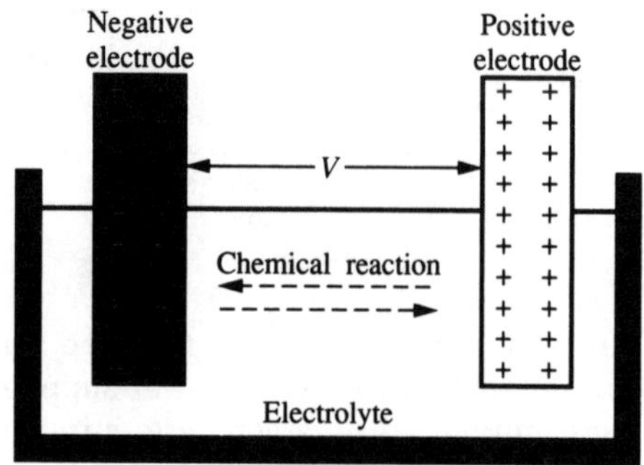

Voltaic cell

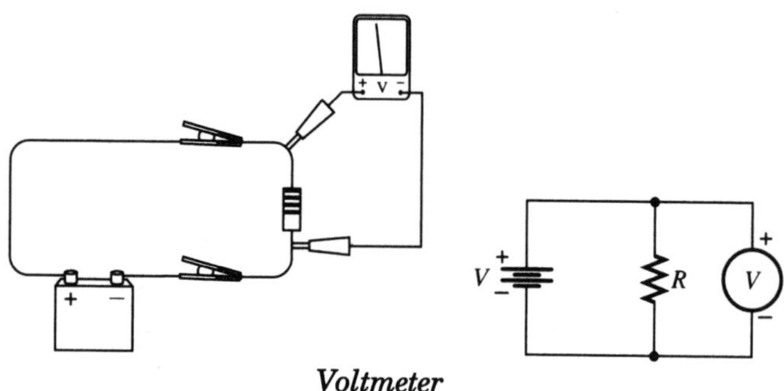

Voltmeter

W Letter symbol for watt.

wall outlet A source of electrical power. A wall outlet consists of spring-loaded female connectors and is enclosed in a metal box. The connectors are permanently connected to the incoming electrical power lines.

warm boot Restarting (booting) the computer without removing and reapplying electrical power. Process is accomplished by a single key or a combination of keys held down at the same time.

watt The unit of measurement for electrical power. One watt is defined as the power required to do work at the rate of one joule per second. Mathematically, $P = IE$, where P is the power (in watts), I is the current (in amps), and E is the voltage (in volts).

wattage rating The maximum power a device can safely handle without being destroyed. The wattage rating of a carbon composition resistor is determined by the physical size of the resistor: the larger the size, the larger the power rating.

watthour The unit of electrical work indicating an expenditure of one watt for one hour.

wattless component The reactive component in an ac circuit.

wattmeter An electrical instrument used to measure the power in a circuit.

wave Any detectable form of energy transmission that has a periodic change in magnitude with time.

waveform 1. A mental or actual graphical representation of an electrical time-varying quantity. An ac waveform is mentally and actually pictured as the graph of a sine wave. 2. A graphical representation of a wave where the vertical axis represents the wave magnitude and the horizontal axis represents time.

waveform synthesis Producing an electrical waveform in any desired shape or duration. Usually constructed by a combination of digital and analog circuits. See music synthesizer.

waveguide A confining envi-

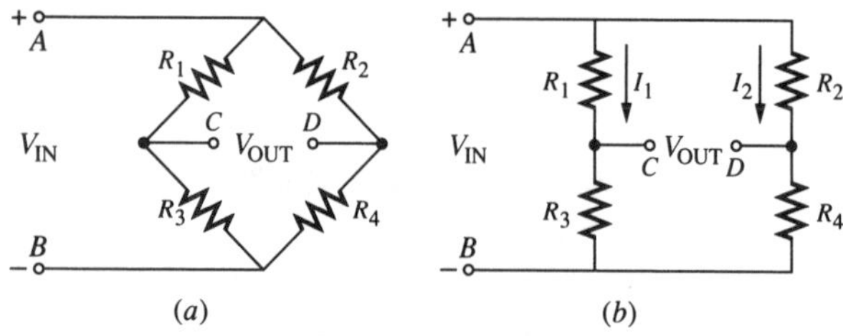

Wheatstone bridge

ronment, usually in the shape of hollow cylinders or rectangles, used to transmit electromagnetic energy in a direction determined by its physical boundaries. Waveguides are used at frequencies where transmission by wire is not efficient for short distances. Waveguides are usually employed at frequencies greater than 1 GHz.

wavelength In a periodic wave the distance between two identical points of the wave.

weber The unit of magnetic flux. One telsa is equal to one weber per square meter.

Wheatstone bridge A balanced bridge circuit that can be used to measure resistors. Named after its inventor.

white noise Random noise where the energy distribution per unit bandwidth is independent of the central frequency of the band.

wideband amplifier An amplifier capable of passing a wide range of frequencies with equal gain.

Wien bridge A classical ac bridge circuit where the value of an inductor or capacitor can be measured as a function of frequency and resistance values used in the bridge.

Wien bridge oscillator A circuit capable of generating its own signal. Uses a lead-lag network with an automatically variable resistor to maintain a stable output amplitude. Frequency of resulting waveform is determined by values of resistor and capacitor in circuit.

winding capacitance The stray capacitance caused by the windings of an inductor or a transformer. The winding capac-

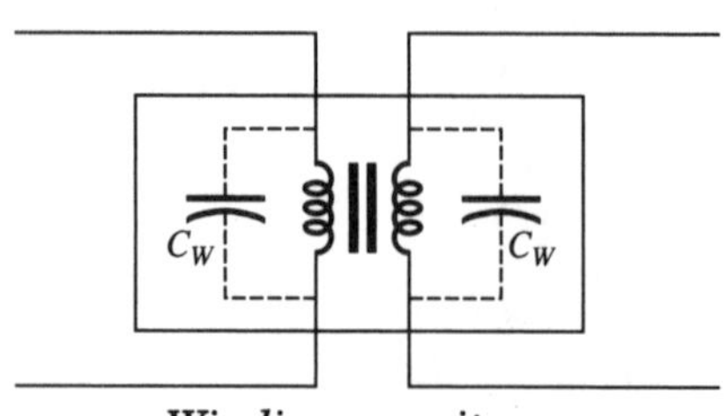

Winding capacitance

itance of a practical inductor or transformer prevents these devices from exhibiting their ideal electrical characteristics.

winding resistance In an inductor or transformer, the resistance inherent in the coil windings. Winding resistance is what causes practical inductors and transformers to not have ideal electrical characteristics.

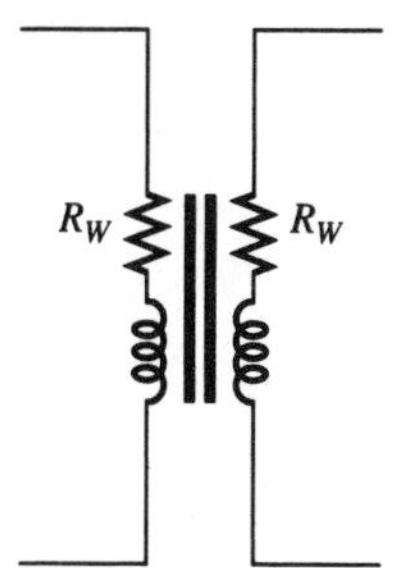

Winding resistance

window In computers, a selected area of the monitor screen. By activating windows, several areas of the monitor screen may be used to present different kinds of information to the viewer just as if several different monitors were being used at the same time.

wire A malleable metallic conductor used to transmit electrical energy from one point to another.

wired AND Logic gates that require a separate external connection done in such a way that the resultant output represents the logical AND function.

wired OR Logic gates that require a separate external connection done in such a way that the resultant output represents the logical OR function.

wirewound resistor A resistance created by a length of high-resistance wire wound around an insulated form producing temperature stability, high wattage ratings, and low resistance values.

wire wrap A technique of making electrical and mechanical connections between circuit components without using soldering. Wire wrapping is accomplished by using a cylindrical wire wrapped tightly (by a special tool) around a small rectangular (or triangular) post. The sharp edges of the post dig into the wire as it is tightly wrapped around the post.

wire wrapping The process of mechanically connecting one circuit element to another for the purpose of an electrical connection with small-diameter wire without the use of solder. The wire is tightly twisted on small rectangular metal connectors causing the wire to be pierced by the sides of the connector. Usually done with a tool called a wire-wrapping tool.

woofer A loud speaker designed to reproduce primarily low audio frequencies up to about 1 kHz.

word In computers, any group of binary digits treated as a unit. Typically, a word consists of 8, 16, 32 or 64 bits.

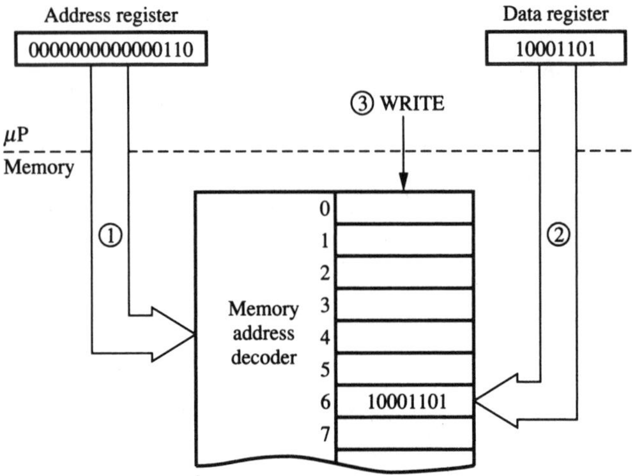

① Address placed on address bus.

② Data placed on data bus.

③ WRITE signal causes data to be stored in address 6_{16}.

Write operation

worm In computers, a program intentionally made for the destruction or modification of other programs to the detriment of the program user whose programs have been penetrated by the worm. A worm does not require the existence of another program in order to exist itself.

write To cause a bit pattern to appear at a specific location in a digital system. As an example to write from memory to a storage disk means to copy the bit pattern from memory to the disk.

write operation In microprocessor terminology, the process of copying information from the microprocessor to a memory location. See read.

wye-connection A circuit that has three elements connected at a common junction to form the letter "Y."

wye-delta A special kind of circuit connection. See wye connection. See delta connection.

R_1 R_2 R_3

Wye

wye-to-delta conversion A method of converting from a wye circuit to a delta circuit.

X Letter symbol for reactance. See reactance.

***X*-axis** On a Cartesian coordinate system, the name of the horizontal axis; the *Y*-axis is the vertical axis.

X-band The range of radio frequencies from approximately 5.2 GHz to 11 GHz.

X_c Symbol for capacitive reactance: $X_c = 1/(2\pi fC)$, where X_c is the capacitive reactance (in ohms), f is the frequency (in hertz), and C is the capacitance (in farads).

xfmr Abbreviation for transformer. See transformer.

X_L Symbol for inductive reactance. $X_L = 2\pi fL$, where X_L is the inductive reactance (in ohms), f is the frequency (in hertz), and F is the inductance (in henrys).

xmtr Abbreviation for transmitter. See transmitter.

XNOR Boolean expression where the result is the inverse of the exclusive OR. The expression is TRUE only when an even number of variables are the same. Thus, $A = \overline{B} + \overline{C}$, where A will be TRUE only when B and C are both TRUE or both FALSE.

XNOR gate An electronic digital circuit with one output and two or more inputs that replicates the Boolean XNOR function.

XOR Abbreviation for exclusive OR. See exclusive OR gate.

XOR gate Abbreviation for exclusive OR gate. An XOR gate has two or more inputs and one output. The output will be TRUE only when ALL of the inputs are FALSE.

X ray Electromagnetic radiation in the wavelength spectrum of from 10^{-7} to 10^{-10} cm. X rays have the ability to penetrate matter and are used in medical diagnosis.

X-ray tube An electronic device used in the production of X rays. The tube is a vacuum tube where high-energy electrons bombard a metal target producing X rays in the process.

xtal Abbreviation for crystal. See crystal.

***X-Y* display** A flat surface drawing usually created by an *XY* plotter. An *X-Y* display uses the Cartesian coordinate system, where data are plotted from two values, their *X*-position (horizontal position) and *Y*-position (vertical position).

***XY* plotter** An electromechanical device for producing line drawings on a flat surface. The device gets its name from the two coordinates it uses to plot a point: the *X*-axis and the *Y*-axis. *XY* plotters are used with computers to create engineering drawings and plot scientific graphs. See *X-Y* display.

Y Letter symbol for admittance. See admittance.

Yagi antenna Antenna consisting of a driven dipole and two parasitic elements called the reflector and the director lined up parallel to each other in the horizontal plane. A Yagi antenna may have more than one dipole and more than two parasitic elements. Named after the Japanese inventors Yagi and Uda. Sometimes called a Yagi-Uda antenna.

***Y*-axis** On a Cartesian coordinate system, the name of the vertical axis where the *X*-axis is the horizontal axis.

Y-connected generator An electrical generator made to operate from a three-wire (with a neutral) or four-wire electrical system in a Y configuration.

Y-connection A connection of electrical components whose schematic representation looks like the letter "Y." Sometimes called a T-network. See wye-connection.

Y-junction In waveguides, a connection of three waveguides

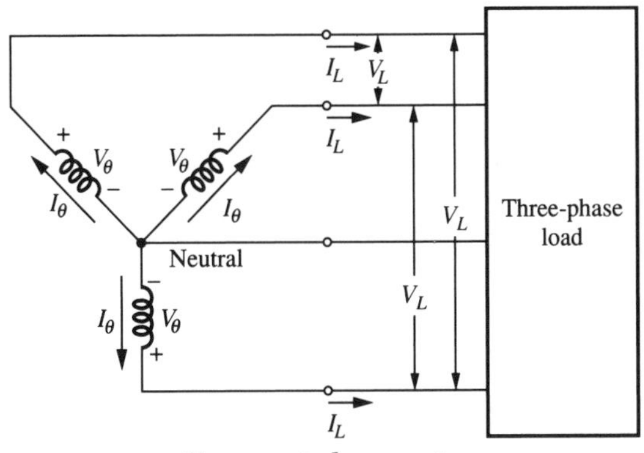

Y-connected generator

such that they form the shape of the letter "Y."

Y-match A method of connecting a lead wire to an unbroken dipole antenna. The ends of the lead wire are spread out to form the letter "Y." This method of connection is used in order to match the impedance of the connection to that of the antenna. Sometimes referred to as a delta match.

yoke A set of coils placed over the neck of a magnetically deflected cathode ray tube to cause electron beams to be deflected vertically and horizontally across the face of the CRT.

Y-parameters Admittance measurements used for the measurement of two-port networks. Historically popular in transistor measurements.

Z

Z Letter symbol for impedance. See impedance.

***Z*-axis** In a three-dimensional coordinate system represented in two dimensions, the *Z*-axis is the axis coming "out" of the flat surface. The *Z*-axis is perpendicular to the plane of the *X* and *Y* axes.

***Z*-axis modulation** Also called intensity modulation. The modulation of the electron beam in a CRT.

zener current Current flow in a pn junction when the junction has been reversed biased to the point where its insulating qualities have broken down.

zener diode A solid-state device that has a predictable and repeatable breakdown voltage in its reverse-biased direction. Zener diodes are used in voltage regulator circuits.

zener voltage In a zener diode, the reverse bias voltage at which the zener current takes place. See zener diode.

zero adjustment A method used to adjust the zero reading of a meter.

zero bias The condition of not having a difference of potential in the biasing voltage of an electrical device. As an example, zero bias in a transistor means that the difference of potential between the base and emitter is zero volts.

zero-crossing detector An electronic circuit capable of producing an output when an input signal changes its polarity from plus to minus or minus to plus (thus "crossing zero").

zero-offset voltage In operational amplifiers, the amount of output voltage when the differential input voltage is zero volts. Ideally, the zero-offset voltage is zero volts.

zero-page addressing In some computers, zero-page addressing allows for shorter, more condensed programming code. This is accomplished by the microprocessor fetching only the

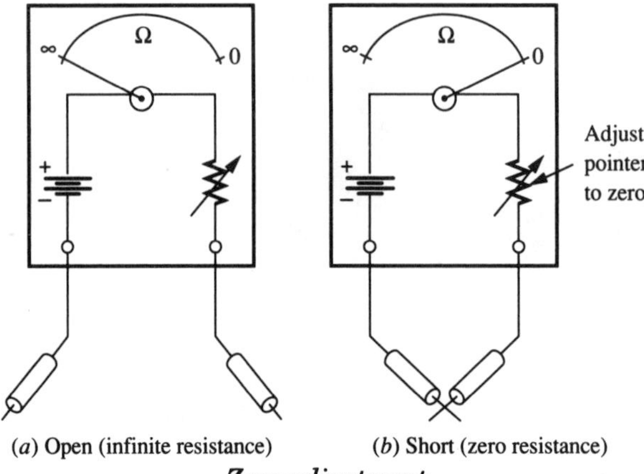

(*a*) Open (infinite resistance) (*b*) Short (zero resistance)

Zero adjustment

lower-order address and assuming that the higher-order address is zero.

Z-parameters Impedance measurements used for the measurement of two-port networks.

Appendix A

Logic Symbols

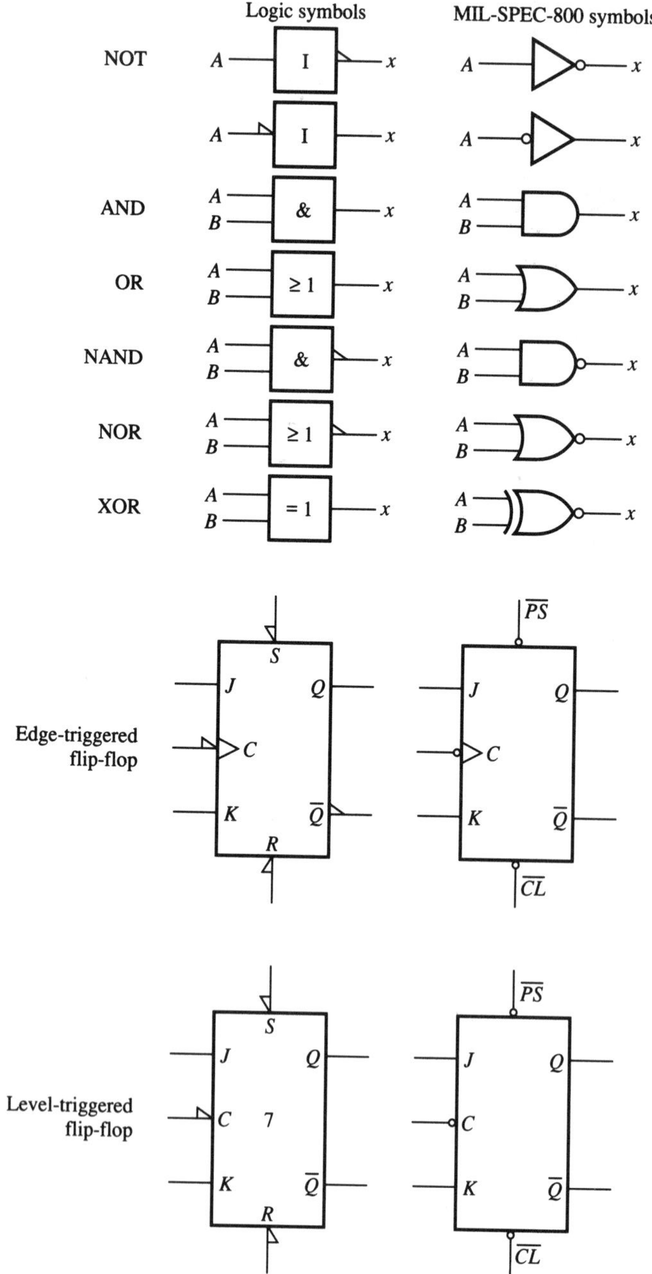
Logic symbols
MIL-SPEC-800 symbols
NOT
AND
OR
NAND
NOR
XOR
A
B
x
I
&
≥ 1
= 1
Edge-triggered flip-flop
Level-triggered flip-flop
S
R
J
K
C
Q
$\overline{Q}$
7
$\overline{PS}$
$\overline{CL}$

Appendix B

Boolean Theorems

Basic OR operation

$A + 0 = A$

$A + 1 = 1$

$A + \overline{A} = 1$

Basic AND operation

$A \cdot 0 = 0$

$A \cdot 1 = A$

$A \cdot \overline{A} = 0$

Basic NOT operation

$\overline{A} = \overline{A}$

$\overline{A} = A$

$\overline{A} = \overline{A}$

Identity laws

$A + A = A$

$A \cdot A = A$

Commutative laws

$A + B = B + A$

$A \cdot B = B \cdot A$

Redundancy law

$A + (A \cdot B) = A$

$A \cdot (A + B) = A$

Associative laws

$A + (B + C) = (A + B) + C$

$A \cdot (B \cdot C) = (A \cdot B) \cdot C$

Distributive laws

$A \cdot (B + C) = (A \cdot B) \cdot (A \cdot C)$

$A + (B \cdot C) = (A + B) \cdot (A + C)$

Special theorems

Nashelsky:

$A + (\overline{A} \cdot B) = A + B$

$A \cdot (\overline{A} + B) = A \cdot B$

DeMorgan:

$(\overline{A + B}) = \overline{A} \cdot \overline{B}$

$(\overline{A \cdot B}) = \overline{A} + \overline{B}$

Appendix C
Schematic Symbols

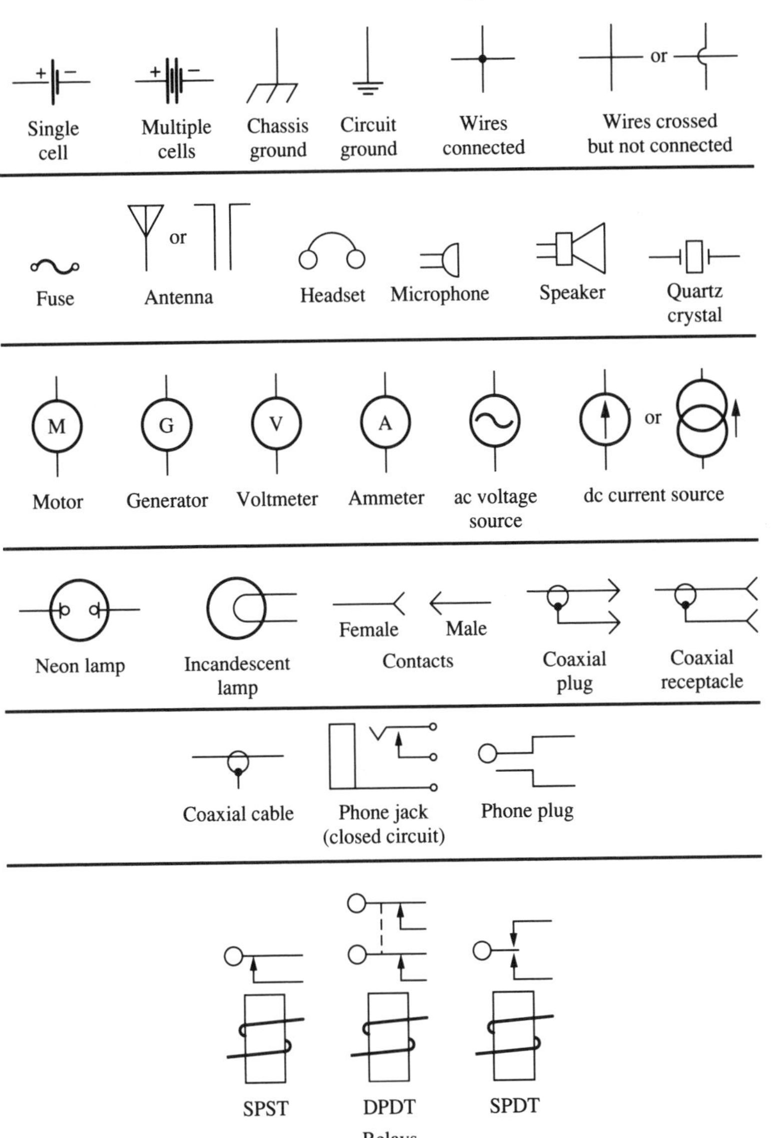
+
−
Single cell
+
−
Multiple cells
Chassis ground
Circuit ground
Wires connected
or
Wires crossed but not connected
Fuse
or
Antenna
Headset
Microphone
Speaker
Quartz crystal
M
Motor
G
Generator
V
Voltmeter
A
Ammeter
ac voltage source
or
dc current source
Neon lamp
Incandescent lamp
Female
Male
Contacts
Coaxial plug
Coaxial receptacle
Coaxial cable
Phone jack (closed circuit)
Phone plug
SPST
DPDT
SPDT
Relays

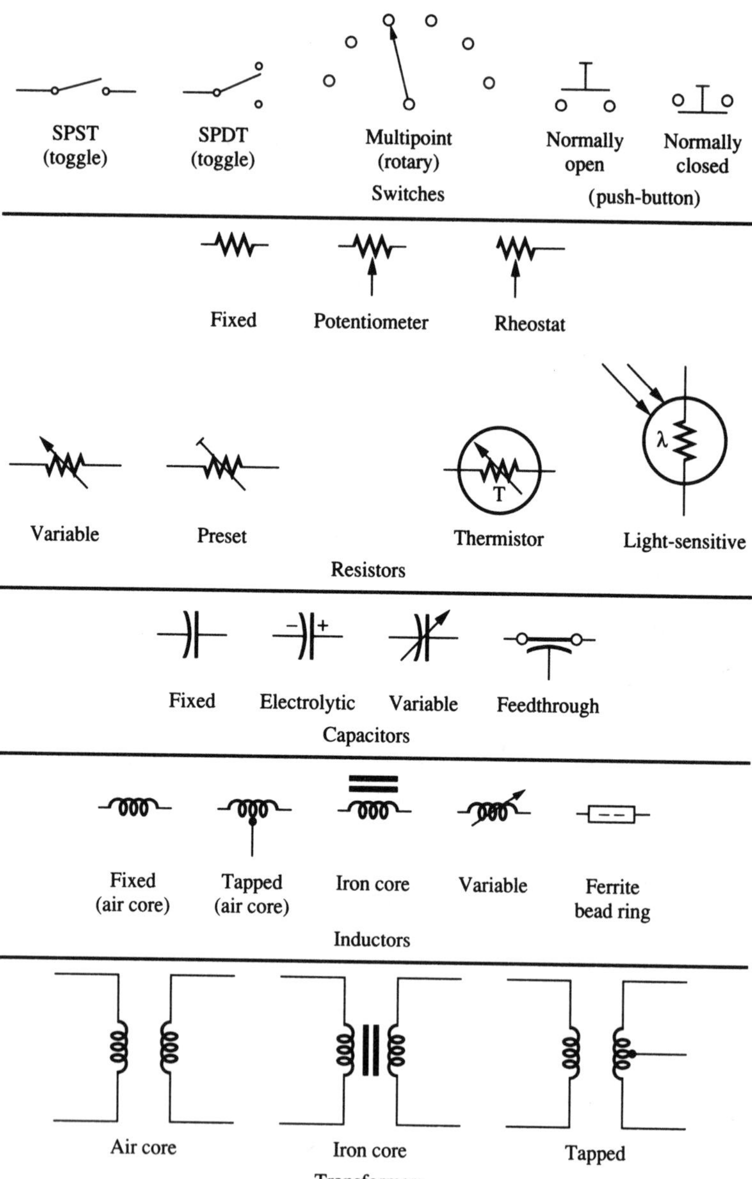
SPST
(toggle)
SPDT
(toggle)
Multipoint
(rotary)
Switches
Normally
open
Normally
closed
(push-button)
Fixed
Potentiometer
Rheostat
λ
T
Variable
Preset
Thermistor
Light-sensitive
Resistors
+
Fixed
Electrolytic
Variable
Feedthrough
Capacitors
Fixed
(air core)
Tapped
(air core)
Iron core
Variable
Ferrite
bead ring
Inductors
Air core
Iron core
Tapped
Transformers

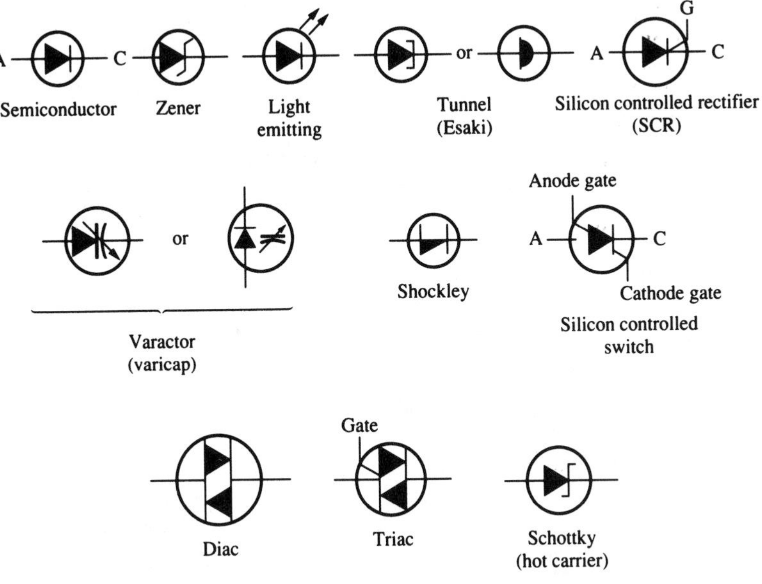

Diodes (drawn with or without circles)

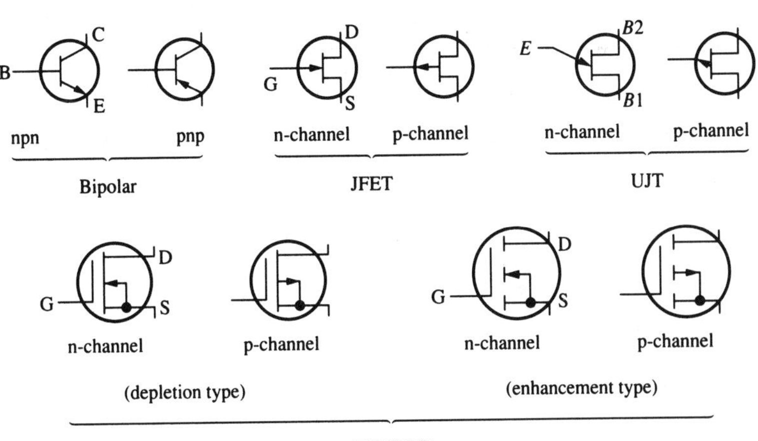

Transistors (drawn with or without circles)

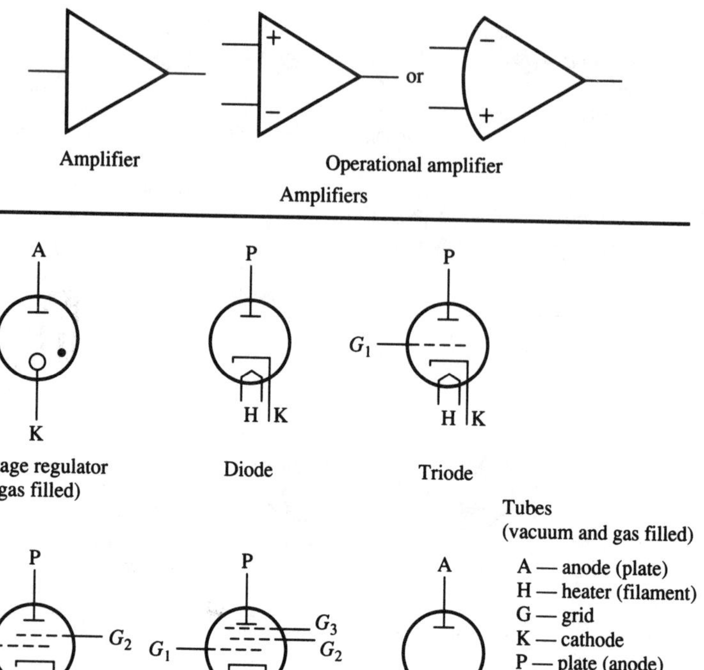

Amplifiers

Tubes
(vacuum and gas filled)

A — anode (plate)
H — heater (filament)
G — grid
K — cathode
P — plate (anode)
• — gas filled

Appendix D

Multiplier Prefixes

The following prefixes are those used in electricity and electronics that indicate multiples and submultiples of units.

Prefix	Symbol	Multiplier	Decimal Value
exa-	E	10^{18}	1 000 000 000 000 000 000
peta-	P	10^{15}	1 000 000 000 000 000
tera-	T	10^{12}	1 000 000 000 000
giga-	G	10^{9}	1 000 000 000
mega-	M	10^{6}	1 000 000
kilo-	k	10^{3}	1 000
deci-	d	10^{-1}	.1
centi-	c	10^{-2}	.01
milli-	m	10^{-3}	.001
micro-	μ	10^{-6}	.000 001
nano-	n	10^{-9}	.000 000 001
pico-	p	10^{-12}	.000 000 000 001
femto-	f	10^{-15}	.000 000 000 000 001
atto-	a	10^{-18}	.000 000 000 000 000 001

Appendix E
Quantity Symbols

The following quantity symbols are those frequently used in the study of electricity and electronics. Where two symbols are separated by three dots, the second is to be used only where there is a specific need to avoid conflict.

Quantity	Symbol	Units
active power	P	watt
admittance	Y	siemen
angle (phase angle)	θ, ϕ	radian, degree
angular frequency	ω	radian per second
apparent power	S . . . P_s	voltampere
capacitance	C	farad
capacitivity (permittivity)	ϵ	farad per meter
conductance	G	siemen
conductivity	Γ, σ	siemen per meter
coupling coefficient	κ	dimensionless
dissipation factor	D	dimensionless
efficiency	η	dimensionless
elastance	S	reciprocal farad, daraf
electric charge	Q	coulomb
electric field strength	E	volt per meter
electric flux	Ψ	coulomb
electrostatic potential	V . . . ϕ	volt
energy	E	joules, watt-hour
force	F	newton
frequency	f . . . ν	hertz
gain (ordinary)	A	dimensionless
gain (logarithmic)	A	bel, neper
illuminance (illumination)	E	lux
impedance	Z	ohm
inductance	L	henry
irradiance	E	watt per square meter
leakage coefficient	σ	dimensionless

Quantity	Symbol	Units
luminance	L	candela per square meter, lambert, nit
luminous intensity	I	candela
luminous flux	Φ	lumen
magnetic field strength	H	ampere per meter
magnetic flux	Φ	weber
magnetic flux density	B	tesla
magnetic flux linkage	Ψ	weber
magnetization	H, M	ampere per meter
magnetomotive force	F, F_m	ampere
period	T	second
permeability	μ	henry per meter
permeance	P, P_m, γ	henry
power	P	watt
power factor	PF	dimensionless
quality factor	Q	dimensionless
reactance	X	ohm
reactive power	Q . . . P_q	var
reluctance	R, R_m	reciprocal henry, ampere-turn per maxwell
reluctivity	ν	meter per henry
resistance	R	ohm
resistivity (specific resistance)	ρ	ohmmeter
susceptance	B	siemen
susceptibility	χ, κ	dimensionless
temperature (customary)	t	degree
temperature (thermodynamic)	T . . . θ	kelvin
time constant	τ . . . T	second
transmittance	τ	dimensionless
turns ratio	N	dimensionless
voltage (potential difference)	V, E . . . U	volt
wavelength	Λ, λ	meter
work	W	joule

Appendix F

Unit Symbols

The following unit symbols are those frequently used in the study of electricity and electronics.

Unit	Symbol	Unit	Symbol
ampere	A	joule	J
ampere-hour	Ah	kelvin	K
ampere-turn	At	lambert	L
baud	Bd	lumen	lm
bel	B	lux	lx
coulomb	C	maxwell	Mx
decibel	dB	meter	m
degree (angle)	. . .°	mho	mho
degree (temperature)		neper	Np
degree Celsius	°C	newton	N
degree Fahrenheit	°F	nit	nt
kelvin	K	oersted	Oe
dyne	dyn	ohm	Ω
electronvolt	eV	radian	rad
farad	F	second	s
foot candle	fc	siemen	S
gauss	G	tesla	T
gilbert	Gb	var	var
henry	H	voltampere	VA
hertz	Hz	watt	W
horsepower	hp	watt-hour	Wh
hour	h	weber	Wb

Appendix G
Wire Sizes

American Wire Gage (AWG) Sizes for Solid Round Copper

AWG #	Area (CM)	Ω/1000 ft at 20 °C	AWG #	Area (CM)	Ω/1000 ft at 20 °C
0000	211,600	0.0490	19	1,288.1	8.051
000	167,810	0.0618	20	1,021.5	10.15
00	133,080	0.0780	21	810.10	12.80
0	105,530	0.0983	22	642.40	16.40
1	83,694	0.1240	23	509.45	20.36
2	66,373	0.1563	24	404.01	25.67
3	52,634	0.1970	25	320.40	32.37
4	41,742	0.2485	26	254.10	40.81
5	33,102	0.3133	27	201.50	51.47
6	26,250	0.3951	28	159.79	64.90
7	20,816	0.4982	29	126.72	81.83
8	16,509	0.6282	30	100.50	103.2
9	13,094	0.7921	31	79.70	130.1
10	10,381	0.9989	32	63.21	164.1
11	8,234.0	1.260	33	50.13	206.9
12	6,529.0	1.588	34	39.75	260.9
13	5,178.4	2.003	35	31.52	329.0
14	4,106.8	2.525	36	25.00	414.8
15	3,256.7	3.184	37	19.83	523.1
16	2,582.9	4.016	38	15.72	659.6
17	2,048.2	5.064	39	12.47	831.8
18	1,624.3	6.385	40	9.89	1049.0

Appendix H

Standard Resistor Values

Resistance Tolerance (±%)

0.1% 0.25% 0.5%	1%	2% 5%	10%	0.1% 0.25% 0.5%	1%	2% 5%	10%	0.1% 0.25% 0.5%	1%	2% 5%	10%	0.1% 0.25% 0.5%	1%	2% 5%	10%	0.1% 0.25% 0.5%	1%	2% 5%	10%	0.1% 0.25% 0.5%	1%	2% 5%	10%
10.0	10.0	10	10	14.7	14.7	—	—	21.5	21.5	—	—	31.6	31.6	—	—	46.4	46.4	—	—	68.1	68.1	68	68
10.1	—	—	—	14.9	—	—	—	21.8	—	—	—	32.0	—	—	—	47.0	—	47	47	69.0	—	—	—
10.2	10.2	—	—	15.0	15.0	15	15	22.1	22.1	22	22	32.4	32.4	—	—	47.5	47.5	—	—	69.8	69.8	—	—
10.4	—	—	—	15.2	—	—	—	22.3	—	—	—	32.8	—	—	—	48.1	—	—	—	70.6	—	—	—
10.5	10.5	—	—	15.4	15.4	—	—	22.6	22.6	—	—	33.2	33.2	33	33	48.7	48.7	—	—	71.5	71.5	—	—
10.6	—	—	—	15.6	—	—	—	22.9	—	—	—	33.6	—	—	—	49.3	—	—	—	72.3	—	—	—
10.7	10.7	—	—	15.8	15.8	—	—	23.2	23.2	—	—	34.0	34.0	—	—	49.9	49.9	—	—	73.2	73.2	—	—
10.9	—	—	—	16.0	—	16	—	23.4	—	—	—	34.4	—	—	—	50.5	—	—	—	74.1	—	—	—
11.0	11.0	11	—	16.2	16.2	—	—	23.7	23.7	—	—	34.8	34.8	—	—	51.1	51.1	51	—	75.0	75.0	75	—
11.1	—	—	—	16.4	—	—	—	24.0	—	24	—	35.2	—	—	—	51.7	—	—	—	75.9	—	—	—
11.3	11.3	—	—	16.5	16.5	—	—	24.3	24.3	—	—	35.7	35.7	—	—	52.3	52.3	—	—	76.8	76.8	—	—
11.4	—	—	—	16.7	—	—	—	24.6	—	—	—	36.1	—	36	—	53.0	—	—	—	77.7	—	—	—
11.5	11.5	—	—	16.9	16.9	—	—	24.9	24.9	—	—	36.5	36.5	—	—	53.6	53.6	—	—	78.7	78.7	—	—
11.7	—	—	—	17.2	—	—	—	25.2	—	—	—	37.0	—	—	—	54.2	—	—	—	79.6	—	—	—
11.8	11.8	—	—	17.4	17.4	—	—	25.5	25.5	—	—	37.4	37.4	—	—	54.9	54.9	—	—	80.6	80.6	—	—
12.0	—	12	12	17.6	—	—	—	25.8	—	—	—	37.9	—	—	—	56.2	—	—	—	81.6	—	—	—
12.1	12.1	—	—	17.8	17.8	—	—	26.1	26.1	—	—	38.3	38.3	—	—	56.6	56.6	56	56	82.5	82.5	82	82
12.3	—	—	—	18.0	—	18	18	26.4	—	—	—	38.8	—	—	—	56.9	—	—	—	83.5	—	—	—
12.4	12.4	—	—	18.2	18.2	—	—	26.7	26.7	—	—	39.2	39.2	39	39	57.6	57.6	—	—	84.5	84.5	—	—
12.6	—	—	—	18.4	—	—	—	27.1	—	27	27	39.7	—	—	—	58.3	—	—	—	85.6	—	—	—
12.7	12.7	—	—	18.7	18.7	—	—	27.4	27.4	—	—	40.2	40.2	—	—	59.0	59.0	—	—	86.6	86.6	—	—
12.9	—	—	—	18.9	—	—	—	27.7	—	—	—	40.7	—	—	—	59.7	—	—	—	87.6	—	—	—
13.0	13.0	13	—	19.1	19.1	—	—	28.0	28.0	—	—	41.2	41.2	—	—	60.4	60.4	—	—	88.7	88.7	91	—
13.2	—	—	—	19.3	—	—	—	28.4	—	—	—	41.7	—	—	—	61.2	—	—	—	89.8	—	—	—
13.3	13.3	—	—	19.6	19.6	—	—	28.7	28.7	—	—	42.2	42.2	—	—	61.9	61.9	62	—	90.9	90.9	—	—

0.1% 0.25% 0.5%	1%	2% 5%	10%	0.1% 0.25% 0.5%	1%	2% 5%	10%	0.1% 0.25% 0.5%	1%	2% 5%	10%	0.1% 0.25% 0.5%	1%	2% 5%	10%	0.1% 0.25% 0.5%	1%	2% 5%	10%	0.1% 0.25% 0.5%	1%	2% 5%	10%
13.5	—	—	—	19.8	—	—	—	29.1	—	—	—	42.7	—	—	—	62.6	—	—	—	92.0	—	—	—
13.7	13.7	—	—	20.0	20.0	20	—	29.4	29.4	—	—	43.2	43.2	43	—	63.4	63.4	—	—	93.1	93.1	—	—
13.8	—	—	—	20.3	—	—	—	29.8	—	—	—	43.7	—	—	—	64.2	—	—	—	94.2	—	—	—
14.0	14.0	—	—	20.5	20.5	—	—	30.1	30.1	30	—	44.2	44.2	—	—	64.9	64.9	—	—	95.3	95.3	—	—
14.2	—	—	—	20.8	—	—	—	30.5	—	—	—	44.8	—	—	—	65.7	—	—	—	96.5	—	—	—
14.3	14.3	—	—	21.0	21.0	—	—	30.9	30.9	—	—	45.3	45.3	—	—	66.5	66.5	—	—	97.6	97.6	—	—
14.5	—	—	—	21.3	—	—	—	31.2	—	—	—	45.9	—	—	—	67.3	—	—	—	98.8	—	—	—

Note: These values are generally available in multiples of 0.1, 1, 10, 100, 1 k, and 1 M.

Appendix I
Conversions

Decimal	BCD (8421)	Octal	Binary	Decimal	BCD (8421)	Octal	Binary	Decimal	BCD (8421)	Octal	Binary
0	0000	0	0	34	00110100	42	100010	68	01101000	104	1000100
1	0001	1	1	35	00110101	43	100011	69	01101001	105	1000101
2	0010	2	10	36	00110110	44	100100	70	01110000	106	1000110
3	0011	3	11	37	00110111	45	100101	71	01110001	107	1000111
4	0100	4	100	38	00111000	46	100110	72	01110010	110	1001000
5	0101	5	101	39	00111001	47	100111	73	01110011	111	1001001
6	0110	6	110	40	01000000	50	101000	74	01110100	112	1001010
7	0111	7	111	41	01000001	51	101001	75	01110101	113	1001011
8	1000	10	1000	42	01000010	52	101010	76	01110110	114	1001100
9	1001	11	1001	43	01000011	53	101011	77	01110111	115	1001101
10	00010000	12	1010	44	01000100	54	101100	78	01111000	116	1001110
11	00010001	13	1011	45	01000101	55	101101	79	01111001	117	1001111
12	00010010	14	1100	46	01000110	56	101110	80	10000000	120	1010000
13	00010011	15	1101	47	01000111	57	101111	81	10000001	121	1010001
14	00010100	16	1110	48	01001000	60	110000	82	10000010	122	1010010
15	00010101	17	1111	49	01001001	61	110001	83	10000011	123	1010011
16	00010110	20	10000	50	01010000	62	110010	84	10000100	124	1010100
17	00010111	21	10001	51	01010001	63	110011	85	10000101	125	1010101
18	00011000	22	10010	52	01010010	64	110100	86	10000110	126	1010110
19	00011001	23	10011	53	01010011	65	110101	87	10000111	127	1010111
20	00100000	24	10100	54	01010100	66	110110	88	10001000	130	1011000
21	00100001	25	10101	55	01010101	67	110111	89	10001001	131	1011001
22	00100010	26	10110	56	01010110	70	111000	90	10010000	132	1011010
23	00100011	27	10111	57	01010111	71	111001	91	10010001	133	1011011
24	00100100	30	11000	58	01011000	72	111010	92	10010010	134	1011100
25	00100101	31	11001	59	01011001	73	111011	93	10010011	135	1011101
26	00100110	32	11010	60	01100000	74	111100	94	10010100	136	1011110
27	00100111	33	11011	61	01100001	75	111101	95	10010101	137	1011111

28	00101000	34	11100	62	01100010	76	111110	96	10010110	140	1100000
29	00101001	35	11101	63	01100011	77	111111	97	10010111	141	1100001
30	00110000	36	11110	64	01100100	100	1000000	98	10011000	142	1100010
31	00110001	37	11111	65	01100101	101	1000001	99	10011001	143	1100011
32	00110010	40	100000	66	01100110	102	1000010				
33	00110011	41	100001	67	01100111	103	1000011